QUANTUM FINANCE

Path Integrals and Hamiltonians for Options and Interest Rates

This book applies the mathematics and concepts of quantum mechanics and quantum field theory to the modelling of interest rates and the theory of options. Particular emphasis is placed on path integrals and Hamiltonians.

Financial mathematics at present is almost completely dominated by stochastic calculus. This book is unique in that it offers a formulation that is completely independent of that approach. As such many new results emerge from the ideas developed by the author.

This pioneering work will be of interest to physicists and mathematicians working in the field of finance, to quantitative analysts in banks and finance firms, and to practitioners in the field of fixed income securities and foreign exchange. The book can also be used as a graduate text for courses in financial physics and financial mathematics.

BELAL E. BAAQUIE earned his B.Sc. from Caltech and Ph.D. in theoretical physics from Cornell University. He has published over 50 papers in leading international journals on quantum field theory and related topics, and since 1997 has regularly published papers on applying quantum field theory to both the theoretical and empirical aspects of finance. He helped to launch the *International Journal of Theoretical and Applied Finance* in 1998 and continues to be one of the editors.

QUANTUM FINANCE

Path Integrals and Hamiltonians for Options and Interest Rates

BELAL E. BAAQUIE

National University of Singapore

CAMBRIDGE
UNIVERSITY PRESS

CAMBRIDGE UNIVERSITY PRESS
Cambridge, New York, Melbourne, Madrid, Cape Town, Singapore, São Paulo

Cambridge University Press
The Edinburgh Building, Cambridge CB2 8RU, UK

Published in the United States of America by Cambridge University Press, New York

www.cambridge.org
Information on this title: www.cambridge.org/9780521840453

First published 2004
This digitally printed version 2007

A catalogue record for this publication is available from the British Library

Library of Congress Cataloguing in Publication data

Baaquie, B. E.
Quantum finance: path integrals and Hamiltonians for options
and interest rates/Belal E. Baaquie.
p.   cm.
Includes bibliographical references and index.
ISBN 0 521 84045 7
1. Stock options – Mathematical models.   2. Interest rates – Mathematical models.   I. Title.
HG6042.B33   2004
332.63′2283′0151539 – dc22   2004045816

ISBN 978-0-521-84045-3 hardback
ISBN 978-0-521-71478-5 paperback

I dedicate this book to my father Mohammad Abdul Baaquie and to the memory of my mother Begum Ajmeri Roanaq Ara Baaquie, for their precious lifelong support and encouragement.

Contents

Foreword

After a few early isolated cases in the 1980s, since the mid-1990s hundreds of papers dealing with economics and finance have invaded the physics preprint server xxx.lanl.gov/cond-mat, initially devoted to condensed matter physics, and now covering subjects as different as computer science, biology or probability theory. The flow of paper posted on this server is still increasing – roughly one per day – addressing a range of problems, from financial data analysis to analytical option-pricing methods, agent-based models of financial markets and statistical models of wealth distribution or company growth. Some papers are genuinely beautiful, others are rediscoveries of results known by economists, and unfortunately some are simply crazy.

A natural temptation is to apply the tools one masters to other fields. In the case of physics and finance, this temptation is extremely strong. The sophisticated tools developed in the last 50 years to deal with statistical mechanics and quantum mechanics problems are often of immediate interest in finance and in economics. Perturbation theory, path integral (Feynman–Kac) methods, random matrix and spin-glass theory are useful for option pricing, portfolio optimization and game theoretical situations, and many new insights have followed from such transfers of knowledge.

Within theoretical physics, quantum field theory has a special status and is regarded by many as the queen of disciplines, that has allowed one to unravel the most intimate intricacies of nature, from quantum electrodynamics to critical phenomena. In the present book, Belal Baaquie tells us how these methods can be applied to finance problems, and in particular to the modelling of interest rates. The interest rate curve can be seen as a string of numbers, one for each maturity, fluctuating in time. The 'one-dimensional' nature of these randomly fluctuating rates imposes subtle correlations between different maturities, that are most naturally described using quantum field theory, which was indeed created to deal with nontrivial correlations between fluctuating fields. The level of complexity of the bond market (reflecting the structure of the interest rate curve) and its derivatives (swaps, caps, floors, swaptions) requires a set of efficient and adapted techniques. My feeling is that the methods of quantum field theory, which naturally grasp complex structures, are particularly well suited for this type of problems. Belal Baaquie's book, based on his original work on the subject, is particularly useful for those who want to learn techniques which will become, in a few years, unavoidable. Many new ideas and results improving our understanding of interest rate markets will undoubtedly follow from an in-depth exploration of the paths suggested in this fascinating (albeit sometimes demanding) opus.

Jean-Philippe Bouchaud
Capital Fund Management and CEA-Saclay

Preface

Financial markets have undergone tremendous growth and dramatic changes in the past two decades, with the volume of daily trading in currency markets hitting over a trillion US dollars and hundreds of billions of dollars in bond and stock markets. Deregulation and globalization have led to large-scale capital flows; this has raised new problems for finance as well as has further spurred competition among banks and financial institutions.

The resulting booms, bubbles and busts of the global financial markets now directly affect the lives of hundreds of millions of people, as was witnessed during the 1998 East Asian financial crisis.

The principles of banking and finance are fairly well established [16,76,87] and the challenge is to apply these principles in an increasingly complicated environment. The immense growth of financial markets, the existence of vast quantities of financial data and the growing complexity of the market, both in volume and sophistication, has made the use of powerful mathematical and computational tools in finance a necessity. In order to meet the needs of customers, complex financial instruments have been created; these instruments demand advanced valuation and risk assessment models and systems that quantify the returns and risks for investors and financial institutions [63, 100].

The widespread use in finance of stochastic calculus and of partial differential equations reflects the traditional presence of probabilists and applied mathematicians in this field. The last few years has seen an increasing interest of theoretical physicists in the problems of applied and theoretical finance. In addition to the vast corpus of literature on the application of stochastic calculus to finance, concepts from theoretical physics have been finding increasing application in both theoretical and applied finance. The influx of ideas from theoretical physics, as expressed for example in [18] and [69], has added a whole collection of new mathematical and computational techniques to finance, from the methods of classical and quantum physics to the use of path integration, statistical mechanics and so on. This book is part of the on-going process of applying ideas from physics to finance.

The long-term goal of this book is to contribute to a quantum theory of finance; towards this end the theoretical tools of quantum physics are applied to problems in finance. The larger question of applying the formalism and insights of (quantum) physics to economics, and which forms a part of the larger subject of econophysics [88, 89], is left for future research.

The mathematical background required of the readership is the following:

- A good grasp of calculus
- Familiarity with linear algebra
- Working knowledge of probability theory

The material covered in this book is primarily meant for physicists and mathematicians conducting research in the field of finance, as well as professional theorists working in the finance industry. Specialists working in the field of derivative instruments, corporate and Treasury Bonds and foreign currencies will hopefully find that the theoretical tools and mathematical ideas introduced in this book broadens their repertoire of quantitative approaches to finance.

This book could also be of interest to researchers from the theoretical sciences who are thinking of pursuing research in the field of finance as well as graduates students with the required mathematical training. An earlier draft of this book was taught as an advanced graduate course to a group of students from financial engineering, physics and mathematics.

Given the diverse nature of the potential audience, fundamental concepts of finance have been reviewed to motivate readers new to the field. The chapters on 'Introduction to finance' and on 'Derivative securities' are meant for physicists and mathematicians unfamiliar with concepts of finance. On the other hand, discussions on quantum mechanics and quantum field theory are meant to introduce specialists working in finance and in mathematics to concepts from quantum theory.

Acknowledgments

I am deeply grateful to Lawrence Ma for introducing me to the subject of theoretical finance; most of my initial interest in mathematical finance is a result of the patient explanations of Lawrence.

I thank Jean-Philippe Bouchaud for instructive and enjoyable discussions, and for making valuable suggestions that have shaped my thinking on finance; the insights that Jean-Philippe brings to the interface of physics and finance have been particularly enlightening.

I would like to thank Toh Choon Peng, Sanjiv Das, George Chacko, Mitch Warachka, Omar Foda, Srikant Marakani, Claudio Coriano, Michael Spalinski, Bertrand Roehner, Bertrand Delamotte, Cui Liang and Frederick Willeboordse for many helpful and stimulating interactions.

I thank the Department of Physics, the Faculty of Science and the National University of Singapore for their steady and unwavering support and Research Grants that were indispensable for sustaining my trans-disciplinary research in physics and finance.

I thank Science and Finance for kindly providing data on Eurodollar futures, and the Laboratoire de Physique Théorique et Hautes Energies, Universités Paris 6 et 7, and in particular François Martin, for their kind hospitality during the completion of this book.

1

Synopsis

Two underlying themes run through this book: first, defining and analyzing the subject of quantitative finance in the conceptual and mathematical framework of quantum theory, with special emphasis on its path-integral formulation, and, second, the introduction of the techniques and methodology of quantum field theory in the study of interest rates.

No attempt is made to apply quantum theory in re-working the fundamental principles of finance. Instead, the term 'quantum' refers to the abstract mathematical constructs of quantum theory that include probability theory, state space, operators, Hamiltonians, commutation equations, Lagrangians, path integrals, quantized fields, bosons, fermions and so on. All these theoretical structures find natural and useful applications in finance.

The path integral and Hamiltonian formulations of (random) quantum processes have been given special emphasis since they are equivalent to, as well as independent of, the formalism of stochastic calculus – which currently is one of the cornerstones of mathematical finance. The starting point for the application of path integrals and Hamiltonians in finance is in stock option pricing. Path integrals are subsequently applied to the modelling of linear and nonlinear theories of interest rates as a two-dimensional quantum field, something that is beyond the scope of stochastic calculus. Path integrals have the additional advantage of providing a framework for efficiently implementing the mathematical procedure of renormalization which is necessary in the study of nonlinear quantum field theories.

The term 'Quantum Finance' represents the synthesis of the concepts, methods and mathematics of quantum theory, with the field of theoretical and applied finance.

To ease the reader's transition to the mathematics of quantum theory, and of path integration in particular, the presentation of new material starts in a few cases with well-established models of finance. New ideas are introduced by first carrying out the relatively easier exercise of recasting well-known results in the formalism of quantum theory, and then going on to derive new results. One unexpected advantage of this approach is that theorists, working in the field of finance – presently focussed on notions drawn from stochastic calculus and partial differential equations – obtain a formalism that completely parallels and mirrors stochastic calculus, and prepares the ground for a (smooth) transition to the mathematics of quantum field theory.

All important equations are derived from first principles of finance and, as far as possible, a complete and self-contained mathematical treatment of the main results is given. A few of the exactly soluble models that appear in finance are closely studied since these serve as exemplars for demonstrating the general principles of quantum finance. In particular, the workings of the path-integral and Hamiltonian formulations are demonstrated by concretely working out, in complete mathematical detail, some of the more instructive models. The models themselves are interesting in their own right, thus providing a realistic context for developing the applications of path integrals to finance.

The book consists of the following three major components:[1]

Fundamental concepts of finance

The standard concepts of finance and equations of option theory are reviewed in this component.

Chapter 2 is an 'Introduction to finance' that is meant for readers who are unfamiliar with the essential ideas of finance. Fundamental concepts and terminology of finance, necessary for understanding the particular set of equations that arise in finance, are introduced. In particular, the concepts of risk and return, time value of money, arbitrage, hedging and, finally, Treasury Bonds and fixed-income securities are briefly discussed.

Chapter 3 on 'Derivative securities' introduces the concept of financial derivatives and discusses the pricing of derivatives. The classic analysis of Black and Scholes is discussed, the mathematics of stochastic calculus briefly reviewed and the connection of stochastic processes with the Langevin equation is elaborated. A derivation from first principles is given of the price of a stock option with stochastic volatility. The material covered in these two chapters is standard, and defines the framework and context for the next two chapters.

Systems with finite number of degrees of freedom

In this part Hamiltonians and path integrals are applied to the study of stock options and stochastic interest rates models. These models are characterized by having

---

[1] The path-integral formulation of problems in finance opens the way for applying powerful computational algorithms; these numerical algorithms are a specialized subject, and are not addressed except for a passing reference in Section 5.16.

finite number of **degrees of freedom,** which is defined to be the **number of independent random variables at each instant of time** t. Examples of such systems are a randomly evolving equity $S(t)$ or the spot interest rate $r(t)$, each of which have one degree of freedom. All quantities computed for quantum systems with a finite number of degrees of freedom are completely finite, and do not need the procedure of renormalization to have a well-defined value.

In Chapter 4 on 'Hamiltonians and stock options', the problem of the pricing of derivative securities is recast as a problem of quantum mechanics, and the Hamiltonians driving the prices of options are derived for both stock prices with constant and stochastic volatility. The martingale condition required for risk-neutral evolution is re-expressed in terms of the Hamiltonian. Potential terms in the Hamiltonian are shown to represent a class of path-dependent options. Barrier options are solved exactly using the appropriate Hamiltonian.

In Chapter 5 on 'Path integrals and stock options', the problem of option pricing is expressed as a Feynman path integral. The Hamiltonians derived in the previous chapter provide a link between the partial differential equations of option pricing and its path-integral realization. A few path integrals are explicitly evaluated to illustrate the mathematics of path integration. The case of stock price with stochastic volatility is solved exactly, as this is a nontrivial problem which is a good exemplar for illustrating the subtleties of path integration.

Certain exact simplifications emerge due to the path-integral representation of stochastic volatility and lead to an efficient Monte Carlo algorithm that is discussed to illustrate the numerical aspects of the path integral.

In Chapter 6 on 'Stochastic interest rates' Hamiltonians and path integrals', some of the important existing stochastic models for the spot and forward interest rates are reviewed. The Fokker–Planck Hamiltonian and path integral are obtained for the spot interest rate, and a path-integral solution of the Vasicek model is presented.

The Heath–Jarrow–Morton (HJM) model for the forward interest rates is recast as a problem of path integration, and well-known results of the HJM model are re-derived using the path integral.

Chapter 6 is a preparation for the main thrust of this book, namely the application of quantum field theory to the modelling of the interest rates.

Quantum field theory of interest rates models

Quantum field theory is a mathematical structure for studying systems that have infinitely many degrees of freedom; there are many new features that emerge for such systems that are beyond the formalism of stochastic calculus, the most important being the concept of renormalization for nonlinear field theories. All the chapters in this part treat the forward interest rates as a quantum field.

In Chapter 7 on 'Quantum field theory of forward interest rates', the formalism of path integration is applied to a randomly evolving curve: the forward interest rates are modelled as a randomly fluctuating curve that is naturally described by quantum field theory. Various linear (Gaussian) models are studied to illustrate the theoretical flexibility of the field theory approach. The concept of psychological future time is shown to provide a natural extension of (Gaussian) field theory models. The martingale condition is solved for Gaussian models, and a field theory derivation is given for the change of numeraire. Nonlinear field theories are shown to arise naturally in modelling positive-valued forward interest rates as well as forward rates with stochastic volatility.

In Chapter 8 on 'Empirical forward interest rates and field theory models', the empirical aspects of the forward rates are discussed in some detail, and it is shown how to calibrate and test field theory models using market data on Eurodollar futures. The most important result of this chapter is that a so-called 'stiff' Gaussian field theory model provides an almost exact fit for the market behaviour of the forward rates. The empirical study provides convincing evidence on the efficacy of the field theory in modelling the behaviour of the forward interest rates.

In Chapter 9 on 'Field theory of Treasury Bonds' derivatives and hedging', the pricing of Treasury Bond futures, bond options and interest caps are studied. The hedging of Treasury Bonds in field theory models of interest rates is discussed, and is shown to be a generalization of the more standard approaches. Exact results for both instantaneous and finite time hedging are derived, and a semi-empirical analysis of the results is carried out.

In Chapter 10 on 'Field theory Hamiltonian of the forward interest rates' the state space and Hamiltonian is derived for linear and nonlinear theories. The Hamiltonian formulation yields an exact solution of the martingale condition for the nonlinear forward rates, as well as for forward rates with stochastic volatility. A Hamiltonian derivation is given of the change of numeraire for nonlinear theories, of bond option price, and of the pricing kernel for the forward interest rates.

All chapters focus on the conceptual and theoretical aspects of the quantum formalism as applied to finance, with material of a more mathematical nature being placed in the Appendices that follow each chapter. In a few instances where the reader might benefit from greater detail the derivations are included in the main text, but in a smaller font size. The Appendix at the end of the book contains mathematical results that are auxiliary to the material covered in the book. The reason for including the Appendices is to present a complete and comprehensive treatment of all the topics discussed, and a reader who intends to carry out some computations would find this material useful. In principle, the Appendices and the derivations in smaller type can be skipped without any loss of continuity.

Part I

Fundamental concepts of finance

2

Introduction to finance

The field of economics is primarily concerned with the various forms of productive activities required to sustain the material and spiritual life of society. **Real assets**, such as capital goods, management and labour force, and so on, are necessary for producing goods and services required for the survival and prosperity of society.

The term **capital** denotes the economic value of the real assets of a society. In most developed economies, economic assets have a monetized form, and capital can be given a monetary value or paper form, called the money form of capital.

Finance is a branch of economics that studies the money (paper) form of capital. Uncertainty and risk are of fundamental importance in finance [87].

The main focus in this book is on financial assets and financial instruments. **Financial assets**, in contrast to real assets, are pieces of paper that entitle its holder to a claim on a fraction of the real assets, and to the income (if any) that is generated by these real assets. For example, a person owning a stock of a company is entitled to yearly dividends (if any), and to a pro rata fraction of the assets if the company liquidates.

What distinguishes finance from other branches of economics is that it is primarily an empirical discipline due to the demands of the finance industry. Vast quantities of financial data are generated every day, in addition to mountains of accumulated historical data. Unlike other branches of economics, the empirical nature of finance makes it closer to the natural sciences, since the financial markets impose the need for practical and transparent quantitative models that can be calibrated and tested.

A financial asset is also called a **security**, and the specific form of a financial asset – be it a stock or a bond – is called a **financial instrument**. A financial asset is at the same time a **financial liability** for the issuing party, since its profit and assets are to be divided amongst all the stockholders. Stocks and bonds are in positive net supply. Derivatives in contrast are in zero net supply since the number of people holding the derivative exactly equals the number of people selling these derivatives – and hence derivatives amount to a zero-sum game. The payoff to the holder of a derivative instrument equals minus the payoff for the seller of the instrument.

An investor can invest in financial assets as well as in real assets, such as real estate, gold or some other commodity [54].

The following are the three primary forms of financial instruments.

- **Equity**, or common **stocks** and **shares** represent a share in the ownership of a company. The possession of a share does not guarantee a return, but only a pro rata fraction of the dividends, usually declared if the company is profitable. The value of a share may increase or decrease over time, depending on the performance of the company, and hence the owner of equity is exposed to the risks faced by the company. The holder of a stock has only a limited liability of losing the initial investment. Hence, the value of a stock is **never negative**, with its minimum value being zero. Equity is a form of asset since the holder of equity is a net owner of capital.
- **Fixed income securities**, also called bonds, are issued by corporations and governments, and promise either a single fixed payment or a stream of fixed payments. Bonds are **instruments of debt**, and the issuer of a bond in effect takes a loan from the buyer of the bond, with the repayment of the debt usually being scheduled over a fixed time interval, called the **maturity** of the bond. There is a great variety of bonds, depending on the different periods of maturity and provisions for the repayment stream. For example, the holder of a five-year coupon US Treasury Bond is promised a stream of interest payments every six months, with the principal being repaid at the end of five years, whereas a holder of a zero coupon US Treasury Bond receives a single cash flow on the maturity of the bond. The risk in the ownership of a fixed-income security is often considered to be less than the ownership of equity since – short of the issuer going bankrupt – the owner of a fixed-income security is guaranteed a return as long as the owner can hold the instrument till maturity. However, due to interest rate risk, credit risk and currency risk for the bonds that are issued in a foreign currency, a bond portfolio can lose as much value, or even more, than a portfolio of equities.
- **Derivative securities** are, as the term indicates, financial assets that are **derived** from other financial assets. The payoff of a derivative security can depend, for example, on the price of a stock or another derivative.

The three primary forms of financial instruments can be combined in an almost endless variety of ways, leading to more complex instruments. For example, a **preferred stock** combines features of equity and debt instruments by entitling the investor to a fraction of the issuer's equity, and at the same time – similar to bonds – to a stream of (fixed) payments.

Theoretical finance takes as its subject the money (paper) form of capital, and is primarily concerned with the problems of the time value of money, risk and return, and the valuation of securities and assets. The creation and arbitrage-free pricing of new financial instruments to suit the myriad needs of investors is of increasing importance. The design, risk-return analysis and hedging of these instruments are issues that are central to finance, and comprise the field of financial engineering.

2.1 Efficient market: random evolution of securities

A financial **market** is where the buyer and seller of a financial asset meet to enact the transaction of buying and selling. If one buys (or agrees to buy) a financial asset, one is said to have a **long position** or is said to be **going long**. On the other hand, if one is selling a financial asset, one is said to be **shorting** the asset, or, equivalently, have a **short position**. If one sells an asset without actually owning it, one is said to be engaged in **short selling**; the repurchase date for short selling is usually some time in the future.

There are various forms in which any market is organized, with the primary ones being the following. A **direct market** is based on a direct search of the buyer and seller, the **brokered market** is one in which the brokers – for fees – match the buyer with the seller, and, lastly, the **auction market** is one in which buyers and sellers interact simultaneously in a centralized market [100]. Financial assets and instruments are traded in specialized markets known as the **financial markets**, which will be discussed in the next section.

The concept of an **efficient market** is of great importance in the understanding of financial markets, and is tied to the concept of the 'fair price' of a security. One expects that for a market in equilibrium, the security will have its fair price, and that investors will consequently not trade in it any further. When in equilibrium, an efficient market is one in which the prices of financial instruments have only small and temporary deviations from their fair price.

Efficient market is closely related to the concept of market information. What differentiates the various players in the market is the amount of market information that is available to each of them. Market information in turn consists of three components, namely: (a) historical data of the prices and returns of financial assets, (b) public domain data regarding all aspects of the financial assets and (c) information known privately to a few market participants. Based on these three categories of information, the concept of 'weak', 'semi-strong' and 'strong' forms of market efficiency can, respectively, be defined [23].

Intuitively speaking, an efficient market in effect means most of the buyers and sellers in the market have equal wealth and information, with no collection of buyers or sellers having any (unfair) advantage over the others. A precise statement of the efficient market hypothesis is the following

For a financial market that is in equilibrium, none of the players, given their endowment and information, want to trade any further. For efficient markets, prices reveal available market information. The inflow of new information comes in randomly – in bits and pieces – causing random responses from the market players, due to the incomplete nature of the incoming information, and results in random changes in the prices of the various financial instruments.

It is worth emphasizing that a far-reaching conclusion of the efficient market hypothesis is that, once the market is in equilibrium, **changes in the prices** of all securities, upto a drift, are **random** [23]. The reason being that in an efficient market the prices of financial instruments have already incorporated all the market information, and resulted in equilibrium prices; any departures of the prices from equilibrium should be uncertain and unpredictable, with changes being equally likely to be above and below the equilibrium price.

Hence changes in the prices of financial instruments should be represented by **random variables**. Suppose the value of an equity at time t is represented by $S(t)$; then the **change** in the value of an equity is random, that is, dS/dt is modelled as a random variable; this in turn implies the security $S(t)$ itself is also a random variable, with its initial (deterministic) condition specified at some time t_0. The extent to which a security $S(t)$ is random is specified by a quantity called the **volatility** of the security, and is usually denoted by σ_S, or simply by σ. The greater the volatility of a security, the greater are the random fluctuations in the price of the security. A volatility of $\sigma = 0$ consequently implies that the security has no uncertainty in its future evolution.

The risk that all investors face is a reflection of the **random evolution** of financial instruments, and is ultimately a reflection of the manner in which (financial) markets incorporate all the relevant features of the underlying real economy.

The efficient market hypothesis does not imply that new information or important events do not move the market; rather, the hypothesis implies that unexpected or unanticipated new information **disturbs** the equilibrium of the market prices of various securities, and systematically moves them to a **new** set of equilibrium prices. Once equilibrium is reached, ordinary information will be available to almost all participants and hence will lead to random changes in the revealed prices of the financial instruments.

Is the efficient market hypothesis empirically testable? As pointed out in [23], there are **two hypotheses** implicit in the existence of an efficient market, namely the hypothesis of efficiency together with the hypothesis that the market is in a particular equilibrium. It is only this **joint hypothesis** – namely of market efficiency and equilibrium – that can be empirically tested and which often leads to spirited academic debates regarding the efficiency of financial markets.

The concept of market equilibrium is similar to the idea of equilibrium for a thermodynamic system. The positions and velocities of individual particles, analogous to the prices of financial instruments, are random even though the system itself is in equilibrium. Furthermore, the efficiency of the market is analogous to the efficiency of a thermodynamic heat engine. No one expects an actual heat engine to have 100% efficiency, and an efficiency of say 60–70% is fairly common. Similarly, even if a financial market is not fully efficient, it is often still justified to apply mathematical models based on this concept.

2.2 Financial markets

The financial markets are organized to trade in various forms of financial instruments. The major segmentation of the financial markets is into the **capital markets** and the **money markets**. Capital markets are structured to trade in the primary forms of financial instruments, namely in instruments of equity, debt and derivatives. Indexes are a part of the capital markets and are equal to the weighted average of a basket of securities of a particular market; given their importance and depth, indexes are treated as entities distinct from the capital markets. Money markets, properly speaking, belong to the debt market, but since money market instruments trade in highly liquid and short-term debt, cash and cash equivalents, foreign currency transactions and so on, a separate market is set up for such transactions.

The following is a breakdown of the main forms of the financial markets:

1 Capital markets
   - Equity market: common stocks; preferred stocks.
   - Debt market: treasury (government) notes and bonds; corporate and municipal bonds; mortgage-backed securities (MBS)
   - Derivative market: options; forwards and futures
2 Indexes
   - Equity indexes: Dow Jones and Standard and Poor's Indexes; Nikkei index; DAX Index; STI Index etc.
   - Debt indexes: bond market indicators
3 Money markets
   - Cash time deposits
   - Treasury bills
   - Certificates of deposit
   - Commercial paper
   - Eurodollar deposits: refers to US$ deposits in non-US banks, or in overseas branches of US banks.

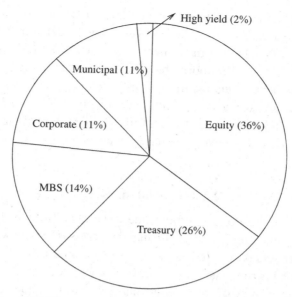

Figure 2.1 The 1993 US capital markets worth US$11.3 trillion. The debt markets consist of corporate, municipal Treasury and high-yield bonds and MBS (mortgage-backed securities) making up 64% of the capital markets, with 36% being in the equity markets.

Data for the 1993 US capital market are given in Figure 2.1. The equity component is only 36% of the capital market; if one takes into account the money market, the share of equity is even lower. The global debt market was worth US$14.08 trillion in 1993 and Figure 2.1 [100] shows the main international borrowers.

The GDP of the USA in 2001 was about US$10 trillion. The size of the credit market in the US for 2002 was about US$29 trillion (with financial sector borrowing making up US$9 trillion). In comparison, the total equity (market capitalization) in the US for 2002 was about US$12 trillion.

Derivatives can be traded in two ways: on regulated exchanges or in unregulated over-the-counter (OTC) markets The size of the derivatives markets are typically measured in terms of the notional value of contracts. Recent estimates of the size of the exchange-traded derivatives market, which includes all contracts traded on the major options and futures exchanges, are in the range of US$13 trillion to $14 trillion in notional amount. OTC derivatives are customized for specific customers. The estimated notional amount of outstanding OTC derivatives as of year end 2000 was US$95.2 trillion, and experts consider even this amount as being most likely on the lower side.

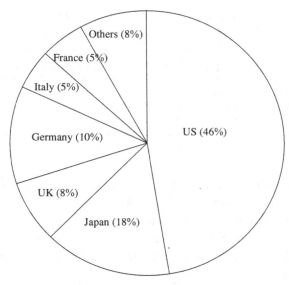

Figure 2.2 Breakdown of the 1993 US$14.8 trillion global debt markets

2.3 Risk and return

For any investor, two considerations are of utmost importance, namely, the return that can be made, and the risk that is inherent in obtaining this return. The trade-off between return and risk is the essence of any investment strategy. Clearly, all investors would like to maximize returns and minimize risk. What constitutes return is quite simple, but the definition of risk is more complex since it involves quantifying the uncertainties that the future holds.

Suppose one buys, at time t, a stock at price $S(t)$, holds it for a duration of time T with the stock price having a terminal value of $S(t + T)$ and during this period earns dividends worth d. The (fractional) rate of return R for the period T is given by

$$R = \frac{S(t + T) + d - S(t)}{S(t)}$$

where R/T is the instantaneous rate of return.

What are risks involved in this investment? The future value of the stock price may either increase on decrease, and it is this uncertainty regarding the future that introduces an element of risk into the investment. There are many possible scenarios for the stock price. One scenario is that there is a boom in the market with stock prices increasing; or there is a downturn and stock prices plummet; or that the market is in the doldrums with only small changes in the stock price.

Table 2.1 *Possible scenarios for the annual change in the price of a security* $S(t)$
with current price of $100

s	Scenario	S	p(s):Likelihood	R(s):Annual Return	Average $\bar{R}$	Risk σ
1	Doldrum	$100	$p(1) = 0.70$	$R(1) = 0.00$		
2	Downturn	$85	$p(2) = 0.20$	$R(2) = -0.15$	-0.025	0.06
3	Boom	$105	$p(3) = 0.10$	$R(3) = 0.05$		

One can assign probabilities for each scenario, and this in turn gives the investor a way of gauging the uncertainties of the market. A typical example of the various scenarios for some security $S(t)$ are shown in Table 2.1.

Label each scenario by a discrete variable s, its probability by $p(s)$, and its return by $R(s)$. The expected return for the investment is the average (mean) value of the return given by

$$\bar{R} = \text{expected return} = \sum_s p(s)R(s) \equiv E[R]$$

where the notation $E[X]$ denotes the expectation value of some random quantity X.

The **risk** inherent in obtaining the expected return is clearly the possibility that the return may deviate from this value. From probability theory it is known that the standard deviation indicates the amount the mean value of any given sample can vary from its expected value, that is

actual return = expected return $\pm$ standard deviation (with some likelihood)

The precise amount by which the actual return will deviate from the expected return – and the likelihood of this deviation – can be obtained only if one knows the probability distribution $p(S)$ of the stock price $S(t)$. Standard deviation, denoted by σ, is the square root of the variance defined by

$$\sigma^2 = \sum_s p(s)\big(R(s) - \bar{R}\big)^2 \equiv E[(R - \bar{R})^2]$$

The risk inherent in any investment is given by σ – the larger risk the greater σ, and vice versa. In the example considered in Table 2.1, at the end of one year the investor with an initial investment of $100 will have an expected amount of cash given by $100 \times [1 + \bar{R}] \pm 6 = \$97.50 \pm 6$.

For some cases, such as a security obeying the Levy distribution, the value of σ is infinite, and a more suitable measure for risk then is what is called 'value at risk' [18].

Instruments such as fixed deposits in a bank and so on are taken to be risk free. The rate of return of a risk-free instrument at time t is the amount earned on an instantaneous deposit in a risk-free bank; the rate is called the spot interest rate, or overnight lending rate, and is denoted by $r(t)$. Hence, for a **risk-free** instrument

$$\sigma_{\text{risk-free}} = 0$$
$$\text{risk-free rate of return} = \text{spot interest rate} = r$$

A risk-neutral investor expects a return equal to the spot interest rate r. However, for risky investments $\sigma > 0$, and clearly to induce investors to take a high risk, there have to be commensurate high rewards. To facilitate the flow of capital towards high-risk investments, the capital market holds out a premium for undertaking high risk with the prospect of high returns. This **risk premium** is the amount by which the rate of return on high-risk investment is above the risk-free rate. For an investment with an average annual rate of return $\bar{R}$, the risk premium – also called the Sharpe ratio – is given by $(\bar{R} - r)/\sigma$.

A speculator would invest in high-risk securities if an analysis shows that the potential return on that investment has a sufficient risk premium. A speculator in this sense is different from a gambler who takes a high risk even in the absence of a risk premium.

A fundamental principle of finance is the **principle of no arbitrage** which states that no risk-free financial instrument can yield a rate of return above that of the risk-free rate. In other words there is no free lunch – if one wants to harvest high returns one has to take the commensurate high risks. The mathematical implications of the principle of no arbitrage is discussed in Section 2.5.

2.4 Time value of money

The money form of capital represents real productive assets of society that can erode over time; furthermore, other factors like inflation, currency devaluations and so on make the value represented by financial assets dependent on time. Financial assets represent the ability to initiate or facilitate economic activities, opportunities which are tied to changing circumstances. For these and many other reasons, the effective value of money is strongly dependent on time.

How does one estimate the time value of money? From economic theory, the sum total of all the endogenous and exogenous effects on the time value of money are contained in the spot interest rate. Money invested in other risky instruments are more complicated to value, as risk premiums are involved that may differ between investors. Ultimately, the time valuation of money involves a discounting of the future value of money to obtain its 'expected' present-day value.

Suppose one has $1 at time t_0, and invests this sum in a risk-free instrument such as a fixed deposit; furthermore, suppose one compounds the investment by reinvesting the returns in the same fixed deposit. Since the rate of return on risk-free instruments is $r(t)$, at some future time t_* the risk-free value of the investment becomes

$$t_0 < t_* : \quad \$1 \text{ at time } t_0 = \$ \, e^{\int_{t_0}^{t_*} dt\, r(t)} \text{ at time } t_*: \text{ augmented value of cash}$$

Equivalently, one discounts the future value of money to obtain its present value

$$\$1 \text{ at time } t_* = \$ \, e^{-\int_{t_0}^{t_*} dt\, r(t)} \text{ at time } t_0: \text{ discounted value of cash}$$

The time value of money essentially means that the correct unit to use for money is **not** $1 – since its effective value is subject to constant variations over time – but instead the **correct units** for measuring risky instruments, such as a stock, or risk-free instruments, such as cash, is the **discounted quantity**. The present-day value of a future cash flow should be discounted by a factor $\exp\{-\int_{t_0}^{t_*} dt\, r(t)\}$, and similarly, the future value of a current cash flow should be augmented by its inverse.

To determine the function $r(t)$ from first principles one has to study the macroeconomic fundamentals of an economy, the supply and demand of money, and so on. The interest rate reflects the marginal utility of consumption, that is, the rate at which people are enticed to forgo current consumption and save (invest) their money for future consumption. It will be seen later, when spot interest rate models are studied, that $r(t)$ is considered to be a stochastic (random) variable. The discounting is then obtained by taking the average of the discounting factor over all possible values of the random function $r(t)$. Hence the discounting factor is given by

$$t_0 < t_* : \$1 \text{ at time } t_* = \$ \, E\left[e^{-\int_{t_0}^{t_*} dt\, r(t)} \right] \text{ at time } t_0 \qquad (2.1)$$
$$: \text{ discounted value of cash}$$

2.5 No arbitrage, martingales and risk-neutral measure

Arbitrage – an idea that is central to finance – is a term for gaining a risk-free (guaranteed) profit by simultaneously entering into two or more financial transactions – be it in the same market or in two or more different markets. Since one has risk-free instruments, such as cash deposits, arbitrage means obtaining guaranteed risk-free returns **above** the risk-less return that one can get from the money market.

For example, suppose that at some instant the share of a company is traded at value US\$1 on the New York stock exchange, and at value S\$1.8 in Singapore, with the currency conversion being US\$1 = S\$1.7. A broker can simultaneously buy 100 shares in New York and sell 100 shares in Singapore making a risk-less profit of S\$10. Transaction costs tend to cancel out arbitrage opportunities for small traders, but for big brokerage houses – which have virtually zero transaction cost – arbitrage is a major source of profits. One can also see that the price of the share in Singapore will tend to move to a value close to S\$1.7 due to the selling of shares by the arbitrageurs.

In an efficient market there are no arbitrage opportunities. Arbitrage is one of the **mechanisms** by which the capital market in practice functions as an efficient market, and determines the equilibrium ('correct') price of any financial instrument. The existence of an efficient market is a sufficient but not a necessary condition for the principle of no arbitrage to hold. In equilibrium no arbitrage opportunities exist. No arbitrage is a robust concept since it expresses the preference of all investors to have more wealth over less wealth. Most models of market equilibrium are based on more restrictive assumptions about investor behaviour.

An important result of theoretical finance is the following: for the price of a financial instrument to be free from any possibility of arbitrage, it is necessary to evolve the discounted value of the financial instrument using a **martingale process** [23, 40, 100]. The real market evolution of a security, for example a stock, does not follow a martingale process since there would then no risk premium for owning such a security. Instead, the martingale evolution of a security is a convenient theoretical construct to price derivative instruments such that their price is then free from arbitrage opportunities.

The concept of a martingale in probability theory (discussed in Appendix A.1) is the mathematical formulation of the concept of a fair game. In an efficient market the **risk-free** evolution of a security is equivalent to its evolution obeying the martingale condition. Since real investors are not risk neutral and demand a risk premium, their evolution requires a change of measure from the risk-neutral one.

Suppose one has a stochastic process given by a collection of $N + 1$ random variables X_i; $1 \le i \le N + 1$, with a joint probability distribution function given by $p(x_1, x_2, \ldots, x_{N+1})$. As discussed in Eq. (A.2), a **martingale process** is defined by the following conditional probability

$$E[X_{n+1}|x_1, x_2, \ldots, x_n] = x_n \quad : \quad \text{martingale} \qquad (2.2)$$

The left-hand side is the expected probability of the random variable X_{n+1}, conditioned on the occurrence of $x_1, x_2, \ldots, x_n$ for random variables $X_1, X_2, \ldots, X_n$. In finance, at time t the random variables are the future prices of a stock $S_1, S_2, \ldots, S_{N+1}$ at the times $t_1, t_2, \ldots, t_{N+1}$ respectively. To apply the martingale condition to the evolution of stock prices, the stock price needs to discounted since the prices of stocks are being compared at two different times. It is shown in Appendix A.6 that for a **complete market** there exists a **unique** risk-neutral measure with respect to which the discounted evolution of all derivatives of an asset obey the martingale condition [40, 42].

Let the value of an equity at future time t be $S(t)$. Assume that there exists a risk-free evolution of the discounted stock price

$$e^{-\int_0^t r(t')dt'} S(t) \tag{2.3}$$

such that it follows a martingale process [42]. From Eq. (A.40) it follows that the conditional probability of the discounted value of the equity at time t, is its present value $S(0)$. In other words

$$S(0) = E\left[e^{-\int_0^t r(t')dt'} S(t)|S(0)\right] \tag{2.4}$$

The result above is of great generality, as it holds for any security; the importance of martingales in financial modelling is discussed [80].

In summary, if there exists a measure such that the evolution of a discounted financial instrument obeys the martingale condition, then it is guaranteed that the prices of all of its derivative instruments are free from arbitrage opportunities. The converse is also true: if a discounted financial instrument's price is free from arbitrage opportunities, then there exists a martingale measure. The existence of a martingale measure is called by some authors [42] the **fundamental theorem of finance**, and is briefly discussed in Appendix A.6.

Most of the models that are analyzed in this book are evolved with a martingale measure, thus ensuring that the price of all (derivative) financial instruments are free from arbitrage opportunities.

2.6 Hedging

Given that the evolution of financial instruments is stochastic, the question naturally arises as to whether one can create a portfolio from risky financial assets that is risk free? In other words, can one cancel the random fluctuations of one instrument with the random fluctuations of another instrument? Can the cancellation be made exact so that the composite instrument becomes risk free? In addition to reducing risk, hedging has another major role: between two portfolio's giving the same return, the one that is hedged has a lower risk, and hence in general is a superior portfolio.

Hedging is the general term for the procedure of **reducing** the random fluctuations of a financial instrument by including it in a portfolio together with other related instruments. A perfectly hedged portfolio is free from all random fluctuations: the random fluctuations in the price of the financial instrument being hedged are exactly cancelled by the compensating fluctuations in other instruments in the portfolio. In practice, however, the best that is usually possible is to have a partially hedged instrument.

Hedging is analogous to buying insurance. The cost of hedging is the transaction costs incurred in buying and selling the needed securities – and, similar to insurance, is the price that one has to pay for reducing risk. High transaction costs make it more costly to hedge, but it is still effective in combating risk. Often, hedging leads to the unwanted result of lowering future payoffs. For example, one can use short positions in futures contracts (futures contracts cost nothing to enter into) in order to hedge a bond. If interest rates increase, the hedge works (gain on futures contracts offset losses on the bond's value); however, if interest rates decrease the bond price increases, but the futures contracts lose money and in doing so lower the net profit. Thus, eliminating fluctuations also eliminates the possibility of some 'good' fluctuations in the process. Options, though not costless to enter, often allow investors to manage risks without having to accept reduced payoffs in the future.

In short, the hedging strategy depends on the objectives of the investor.

There is in general no guarantee that all the fluctuations in the price of a financial instrument can be hedged. For a complete market there exist, in principle, assets that can be used to hedge every risk of a specific instrument. In practice whether an instrument can be perfectly hedged or not depends on the other instruments that are actually available in the market; a major impetus for the development of derivative instruments stems from the need to hedge commonly used financial instruments.

To hedge a financial instrument, one needs to have at least a second instrument so that a cancellation between the fluctuations of the two instruments can be attempted. The second instrument clearly has to depend on the instrument one intends to hedge, since only then can one expect a connection between their random fluctuations. For example, to hedge a primary instrument, what is often required is a **derivative instrument**, and vice versa. Since the derivative instrument is driven by the same random process as the primary instrument, the derivative instrument has the important property that its evolution is perfectly **correlated** with the fundamental underlying instrument, and hence allowing for perfect hedging.

Consider the case of a security, say a common stock, that is represented by the stochastic variable $S(t)$. Suppose a reduction in the risk of holding a stock is sought by attenuating the fluctuations in the value of $S(t)$; one needs to consequently hold a second instrument, a derivative of $S(t)$ – denoted by $D(S)$ – such that taken together the portfolio will have fewer fluctuations. Suppose that $S(t)$ can be perfectly hedged, and denote the hedged portfolio by $\Pi(S, t)$. The portfolio for example can consist of the investor going long (buying) on a single derivative $D(S)$, and short selling $\Delta(S)$ worth of the underlying stock S. The portfolio at some instant t is then given by

$$\Pi(S, t) = D(S) - \Delta(S)S$$

Since value of the security $S(t)$ is known at time t, the portfolio $\Pi(S, t)$ is perfectly hedged if its time evolution has no random fluctuations, and is in effect a deterministic function. In other words

$$\frac{d\Pi(S, t)}{dt} : \text{no randomness} \rightarrow \text{perfectly hedged}$$

Since there are no random fluctuations in the value of $d\Pi(S, t)/dt$ it is a risk-free security; the principle of no arbitrage then requires that the rate of return on the perfectly hedged portfolio must be equal to risk-free spot rate $r(t)$. Hence

$$\frac{d\Pi(S, t)}{dt} = r(t)\Pi(S, t)$$

This, in short, is the procedure for hedging a financial asset.

In practice there are many conditions that need to be met for hedging to be possible.

- The market must trade in the derivative instrument $D(S)$; otherwise one cannot create a hedged portfolio. There are many financial instruments that cannot be hedged because the appropriate derivative instruments are not traded in the market, as, for example, is the case with the volatility of a security.
- It needs to be ascertained whether the hedging parameter $\Delta(S)$ exists, and what is its functional dependence on the stock price S. For this the precise relation of the derivative $D(S)$ with the stock price $S(t)$ needs to be known, as well as the detailed description of the (random) dynamics of $S(t)$.
- Since the portfolio $\Pi(S, t)$ depends on time, hedging needs to be done continuously; for this to be possible the market has to have enough liquidity and this in turn determines the transaction costs involved in hedging.

The concept of hedging an equity is discussed in Chapter 3 on derivatives, and in Chapter 9 where the hedging of Treasury Bonds is discussed in some detail.

2.7 Forward interest rates: fixed-income securities

Forward interest rates and fixed-income securities are fundamental to the debt market [58].

An instantaneous loan at time t costs the borrower a spot interest rate $r(t)$, and is usually quoted as an annual percentage; spot interest rates typically vary from 0.1% to 20% per year.

It is often the case that a borrower may need to borrow money at some specific time in the future, for example to buy and sell a commodity a year in the future; such a borrower would like to lock-in the interest rate needed for the expected transaction. The capital market has an instrument for such a borrower called the **forward interest rates**, or **forward rates**, and is denoted by $f(t, x)$. From a mathematical point of view, both the spot interest rate and the forward rates are the instantaneous cost of borrowing money, that is of borrowing money for an infinitesimal time ϵ.

The forward rate $f(t, x)$ is the instantaneous interest rate agreed upon (in the form of a contract) at an earlier time $t < x$, for a borrowing between future times x and $x + dx$. The forward rates constitute the **term structure of the interest rates**, and is related to the interest rate **yield curve**.

From its definition, that the spot interest rate $r(t)$ for an overnight loan at some time t, is given by

$$r(t) = f(t, t) \qquad (2.5)$$

Bonds are financial instruments of debt that are issued by governments and corporations to raise money from the capital market. Bonds entail a financial obligation on the part of the issuer to pay out a predetermined and fixed set of cash flows, and hence the generic term **fixed-income securities** is used for the various categories of bonds.

A **Treasury Bond** is an instrument for which there is no risk of default in receiving the payments, whereas for corporate, municipal bonds and sovereign bonds of certain countries – such as Russia, Argentina, and so on – there is in principle such a risk. Due to the risk-free nature of the US Treasury Bond the US government is able to engage in large-scale international borrowing at the lowest possible interest rates.

A **zero coupon** Treasury Bond is a risk-free financial instrument which has a single cash flow consisting of a fixed payoff of say $1 at some future time T; its price at time $t < T$ is denoted by $P(t, T)$, with $P(T, T) = 1$.

From the time value of money, for a bond maturing at time T its value $P(t, T)$ before maturity is given by discounting $P(T, T) = 1$ to the time t by the spot interest rate. For the general case when interest rates are considered to be stochastic, Eq. (2.1) gives

$$P(t, T) = E\left[e^{-\int_t^T dt' r(t')} P(T, T)\right]$$
$$= E\left[e^{-\int_t^T dt' r(t')}\right] \qquad (2.6)$$
$$: \text{discounted value of } P(T, T) = 1$$

where the expectation value is taken with respect to the stochastic process obeyed by $r(t)$. Eq. (2.6) shows that the Treasury Bond is a function of only the initial value $r(t)$ of the spot rate.

A **coupon** Treasury Bond $\mathcal{B}(t, T)$ has a series of predetermined cash flows that consist of coupons worth c_i paid out at increasing times T_i, with the principal worth L being paid on maturity at time T. Using the principle that two financial instruments are identical if they have the same cash flows, it can be shown [58] that $\mathcal{B}(t, T)$ is given in terms of the zero coupon bonds as

$$\mathcal{B}(t, T) = \sum_{i=1}^{K} c_i P(t, T_i) + LP(t, T) \qquad (2.7)$$

From above it can be seen that a coupon bond is equivalent to a portfolio of zero coupon bonds. Hence, any model of the zero coupon bonds automatically provides a model for the coupon bonds as well.

Municipal, corporate and high-yield bonds are more complex to model due to taxation rules, liquidity, and so on, have a finite likelihood of default, and hence carry an element of risk not present in Treasury Bonds. Risky bonds consequently pay a risk premium over and above that of Treasury Bonds.

The price of a zero coupon Treasury Bond $P(t, T)$ can be written in terms of the forward rates, which recall are defined only for instantaneous future borrowing. Since a zero coupon bond is a loan taken by the issuer for a **finite** duration, one has to iterate the discounting by the forward rates to obtain the present value of the Treasury Bond.

At maturity $P(T, T) = \$1$; hence, $P(t, T)$ is obtained by successively discounting $\$1$ from future time T to the present time t. For this purpose, discretize time into a set of instants with time interval ϵ; the set of forward rates $f(t, x_n)$ are then defined for future times $x_n = t + n\epsilon$; $n = 0, 1, \ldots [(T - t)/\epsilon]$.

The discounting of an instantaneous loan from future time x_n to time $x_{n-\epsilon}$ is given by $e^{-\epsilon f(t, x_n)}$. Successively discounting the deterministic payoff of $\$1$ at time T to present time t, gives

$$P(t, T) = e^{-\epsilon f(t, x_0)} e^{-\epsilon f(t, x_1)} \ldots e^{-\epsilon f(t, x_{N-1})} e^{-\epsilon f(t, x_N)} 1$$

Taking the limit of $\epsilon \to 0$ yields

$$P(t, T) = e^{-\int_t^T dx f(t, x)} \qquad (2.8)$$

$$\Rightarrow -\frac{\partial}{\partial T} \ln P(t, T) = f(t, T) \qquad (2.9)$$

The expression obtained for the Treasury Bonds in terms of the forward rates is an identity, and can be taken as the **definition** of the forward rates. Moreover, from

Eqs. (2.8) and (2.9), once the value of $P(t, T)$ is known, the value of the forward rate $f(t, T)$ is also known, and vice versa.

The martingale condition, given in Eq. (2.4), when applied to the evolution of the forward rates states the following. Suppose that a zero coupon Treasury Bond that matures at time T has a price $P(t_0, T)$ for some $t_0 < T$, and a price of $P(t_*, T)$ at time $t_0 < t_* < T$. To satisfy the martingale condition the price of the bond at t_*, evolved backward to time t_0 – and continuously discounted by the risk-free spot rate $r(t)$ – must be equal to the price of the bond at time t_0. The martingale condition for the zero coupon bond, from Eq. (2.4), has the following very general and **model-independent** expression given by

$$P(t_0, T) = E_{[t_0, t_*]} \left[e^{-\int_{t_0}^{t_*} r(t)dt} P(t_*, T) \mid P(t_0, T) \right] \qquad (2.10)$$

where the notation $E_{[t_0, t_*]}[X]$ denotes the average of the stochastic variable X over the time interval $(t_0, t_*]$.

There are two definitions of the zero coupon bonds, one in terms of the spot rate given in Eq. (2.6) and the other in terms of the forward rates given in Eq. (2.8). The two expressions are very different.

One can in principle take the spot interest rate as being fundamental, as is the case in Eq. (2.6), and the forward rates as derived quantities. The advantage of this approach is its simplicity, and it is adequate for addressing questions directly related to the behaviour of only the spot rate. The disadvantage of this approach is that the spot rate must be consistent with observed bond prices of many maturities.

Writing $P(t, T)$ in terms of the forward interest rates as given in Eq. (2.8) is a more general expression than the one given by the spot rate as in Eq. (2.6). The reason being that the forward rates can in principle be directly determined from the market. In the modelling of the forward rates across all maturities, an entire initial term structure is used as an input. Hence, the model will not generate arbitrage opportunities based on observed bond prices.

Of course, the trade-off is that forward rates $f(t, x)$ depend on two variables, namely t, x whereas the spot interest rate $r(t)$ depends only on time t. In spite of its greater complexity, considering the forward interest rates as fundamental, and regarding the spot rate as just one point of the forward rates curve, is nevertheless a very productive approach to the modelling of forward rates.

2.8 Summary

Some of the key ideas of finance were reviewed, and which will repeatedly appear in later discussions.

The random evolution of financial instruments forms the basis for the entire exercise of modelling the evolution of these instruments using random variables. It is the random time evolution of financial instruments that provides the crucial and far-reaching connection with quantum theory, and with path integration.

The concept of risk is at the center of the notion of hedging. The generalization of the concept of risk for the infinite-dimensional case will play a key role in defining the quantum field theory of hedging for Treasury Bonds.

The time value of money leads one naturally to the concept of the martingale measure, which is defined by demanding that the discounted future value of a traded instrument be equal to its present-day value. A lot of effort will be expended, specially for the case of nonlinear forward interest rates, in finding the appropriate martingale measure.

The concepts of hedging and options form a cornerstone of the theory of financial derivatives, and the pricing of options is the focus of the first half of this book.

The second half of the book addresses the subject of forward interest rates and fixed-income securities. In particular, the study of the forward rates naturally leads to its formulation as a quantum field, which is the main thrust of this book.

3

Derivative securities

Derivative securities, or derivatives in short, are important forms of financial instruments that are traded in the financial markets. As the name implies, derivatives are **derived** from underlying financial instruments: the cash flows of a derivative depend on the prices of the underlying instruments.

Derivatives have many uses from being an ingredient in the hedging of a portfolio, to their use as instruments for speculation.

Given the uncertainties of the financial markets, there is a strong demand from the market for predicting the future behaviour of securities. Derivative instruments are a response to this need, and contain information for estimating the behaviour of a security in the future. There are three general categories of derivatives, namely forwards, futures and options.

3.1 Forward and futures contracts

Suppose a corporation needs to import steel one year in the future, denoted by T. Since the price of steel can vary over time, the corporation would like to guard against the risk of the price of steel increasing by locking-in the price of steel today, denoted by t.

Let the price of steel per ton at time t be denoted by $S(t)$. The **forward contract** is a contract between a buyer of steel – who is said to have a long position, and a seller – who is said to have a short position. The seller agrees, at time t, to provide steel at future time T, at the forward price $F(t, T)$ that reflects the current prevailing price and interest rates. The contract is entered into at time t, and there is no initial cash transaction; the value of the forward contract is chosen such that its initial value is **zero**; the value of the forward contract fluctuates till its maturity at time T. On maturing (at time T) the value of the forward contract is

$$S(T) - F(t, T) \quad \text{if long}$$
$$F(t, T) - S(T) \quad \text{if short}$$

There is a single cash flow at the maturity of contract worth $F(t, T)$; this value can be either positive or negative depending of the price of steel at time T.

A **futures contract**, denoted by $\mathcal{F}(t, T)$, is similar to a forward contract in the sense that it is an agreement between a buyer and seller, entered into at some present time t, for the delivery of a specified quantity, of say steel, at a fixed time T in the future. Futures contracts are highly regulated and have a standard format. There is always a third party, usually a **clearing house**, that acts as a middle man in the contract, and imposes margin payments on both the seller and buyer to increase liquidity, and reduce owner party risk, creating a series of cash flows from the time the contract is initiated until it matures.

One needs to draw a distinction, at time t, between the value and the price of a futures contract.[1] On initiating the contract, neither party pays any cash amount and hence the value of the contract is zero. However, a notional fair price is assigned to the futures contract at time of writing the contract, namely $\mathcal{F}(t_0, T) \neq 0$. An **initial margin** is paid on initiating the contract. There is a daily **maintenance margin** if the price moves from its initial value, and is called marking-to-market: if the price of steel deviates overnight, depending on the direction of the movement, money is either paid into the margin account, or withdrawn from it. Since the price of an asset fluctuates randomly, the value of the futures contract also fluctuates randomly.

When the contract matures the value of the futures contract converges to the value of $S(T)$ due to pressure from arbitrage. If the value of $\mathcal{F}(T, T) > S(T)$, the seller can short the futures contract and buy the stipulated amount from the market at price $S(T)$, thus making a risk-less profit; on the other hand, if $\mathcal{F}(T, T) < S(T)$, companies interested in buying steel will arbitrage by entering into a long futures contract and acquire the asset below market price. Hence

$$\mathcal{F}(T, T) = S(T)$$

During its duration, the futures contract $\mathcal{F}(t, T)$ can be above or below $S(t)$, the spot price for steel as shown in Figure 3.1.

Both the forward and futures contracts are designed to meet the needs of the corporation. Forward contracts are customized to serve the needs of the buyer and seller, whereas futures contracts are standardized, being traded in the exchange market, and hence are more liquid. An important difference between the two is that for the forward contract, there is only one cash flow at the maturity of the contract, whereas for the futures contract there is a series of cash flows generated by the procedure of marking-to-market.

---

[1] In general, there is no need to draw a distinction between the value and price of a financial instrument, and these two terms will usually be used interchangeably.

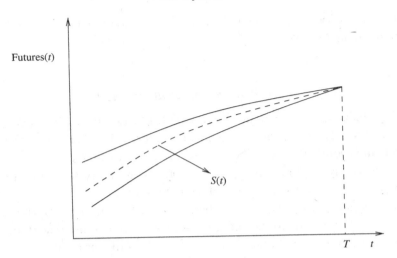

Figure 3.1 Possible futures prices. The dashed line is the price of the security.

3.2 Options

Options have been widely studied in finance [15, 51, 77], and are a fertile ground for the application of quantitative methods [27, 92, 105].

Options are derivatives that can written on any security, including other derivative instruments. In the case of the forward and futures derivatives, the seller is obliged by the contract to take delivery of the asset in question. In contrast, an investor may be more interested in the profit than can be made by entering into a contract, rather than actually possessing the asset.

An option C is a contract to buy or sell (called a call or a put) that is entered into by a buyer and a seller. For a European **call option** the seller of the option is obliged to provide the stock of a company S at some pre-determined price K and at some fixed time in the future; the buyer of the option, on the other hand has the right to either **exercise** or **not exercise** the option. If the price of the stock on maturity is less than K, then clearly the buyer of a call option should not exercise the option. If, however, the price of the stock is greater than K, then the buyer makes a profit by exercising the option. Conversely, the holder of a **put option** has the right to sell or not sell the security at a pre determined price to the seller of the put option.

In general a European **option** is a contract with a fixed maturity, and in which the buyer has the option to either buy or sell a security to the seller of the option at some pre-determined (but not necessarily fixed) strike price [51]. The precise form of the strike price is called the **payoff function** of the option. There are a great variety of options, and these can be broadly classed into **path-independent** and **path-dependent** options.

Most of the options are traded in the derivatives market, which is a growing and a highly diversified market.

3.2.1 Path-independent options

Path-independent options are defined by a payoff function that only depends of the value of the underlying security at the time of maturity. In other words, the payoff function is **independent** of how the security arrives at its final price.

The most widely used path-independent options are the European options, and these come in two varieties, the call and the put options.

Consider an underlying security S. The price of a European **call option** on $S(t)$ is denoted by $C(t) = C(t, S(t))$, and gives the owner of the instrument the option to buy the security at some future time $T > t$ for the strike price of K.

At time $t = T$, when the option matures the value of the call option $C(t, S(t))$ is clearly given by

$$C(T, S(T)) = \begin{cases} S(T) - K, & S(T) > K \\ 0, & S(T) < K \end{cases}$$
$$= g(S)$$

where $g(S)$ is the payoff function.

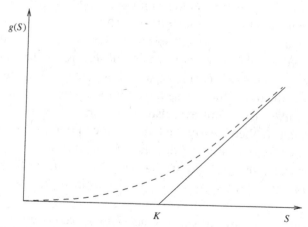

Figure 3.2 Payoff for call option. The dashed line is possible values of the option before maturity.

The price of an European put option, denoted by $P(t)$ is the same as above, except that the holder now has the option to sell a security S at a price of K. Suppose the spot interest rate is given by r, and is a constant. A simple no-arbitrage argument [51] shows that

$$C(t) + Ke^{-r(T-t)} = P(t) + S(t); \quad t \leq T$$

and is called put–call parity.

Clearly, the European call and put options are path independent since the payoff function depends only on the final price of the security.

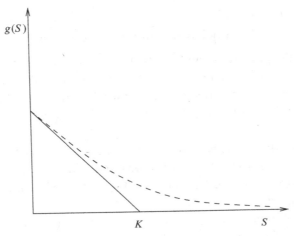

Figure 3.3 Payoff for put option. The dashed line is possible values of the option before maturity.

3.2.2 Path-dependent options

Path-dependent options [51] are defined by payoff functions that depend on the (entire) path that the security takes before the option expires. The most well-known such option is the American option, which is the same as the European option **except** that it can be exercised at any time before the expiry of the contract.

The American option is clearly path dependent, since the choice of an early exercise depends on the value of the security for the entire duration before expiry of the option. No-arbitrage arguments [51] show that for a security that does not pay a dividend, the American call option has the same value as a European call option. On the other hand, an American put option in general has a higher price than a European put option.

Another path-dependent option is the Asian option. It has a payoff function that depends on the average value of the security during the whole period of its duration, namely from the time it is written at time t till the time it expires at T.

A class of path-dependent options are the barrier options, for which a pre-specified barrier is set for the value of the stock price. The knock-out barrier European call option is the same as the European one, except that the barrier option becomes valueless the moment the security price $S(t)$ exceeds the barrier.

There is also a knock-in barrier option similarly defined, and various combinations of knock-in and knock-out options [51].

More exotic options such as the look-back option, quanto option, basket option, hybrid option, dual-strike option and so on are designed to serve the specific needs of investors. There are also OTC (over-the-counter) options that are customized for the specific needs of investors [27,51].

3.3 Stochastic differential equation

Let C stand for the price of a typical option. The fundamental problem of option pricing is the following: given the payoff function $g(S)$ of the option maturing at time T, what should be the price of the option C, at time $t < T$, if the price of the security is $S(t)$? Clearly $C = C(t, S(t))$ with the final value of $C(T, S(T)) = g(S(T))$.

If the payoff function depends only on the value of S at time T, then the pricing of the option $C(t, S(t))$ is a **final value problem** since the final value of C at $t = T$, namely $g(S)$, has been specified, and the value of C at an earlier time t needs to be evaluated.

The present-day price of the option depends on the future value of the security; clearly, the price of the option C will be determined by how the security $S(t)$ evolves to its future value of $S(T)$. In theoretical finance it is common to model the stock price $S(t)$ as a (random) **stochastic process** that is evolved by a **stochastic differential equation** [81,92] given by[2]

$$\frac{dS(t)}{dt} = \phi S(t) + \sigma S R(t) \tag{3.2}$$

where ϕ is the expected return on the security S, σ is its volatility and R is a Gaussian white noise with zero mean. Following the Black–Scholes analysis [15] consider, for now, σ to be a constant. σ is a measure of the randomness in the evolution of the stock price; for the special case of $\sigma = 0$, the stock price evolves deterministically with its future value given by $S(t) = e^{\phi t} S(0)$.

Gaussian white noise is discussed in Appendix A.4. Since white noise is assumed to be independent for each time t, the Dirac delta function correlator is given by[3]

$$E[R(t)] = 0; \quad E[R(t)R(t')] \equiv < R(t)R(t') > = \delta(t - t'). \tag{3.3}$$

---

[2] A stochastic differential equation is known in physics as the Langevin equation [106], and in probability theory as an Ito–Wiener process [65, 104].
[3] The Dirac delta-function is discussed in Appendix A.2, and a brief discussion on the Langevin equation is given in Appendix A.5.

Both the notations $E[X]$ and $<X>$ are used for denoting the expectation value of a random variable X.

White noise $R(t)$ has the following important property. On discretizing time $t = n\epsilon$, $R(t) \to R_n$, the probability distribution function of the white noise is given by

$$P(R_n) = \sqrt{\frac{\epsilon}{2\pi}} e^{-\frac{\epsilon}{2}R_n^2} \tag{3.4}$$

For random variable $R(t)$ it is shown in Appendix A.4, using the equation above, that

$$R_n^2 = \frac{1}{\epsilon} + \text{ random terms of } 0(1) \tag{3.5}$$

In other words, to leading order in ϵ, the square of white noise (random variable), namely $R^2(t)$, is in fact **deterministic**. This property of white noise leads to a number of important results, and goes under the name of Ito calculus in probability theory.

3.4 Ito calculus

The application of stochastic calculus to finance is discussed in great detail in [65], and a brief discussion is given to relate Ito calculus to the Langevin equation. Due to the singular nature of white noise $R(t)$, functions of white noise, such as the security $S(t)$ and the option $C(t)$, have new features. In particular, the infinitesimal behaviour of such functions, as seen in their Taylor expansions, acquire new terms.

Let f be some arbitrary function of white noise $R(t)$. From the definition of a derivative

$$\frac{df}{dt} = \lim_{\epsilon \to 0} \frac{f(t+\epsilon, S(t+\epsilon)) - f(t, S(t))}{\epsilon}$$

or, using Taylors expansion

$$\frac{df}{dt} = \frac{\partial f}{\partial t} + \frac{\partial f}{\partial S}\frac{dS}{dt} + \frac{\epsilon}{2}\frac{\partial^2 f}{\partial S^2}\left[\frac{dS}{dt}\right]^2 + 0(\epsilon^{1/2}) \tag{3.6}$$

The last term in Taylors expansion is order ϵ for smooth functions, and goes to zero. However, due to the singular nature of white noise

$$\left[\frac{dS}{dt}\right]^2 = \sigma^2 S^2 R^2 + 0(1)$$

$$= \frac{1}{\epsilon}\sigma^2 S^2 + 0(1) \qquad \cdot \tag{3.7}$$

Hence, from Eqs. (3.2), (3.6) and (3.7), for $\epsilon \to 0$

$$\frac{df}{dt} = \frac{\partial f}{\partial t} + \frac{\partial f}{\partial S}\frac{dS}{dt} + \frac{\sigma^2}{2}S^2\frac{\partial^2 f}{\partial S^2}$$

$$= \frac{\partial f}{\partial t} + \frac{1}{2}\sigma^2 S^2 \frac{\partial^2 f}{\partial S^2} + \phi S\frac{\partial f}{\partial S} + \sigma S\frac{\partial f}{\partial S}R \qquad (3.8)$$

Suppose $g(t, R(t)) \equiv g_t$ is another function of the white noise $R(t)$. The abbreviated notation $\delta g_t \equiv g_{t+\epsilon} - g_t$ yields

$$\frac{d(fg)}{dt} = \lim_{\epsilon \to 0}\frac{1}{\epsilon}[f_{t+\epsilon}g_{t+\epsilon} - f_t g_t]$$

$$= \lim_{\epsilon \to 0}\frac{1}{\epsilon}[\delta f_t g_t + f_t \delta g_t + \delta f_t \delta g_t]$$

Usually the last term $\delta f_t \delta g_t$ is of order ϵ^2 and goes to zero. However, due to the singular nature of white noise

$$\frac{d(fg)}{dt} = \frac{df}{dt}g + f\frac{dg}{dt} + \frac{df}{\sqrt{dt}}\frac{dg}{\sqrt{dt}} \quad : \text{Ito's chain rule} \qquad (3.9)$$

Since Eq. (3.8) is of central importance for the theory of security derivatives a derivation is given based on Ito calculus. Rewrite Eq. (3.2) in terms differentials as

$$dS = \phi S dt + \sigma S dz \; ; \quad dz = R dt \qquad (3.10)$$

where dz is a Wiener process. Since from Eq. (3.5) $R^2(t) = 1/dt$

$$(dz)^2 = R_t^2(dt)^2 = dt + 0(dt^{3/2})$$

and hence

$$(dS)^2 = \sigma^2 S^2 dt + 0(dt^{3/2})$$

From the equations for dS and $(dS)^2$ given above

$$df = \frac{\partial f}{\partial t}dt + \frac{\partial f}{\partial S}dS + \frac{1}{2}\frac{\partial^2 f}{\partial S^2}(dS)^2 + 0(dt^{3/2})$$

$$= \left(\frac{\partial f}{\partial t} + \frac{1}{2}\sigma^2 S^2 \frac{\partial^2 f}{\partial S^2}\right)dt + \sigma S\frac{\partial f}{\partial S}dz + \phi S\frac{\partial f}{\partial S}dt$$

and Eq. (3.8) is recovered using $dz/dt = R$. Similar to Eq. (3.9), in terms of infinitesimals, the Ito chain rule is given by

$$d(fg) = dfg + fdg + dfdg$$

Stock price a lognormal random variable

To illustrate stochastic calculus, the stochastic differential equation Eq. (3.2) is integrated. Consider the change of variable and the subsequent integration

$$x(t) = \ln[S(t)] \; ; \quad \Rightarrow \frac{dx}{dt} = \phi - \frac{\sigma^2}{2} + \sigma R(t) \tag{3.12}$$

$$\Rightarrow x(T) = x(t) + \left(\phi - \frac{\sigma^2}{2}\right)(T - t) + \sigma \int_t^T dt' R(t') \tag{3.13}$$

The random variable $\int_t^T dt' R(t')$ is a sum of normal random variables and is shown in Eq. (A.29) to be equal to a normal $N(0, \sqrt{T - t})$ random variable. Hence

$$S(T) = S(t) e^{(\phi - \frac{\sigma^2}{2})(T-t) + (\sigma\sqrt{T-t})Z} \text{ with } Z = N(0, 1) \tag{3.14}$$

The stock price evolves randomly from its given value of $S(t)$ at time t to a whole range of possible values $S(T)$ at time T. Since the random variable $x(T)$ is a normal (Gaussian) random variable, the security $S(T)$ is a lognormal random variable. Campbell *et al.* [23] discuss the results of empirical studies on the validity of modelling security S as a lognormal random variable.

Geometric Mean of Stock Price

The probability distribution of the (path-dependent) geometric mean of the stock price can be exactly evaluated. For $\tau = T - t$ and $m = x(t) + \frac{1}{2}(\phi - \frac{\sigma^2}{2})\tau$, Eq. (3.13) yields

$$S_{\text{geometric mean}} = e^G$$

$$G \equiv \frac{1}{\tau} \int_t^T dt' x(t') = m + \frac{\sigma}{\tau} \int_t^T dt' \int_t^{t'} dt'' R(t'')$$

$$= m + \frac{\sigma}{\tau} \int_t^T dt'(T - t')R(t')$$

From Eq. (A.29) the integral of white noise is a Gaussian random variable, which is completely specified by its means and variance. Hence, using $E[G] = m$ and Eq. (3.3) for $E[R(t)R(t')] = \delta(t - t')$ yields

$$E[(G - m)^2] = \left(\frac{\sigma}{\tau}\right)^2 \int_t^T dt'(T - t') \int_t^T dt''(T - t'')E[R(t')R(t'')]$$

$$= \left(\frac{\sigma}{\tau}\right)^2 \int_t^T dt'(T - t')^2 = \frac{\sigma^2 \tau}{3}$$

Hence

$$G = N\left(m, \frac{\sigma^2 \tau}{3}\right) \tag{3.15}$$

The geometric mean of the stock price is lognormal with the same mean as the stock price, but with its volatility being one-third of the stock price's volatility.

3.5 Black–Scholes equation: hedged portfolio

How can one obtain the price of the option C? The option price has to primarily obey the condition of no arbitrage. Black and Scholes made the fundamental observation that if one could **perfectly hedge** an option, then one could price it as well. The reason being that a perfectly hedged portfolio has no uncertainty, and hence has a risk-free rate of return given by the spot interest rate r. In order to form a perfectly hedged portfolio, the time evolution of the option has to be analyzed.

The fundamental idea of Black and Scholes [15,76] is to form a hedged portfolio such that, instantaneously, the change of the portfolio is **independent** of the white noise R. Such a portfolio is perfectly hedged since it has no randomness.

Consider the portfolio

$$\Pi = C - \frac{\partial C}{\partial S} S \qquad (3.16)$$

Π is a portfolio in which an investor holds an option C and short sells $\partial C/\partial S$ amount of security S. Hence, from Eqs. (3.8) and (3.2)[4]

$$\frac{d\Pi}{dt} = \frac{dC}{dt} - \frac{\partial C}{\partial S}\frac{dS}{dt}$$
$$= \frac{\partial C}{\partial t} + \frac{1}{2}\sigma^2 S^2 \frac{\partial^2 C}{\partial S^2} \qquad (3.17)$$

At time t, since $S(t)$ is known, the price $C(t, S(t))$ is deterministic and hence from above the change in the value of the portfolio Π is **deterministic**. Since the random term coming from dS/dt has been removed, due to the choice of the portfolio, $d\Pi/dt$ is consequently free from the risk that comes from the stochastic nature of the security. This technique of cancelling the random fluctuations of one security (in this case of C) by another security (in this case S) is a key feature of hedging.

Since the rate of (change) return on Π is deterministic, it must equal the risk-free return given by the short-term risk-free interest rate r, since otherwise one could arbitrage [51,60]. Hence, based on the absence of arbitrage opportunities the price

---

[4] The term $(\partial^2 C/\partial S\partial t)S$ is considered to be negligible.

of the portfolio has the following evolution

$$\frac{d\Pi}{dt} = r\Pi$$

which yields from Eq. (3.17) the famous Black–Scholes equation [15, 75, 77]

$$\frac{\partial C}{\partial t} + rS\frac{\partial C}{\partial S} + \frac{1}{2}\sigma^2 S^2 \frac{\partial^2 C}{\partial S^2} = rC \tag{3.18}$$

The parameter ϕ of Eq. (3.2) has dropped out of Eq. (3.18) showing that a risk-neutral portfolio Π is independent of the investor's expectation as reflected in ϕ; or, equivalently, the pricing of the security **derivative** is based on a risk-free process that is independent of the investor's risk preferences.

The hedging of an instrument in effect implies that for the hedged portfolio, there exists a risk-neutral evolution of the security $S(t)$ – also called the risk-neutral or risk-free measure.

The Black–Scholes framework for the pricing of options hinges on the concept of a risk-less, hedged portfolio – something that can never be achieved in practice. Generalizations of the Black–Scholes have been made [20, 27, 28], as well as other approaches to option pricing are considered in [17, 60, 64, 93, 94, 98]. The work by Bouchaud and Potters [18] takes a more realistic approach by pricing options using imperfectly hedged portfolios that always have a finite amount of residual risk. A similar approach is taken in the quantum field theory of hedging of Treasury Bonds in Chapter 9 in that there is a finite amount of risk for the hedged bond portfolio, called the residual variance, that cannot be removed.

3.5.1 Assumptions in the derivation of Black–Scholes

The following assumptions were made in the derivation of the Black–Scholes equation.

- The portfolio satisfies the no-arbitrage condition.
- To form the hedged portfolio Π the stock is infinitely divisible, and that short selling of the stock is possible.
- The stock price has a continuous-time evolution. If the stock price follows a more general stochastic process that includes discontinuous jumps, it can be shown that the portfolio cannot be perfectly hedged, and the Black–Scholes analysis is no longer applicable [18, 75].
- The spot interest rate r is constant (this can be generalized to a stochastic spot interest rate).

- The portfolio Π can be re-balanced continuously.
- There is no transaction cost.

The conditions above are not fully met in the financial markets. In particular, transactions costs are significant. In spite of this, the market uses the Black–Scholes option pricing as the industry standard, and which forms the basis for the pricing of more complex options.

3.5.2 Risk-neutral solution of the Black–Scholes equation

There are several ways of solving the Black–Scholes equation. One can directly solve the Black–Scholes equation as a partial differential equation; other approaches will be given in Sections 4.6.1 and 5.2 using techniques based on the Hamiltonian and path integration respectively.

An elegant and simple solution is obtained by using the principle of risk-neutral valuation. Since this principle is used extensively in pricing Treasury Bond options, the price of a European call option is solved for the purpose of illustrating risk-neutral valuation. Recall from Eq. (3.14) that, for remaining time $\tau = T - t$

$$S(T) = e^{x(T)} \;\; ; \;\; x(T) = N\left(\ln S(t) + \left(\phi - \frac{\sigma^2}{2} \right) \tau, \sigma\sqrt{\tau} \right) \tag{3.19}$$

The principle of risk-neutral valuation implies that the present value of the European call option is the expected final value $E[\max(S - K, 0)]$, discounted by the risk-free interest rate. This risk-neutral probability distribution, also called the **martingale measure**, satisfies the martingale condition Eq. (A.40) given by

$$S(t) = E[e^{-r\tau} S(T) | S(t)] = \int_{-\infty}^{+\infty} e^x \, P_m(x) dx \tag{3.20}$$

Using the lognormal probability distribution for the stock price given in Eq. (3.19) to solve the martingale condition yields that $\phi = r$. Hence the martingale probability distribution, for $\ln S(t) = x(t)$ and $x(T) = x$, is given by

$$P_m(x) = \frac{1}{\sqrt{2\pi\sigma^2\tau}} e^{-\frac{1}{2\sigma^2\tau}[x - x(t) - (r - \frac{\sigma^2}{2})\tau]^2} \tag{3.21}$$

ϕ in Eq. (3.19) has been replaced by r to obtain the probability distribution function $P_m(x)$, in accordance with the principle of risk-neutral valuation.

The present-day price of the option is its future value, discounted to the present using the martingale measure. Hence

$$C = e^{-r\tau} E[\max(S - K, 0)] = e^{-r\tau} \int_{\ln K}^{+\infty} (e^x - K) P_m(x) dx \qquad (3.22)$$

The value of the integral (3.22) is derived in Eq. (3.25) below, and the price of the European call option is given by [51]

$$C(\tau, S, K, r) = SN(d_+) - Ke^{-r\tau} N(d_-) \qquad (3.23)$$

where the cumulative distribution for the normal random variable $N(x)$, from Eq. (A.22), is defined by

$$N(x) = \frac{1}{\sqrt{2\pi}} \int_{-\infty}^{x} e^{-\frac{1}{2}z^2} dz \; ; \; d_{\pm} = \frac{\ln\left(\frac{S}{K}\right) + \left(r \pm \frac{\sigma^2}{2}\right)\tau}{\sigma\sqrt{\tau}} \qquad (3.24)$$

How well does the option price given in Eq. (3.23) describe the actual observed option price? The answer is: not very well. The empirical analysis of the Black–Scholes option pricing equation is a vast subject, and is discussed in [27]. There are two important **limitations** of the Black–Scholes option price given in Eq. (3.23).

- The volatility parameter σ needs to be estimated in principle σ is independent of the strike price K, but this is not consistent with the data. Hence, instead of predicting the option price using Eq. (3.23), practitioners and traders fit the observed price of an option, for each strike price K, by adjusting the volatility [53]. This gives a strike-price dependent volatility $\sigma(K)$, called the implied volatility, that is not a constant as predicted by the Black–Scholes analysis, and which in turn is used by the traders as a guide for their trading strategy.
- For the Gaussian distribution, given in Eq. (3.21), the probability that the stock price will have values $x(T)$ different from its value initial value $x(t)$ falls off sharply[5] once the difference is greater than $\sigma\sqrt{T-t}$. However, in practise it is observed that the difference is much greater than is predicted by the Gaussian distribution, and is known as the 'fat tail' phenomenon. Other non-Gaussian distributions, such as the Levy probability distribution, have been suggested to explain this feature of the stock price's evolution [18,68,69].

Black–Scholes price for the European call option

Due to its widespread usage, an explicit derivation is given of the call option price. Let $S(T) = e^x$, $x_0 = \ln S + \tau(r - \frac{\sigma^2}{2})$. The option price, from Eqs. (3.22) and (3.21),

---

[5] The drift term is being ignored as it makes no difference to the argument.

is given by

$$C = e^{-r\tau} \int_{-\infty}^{+\infty} \frac{dx}{\sqrt{2\pi\tau\sigma^2}} (e^x - K)_+ e^{-\frac{1}{2\tau\sigma^2}(x-x_0)^2}$$

$$= e^{-r\tau} \int_{-\infty}^{+\infty} \frac{dx}{\sqrt{2\pi\tau\sigma^2}} (e^{x+x_0} - K)_+ e^{-\frac{1}{2\tau\sigma^2}x^2}$$

$$= e^{-r\tau} \int_{\ln K - x_0}^{+\infty} \frac{dx}{\sqrt{2\pi\tau\sigma^2}} (e^{x+x_0} - K) e^{-\frac{1}{2\tau\sigma^2}x^2}$$

$$= S \left[\int_{\ln K - x_0}^{+\infty} \frac{dx}{\sqrt{2\pi\tau\sigma^2}} e^{-\frac{1}{2\tau\sigma^2}(x+\tau\sigma^2)^2} \right] - e^{-r\tau} K N(d_-)$$

$$= S N(d_+) - e^{-r\tau} K N(d_-) \tag{3.25}$$

3.6 Stock price with stochastic volatility

Consider the more complex case for which the security and its volatility are both stochastic. As in the case of constant volatility, an attempt will be made to form a (risk-less) hedged portfolio for pricing the stock option. However, since the volatility of a stock is not traded in the capital market, there are not enough financial instruments to perfectly hedge the volatility of the stock. The market is hence incomplete and there is no unique risk-neutral measure for evolving stochastic volatility, but, rather, the evolution depends on the risk preferences of the investors.

Various stochastic processes for the volatility of a stock price have been considered. For example, Hull and White [49,50], Heston [45] and others [13,59,78, 85,96] have considered the following process

$$\frac{dV}{dt} = a + bV + \xi V^{1/2} Q \; ; \; \sigma^2 \equiv V$$

where Q is white noise. Baaquie [5], Hull and White [50] and others have considered

$$\frac{dV}{dt} = \mu V + \xi V Q$$

while Stein and Stein [99] consider

$$d\sigma = -\delta(\sigma - \theta)dt + kdz \tag{3.26}$$

where δ and θ are constants representing the mean reversion strength and the mean value of the volatility respectively.

All the processes above except for $(3.26)^6$ are special cases of the following general form [51]

$$\frac{dV}{dt} = \lambda + \mu V + \xi V^\alpha Q \tag{3.27}$$

The choice of λ and μ is restricted by the condition that $V > 0$. The complete coupled process is

$$\frac{dS}{dt} = \phi S dt + S\sqrt{V} R_1 \tag{3.28}$$

$$\frac{dV}{dt} = \lambda + \mu V + \xi V^\alpha R_2 \; ; \; V \equiv \sigma^2 \tag{3.29}$$

where R_1 and R_2 are Gaussian white noises with correlation $-1 \leq \rho \leq 1$

$$<R_1(t)R_1(t') >=< R_2(t)R_2(t') >= \delta(t - t') = \frac{1}{\rho} < R_1(t)R_2(t')> \tag{3.30}$$

and ϕ, λ, μ and ξ are constants. On discretizing time yields, to leading order in ϵ, the following

$$R_1^2(t) = \frac{1}{\epsilon} = R_2^2(t) \; ; \; R_1 R_2(t) = \frac{\rho}{\epsilon} \tag{3.31}$$

The generalization of Eq. (3.8) is obtained by considering an arbitrary function f of the white noise R_1, R_2 for the case of stochastic volatility. This yields

$$\frac{df}{dt} = \frac{\partial f}{\partial t} + \phi S \frac{\partial f}{\partial S} + (\lambda + \mu V)\frac{\partial f}{\partial V} + \frac{\sigma^2 S^2}{2}\frac{\partial^2 f}{\partial S^2} + \rho V^{1/2+\alpha}\xi\frac{\partial^2 f}{\partial S \partial V}$$

$$+ \frac{\xi^2 V^{2\alpha}}{2}\frac{\partial^2 f}{\partial V^2} + \sigma S \frac{\partial f}{\partial S}R_1 + \xi V^\alpha \frac{\partial f}{\partial V}R_2$$

$$= \Theta + \Xi R_1 + \Psi R_2 \tag{3.32}$$

where Eq. (3.32) has been written in a form that separates the stochastic and non-stochastic terms.

3.7 Merton–Garman equation

To obtain the price of the option, similar to the Black–Scholes case, consider two different options, C_1 and C_2 on the same underlying security with strike prices and maturities given by K_1, K_2, T_1 and T_2 respectively. Form a portfolio

$$\Pi = C_1 + \Gamma_1 C_2 + \Gamma_2 S$$

---

[6] This process can be included if a term of the form $\gamma V^{1/2}$ is added to the drift term.

so that

$$\frac{d\Pi}{dt} = \Theta_1 + \Gamma_1\Theta_2 + \Gamma_2\phi + (\Xi_1 + \Gamma_1\Xi_2 + \Gamma_2\sigma S)R_1 + (\Psi_1 + \Gamma_1\Psi_2)R_2$$

As in the case of constant volatility, to ensure perfect hedging the stochastic terms have to be eliminated from the change in the value of the portfolio. Hence, fix Γ_1 and Γ_2 by setting the random terms in the hedged portfolio to zero, and obtain

$$\Xi_1 + \Gamma_1\Xi_2 + \Gamma_2\sigma S = 0$$
$$\Psi_1 + \Gamma_1\Psi_2 = 0$$

which yields

$$\Gamma_1 = -\frac{\Psi_1}{\Psi_2} = -\frac{\partial C_1/\partial V}{\partial C_2/\partial V}$$

$$\Gamma_2 = \frac{\Psi_1}{\Psi_2}\frac{\partial C_2}{\partial S} - \frac{\partial C_1}{\partial S} = \frac{\partial C_1/\partial V}{\partial C_2/\partial V}\frac{\partial C_2}{\partial S} - \frac{\partial C_1}{\partial S}$$

Since the portfolio is now risk-less, it must increase at the risk-free interest rate by the principle of no arbitrage. In other words

$$\frac{d\Pi}{dt} = r\Pi$$

Simplifying Π, and after a **separation of variables**, yields

$$\frac{1}{\partial C_1/\partial V}\left(\frac{\partial C_1}{\partial t} + (\lambda + \mu V)\frac{\partial C_1}{\partial V} + rS\frac{\partial C_1}{\partial S} + \frac{VS^2}{2}\frac{\partial^2 C_1}{\partial S^2}\right.$$
$$\left. + \rho V^{1/2+\alpha}\xi\frac{\partial^2 C_1}{\partial S\partial V} + \frac{\xi^2 V^{2\alpha}}{2}\frac{\partial^2 C_1}{\partial V^2} - rC_1\right)$$
$$= \frac{1}{\partial C_2/\partial V}\left(\frac{\partial C_2}{\partial t} + (\lambda + \mu V)\frac{\partial C_2}{\partial V} + rS\frac{\partial C_2}{\partial S} + \frac{VS^2}{2}\frac{\partial^2 C_2}{\partial S^2}\right. \tag{3.33}$$
$$\left. + \rho V^{1/2+\alpha}\xi\frac{\partial^2 C_2}{\partial S\partial V} + \frac{\xi^2 V^{2\alpha}}{2}\frac{\partial^2 C_2}{\partial V^2} - rC_2\right)$$
$$\equiv \beta(S, V, t, r)$$

β is not a function of K_1, K_2, T_1 or T_2 since the first expression is dependent only on K_1 and T_1 while the second depends only on K_2 and T_2.

The term β is referred to as the market price of volatility risk. This is because the higher is the value of β, the more averse are the investors to volatility risk. The reason this parameter is needed to price options with stochastic volatility, and not for Black–Scholes pricing formula, is that volatility is not traded in the market, and hence there are not enough instruments to perfectly hedge against volatility. The investors' risk preferences, as expected, have to be taken into account when considering stochastic volatility or, in other words, risk-neutral valuation can be applied directly to volatility, but is no longer unique.

The parameter β is difficult to estimate empirically and there is some evidence that it is non-zero [66]. In the Cox, Ingersoll and Ross [28] model, consumption growth has constant correlation with the spot-asset return, and this gives rise to a risk premium that is proportional to volatility. This model is assumed for simplicity since the only the effect it has is of redefining λ in the above equation.

Henceforth, it is assumed that the market price of risk has been included in the Merton–Garman equation by redefining λ. Therefore, the Merton–Garman equation [35, 77] is

$$\frac{\partial C}{\partial t} + rS\frac{\partial C}{\partial S} + (\lambda + \mu V)\frac{\partial C}{\partial V} + \frac{1}{2}VS^2\frac{\partial^2 C}{\partial S^2} + \rho \xi V^{1/2+\alpha} S\frac{\partial^2 C}{\partial S\partial V}$$

$$+ \xi^2 V^{2\alpha}\frac{\partial^2 C}{\partial V^2} = rC \tag{3.34}$$

3.8 Summary

The main forms of derivative instruments were discussed, namely forward, futures and options contracts. Various forms of options were briefly discussed and Ito calculus was reviewed. The concept of a hedged portfolio was introduced to derive the Black–Scholes equation using the Langevin formulation of stochastic differential equations. The ideas developed in the analysis of the Black–Scholes equation were applied to the case of a stock price having stochastic volatility, leading to the Merton–Garman equation for the pricing of an option.

3.9 Appendix: Solution for stochastic volatility with $\rho = 0$

Although the stock price process as given in Eq. (3.28) depends on the volatility V, the process for volatility given in Eq. (3.29) is independent of the stock price S for $\rho = 0$; hence for this case the evolution of volatility can be treated independent of the stock price.

A theorem of Merton [70, 75, 96] states that the solution for a stochastic volatility process is the Black–Scholes price, but with the volatility variable replaced by the average volatility. If the volatility follows, independent of S, the generic process $V(t)$ (where V may be stochastic), the option price is given by the following generalization of Eq. (3.23) ($\tau = T - t$)

$$C = \int_0^\infty [SN(d_+(\bar{V})) - Ke^{-r\tau}N(d_-(\bar{V}))]P_M(\bar{V})\frac{d\bar{V}}{\bar{V}} \tag{3.35}$$

$$\bar{V} = \frac{1}{\tau}\int_t^T dt'V(t')$$

where P_M is the probability distribution function for the mean of the volatility $\bar{V} =$, and $d_{\pm}(\bar{V})$, as in Eq. (3.24), is given by

$$d_{\pm} = \frac{\ln(S/K) + \tau(r \pm \frac{1}{2}\bar{V})}{\sqrt{\bar{V}\tau}}$$

and K is the strike price.

An independent derivation of Eq. (3.35), together with a formal expression for P_M, is given in Eq. (5.115) using the path-integral formulation of option pricing for stochastic volatility.

Consider two simple examples to illustrate this result [70], starting with a deterministic process

$$V = V_0 e^{\mu t}, \ 0 \le t \le T$$

In this case, the probability distribution function of the mean of the volatility is given by $(\tau = T)$

$$V_M = \delta\left(V - V_0 \frac{e^{\mu\tau} - 1}{\mu\tau}\right)$$

giving the Black–Scholes result with σ replaced by $\sqrt{V_0 \frac{e^{\mu\tau}-1}{\mu\tau}}$.

For a stochastic volatility process, choose[7] $\lambda = \mu = \alpha = 0$ in Eq. (3.29) to obtain

$$\frac{dV}{dt} = \xi R(t), \ V(0) = V_0, \ 0 \le t \le T$$

where $R(t)$ represents white noise. The distribution of the mean of V during the time interval $(0, T)$ is given, from Eq. (3.15), by

$$P_M = N\left(V_0, \frac{\xi^2\tau}{3}\right)$$

Hence, the option price is given by

$$f = \sqrt{\frac{3}{2\pi\xi^2\tau}} \int_0^\infty [SN(d_+(V)) - Ke^{-r\tau}N(d_-(V))] \exp\left(-\frac{3(V - V_0)^2}{2\xi^2\tau}\right) \frac{dV}{V}.$$

[7] This is not a realistic process as $P(V < 0) > 0$, while V is obviously non-negative. However, it might be a reasonable approximation for relatively short times for which $P(V < 0)$ is negligible.

Part II

Systems with finite number of degrees of freedom

4

Hamiltonians and stock options

In this chapter the concept of the Hamiltonian is introduced in the pricing of options. Hamiltonians occur naturally in finance; to demonstrate this the analysis of the Black–Scholes equation is recast in the formalism of quantum mechanics. It is then shown how the Hamiltonian plays a central role in the general theory of option pricing.

The Hamiltonian formulation provides new tools for obtaining solutions for option pricing; two key concepts related to the Hamiltonian are (a) eigenfunctions and (b) potentials. Knowledge of all the eigenfunctions of a Hamiltonian yields an exact solution for a large class of path-dependent and path-independent options. For example, barrier options can be modelled by placing constraints on the eigenfunctions of the Hamiltonian. The potentials are a means for defining new financial instruments, and for modelling path-dependent options.

4.1 Essentials of quantum mechanics

It is shown in this chapter that option pricing in finance has a mathematical description that is identical to a quantum system; hence the key features of quantum theory are briefly reviewed.

Quantum theory is a vast subject that forms the bedrock of contemporary physics, chemistry and biology [39]. Only those aspects of quantum mechanics are reviewed that are relevant for the analysis of option pricing.

In classical mechanics the position of a particle at time t, denoted by x_t, is a deterministic function of t, and is given by Newton's law of motion. Classical mechanics is analogous to the case of the evolution of a stock price with zero volatility ($\sigma = 0$) that yields a deterministic evolution of the stock price. In contrast, in quantum mechanics, the particle's evolution is random, analogous to the case of the evolution of a stock price having non-zero volatility ($\sigma \neq 0$).

The quantum particle's **position** at each instant t, namely x_t, is called a **degree of freedom** in physics; the role of time t is that of a parameter that labels the independent random variables x_t for different instants. The allowed configurations for the particle's random position x_t are all of the points on the real line $\mathcal{R}$. The probability distribution of the quantum particle's position x, at some fixed time t, is given by $|\psi(t, x)|^2 \equiv \psi^*(t, x)\psi(t, x)$, where $*$ stands for complex conjugation.

Note the system that is being considered has only **one** degree of freedom, namely x_t, since a degree of freedom is a **random variable** that the system has at a given instant; if one collects all the degrees of freedom (random variables) over time, one would obtain a collection of random variables $\{x_t\}$ that is called a stochastic process in probability theory.

The function ψ is called the state vector of the quantum system, and is an element of a linear vector space – called a **state space** and denoted by $\mathcal{V}$ – that consists of functions of the allowed configurations.[1] The dual space of $\mathcal{V}$ – denoted by $\mathcal{V}_{\text{dual}}$ – consists of all linear mappings from $\mathcal{V}$ to the complex numbers, and is also a linear vector space.

In Dirac's bracket notation for the state vectors, an element of $\mathcal{V}$ is denoted by the ket vector $|g>$ and an element of $\mathcal{V}_{\text{dual}}$ by the dual bra vector $<p|$. The scalar product is defined for any two vectors from the state space and its dual, and is given by the bracket $<p|g> = <g|p>^*$, which is a complex number. Both $\mathcal{V}$ and its dual $\mathcal{V}_{\text{Dual}}$ will be referred to as the state space of the system.

In quantum mechanics, physically measurable quantities such as energy, position and so on are represented by Hermitian operators that map the linear vector space on to itself. For Hermitian operators, the state space $\mathcal{V}$ and its dual $\mathcal{V}_{\text{Dual}}$ are isomorphic.

In Appendix 4.11 the simplest possible quantum system is analyzed, namely a quantum system that has only two possible states, called a two-state system.

In summary the fundamental mathematical structure of quantum mechanics has two **independent** ingredients (a) a linear vector state space and its dual, namely $\mathcal{V}$ and $\mathcal{V}_{\text{dual}}$ and (b) (linear) operators that are linear mappings of the state space $\mathcal{V}$ on to itself. Hence

- A system in quantum mechanics is described by a state vector $|\psi>$ that is an element of a state space $\mathcal{V}$.
- All properties of a quantum system are represented by linear operators acting on the state vector $|\psi>$ of the system.

---

[1] For the case of a quantum particle moving on $\mathcal{R}$, the space $\mathcal{V}$ consists of all possible functions $\psi(x)$, with $x \in \mathcal{R}$ such that $\int_{\mathcal{R}} dx|\psi(x)|^2 = 1$.

4.2 State space: completeness equation

The state space in quantum mechanics, and in option pricing, is one of the fundamental ingredients in the description of a quantum system. A complete description
of the state space consists of enumerating a collection of (basis) vectors so that any
arbitrary vector can be represented as a linear combination of these basis states.
The completeness equation is a statement that one has a complete set of (linearly
independent) basis vectors. Although the analysis of this section may seem formal and mathematical, it is of central importance in many of the derivations and
calculations.

An important conclusion of Appendix 4.11 is to concretely illustrate, in
Eqs. (4.55) and (4.58), the concept of the completeness equation for a two-state
system. For all the applications that will be studied, the 'particle' moves on a continuous line $\mathcal{R}$; each point on the continuous line is a possible state for the system,
and hence the particle requires continuously infinitely many independent basis vectors for its description. The completeness equation of a two state consequently is
generalized to an N-state system, and then the limit of $N \to \infty$ is taken.

Consider an electron moving in space, with its position denoted by x, but with
the restriction that it can only hop on a lattice of discrete points with lattice spacing
of distance a; the lattice points are given by $x = na$. The basis states are labelled
by $|n>$, and can be represented by an infinite column vector with the only non-
zero entry being unity in the nth position. Hence

$$n = 0, \pm 1, \pm 2, \ldots \pm \infty$$

$$|n> = \begin{bmatrix} \ldots \\ 0 \\ 1 \\ 0 \\ \ldots \end{bmatrix} \quad : n\text{th position}\,; \quad <n| = [\ldots 0\ 1\ 0 \ldots]$$

$$<m|n> = \delta_{n-m} \equiv \begin{cases} 1\ n = m \\ 0\ n \neq m \end{cases}$$

$$\sum_{n=-\infty}^{+\infty} |n><n| = \mathcal{J} \quad : \text{completeness equation}$$

where $\mathcal{J}$ above is the infinite-dimensional unit matrix. The completeness is also
referred to as the **resolution of the identity** since only a complete set of basis
states can, taken together, construct the identity operator on state space.

The allowed configurations for the particle are all the various positions $x \in \mathcal{R}$,
and hence the limit of $a \to 0$ needs to be taken. The state vector for the particle
is given by the 'ket vector' $|x>$, with its dual given by the 'bra vector' $<x|$. In

terms of the underlying lattice $(x = na)$

$$|x> = \lim_{a \to 0} \frac{1}{\sqrt{a}} |n> \ ; \ -\infty \le x \le \infty$$

with the scalar product, from Eq. (A.13), given by the Dirac delta function

$$<x|x'> = \delta(x - x') \equiv \begin{cases} \infty & x = x' \\ 0 & x \ne x' \end{cases}$$

The completeness equation is given by

$$\sum_{n=-\infty}^{+\infty} |n><n| \to a \sum_{n=-\infty}^{+\infty} |x><x|$$

$$\Rightarrow \int_{-\infty}^{\infty} dx|x><x| = \mathcal{J} \ : \ \text{completeness equation}$$

where $\mathcal{J}$ is the identity operator on (function) state space.

A more direct derivation of the completeness equation is to consider the scalar product of two functions, namely

$$<\psi|g> \equiv \int dx \psi^*(x) g(x)$$

$$= <\psi| \left\{ \int_{-\infty}^{\infty} dx|x><x| \right\} |g>$$

and this yields the completeness equation

$$\mathcal{J} = \int_{-\infty}^{\infty} dx|x><x| \tag{4.2}$$

The completeness equation given by Eq. (4.2) is a key equation in the analysis of the state space. For the case of two quantum particles with positions x, y, the completeness equation is given by

$$\mathcal{J} = \int_{-\infty}^{\infty} dxdy|x, y><x, y| \tag{4.3}$$

where $|x, y> \equiv |x> \otimes |y>$. The generalization to many quantum particles is straightforward.

The bra and ket vectors $<x|$ and $|x>$ are the basis vectors of the V_{dual} and V respectively. An element of the state space V is the ket vector $|\psi>$, which can be thought of as an infinite-dimensional vector with components given by $\psi(x) = <x|\psi>$. The vector $|\psi>$ in quantum mechanics can be mapped to a unique **dual** vector denoted by $<\psi| \in V_{\text{dual}}$. In components $\psi^*(x) = <\psi|x>$. The vector $|\psi>$ and its dual $<\psi|$ have the important property that they define the 'length' $< \psi | \psi >$ of the vector. The completeness equation Eq. (4.2) yields the following[2]

$$< \psi | \psi > = < \psi \int_{-\infty}^{\infty} | x > < x | \psi >$$

$$= \int_{-\infty}^{\infty} \psi(x)^* \psi(x) \geq 0$$

4.3 Operators: Hamiltonian

An operator is defined as a linear mapping of the state space V on to itself, and is an element of the tensor product space $V \otimes V_{\text{dual}}$. For a two-state system discussed in Appendix 4.11, operators are 2×2 matrices. Consider a state space that consists of all functions of single (real) variable x, namely $V = \{\psi(x) | x \in \Re\}$, where $< x | \psi >= \psi(x)$; operators on this state space are infinite-dimensional generalizations of $N \times N$ matrices, with $N \to \infty$.

One of the most important operators is the co-ordinate operator $\hat{x}$ that simply multiplies $\psi(x) \in V$ by x, that is $\hat{x}\psi(x) \equiv x\psi(x)$. Another important operator is the differential operator $\partial/\partial x$ that maps $\psi(x) \in V$ to its derivative $\partial\psi(x)/\partial x$. All the operators that will be studied are functions of a combination of the operators $\hat{x}$ and $\partial/\partial x$.

Similar to a $N \times N$ matrix M that is fully specified by its matrix elements $M_{ij}, i, j = 1, \ldots, N$, an operator is also specified by its matrix elements. For the operators $\hat{x}$ and $\partial/\partial x$, in the notation of Dirac

$$\hat{x}\psi(x) = x\psi(x)$$
$$\Rightarrow < x|\hat{x}|\psi > = x < x|\psi >= x\psi(x)$$
$$< x|\frac{\partial}{\partial x}|\psi > = \frac{\partial\psi(x)}{\partial x}$$

In other words, the matrix element $< x|\hat{x}|\psi >$ of the operator $\hat{x}$ is given by $x\psi(x)$. Choosing the function $|\psi >= |x' >$ yields

$$< x|\hat{x}|x' > = x < x|x' >= x\delta(x - x')$$

Pursuing the analogy with matrices further, it is known that a matrix M has a Hermitian conjugate defined by $M_{ij}^\dagger \equiv M_{ji}^*$. Similar to a matrix, the Hermitian

---

[2] In quantum mechanics, only the subspace of V consisting of state vectors that have unit norm, defined by $< \psi | \psi >= 1$ is allowed, and is called a Hilbert space. In finance the state space is larger than a Hilbert space since many financial instruments are represented by state vectors, such as the price of a stock given by e^x, that do not have a finite length.

conjugate of an arbitrary operator $\mathcal{O}$ is defined by[3]

$$< f|\mathcal{O}^\dagger|g > \equiv < g|\mathcal{O}|f >^*$$ (4.4)

Furthermore, similar to matrices, the Hermitian adjoint of a sum of operators is given by $(A + B + \ldots)^\dagger = A^\dagger + B^\dagger \ldots$, and that of a product of operators is given by $(AB \ldots)^\dagger = \ldots B^\dagger A^\dagger$.

Hermitian adjoint of $\hat{x}$ and $\partial/\partial x$

The completeness Eq. (4.2) and Eq. (4.4) yield the following expression for $\hat{x}^\dagger$

$$< f|\hat{x}^\dagger|g > \equiv < g|\hat{x}|f >^* = \left[\int_{-\infty}^{\infty} dx < g|x >< x|\hat{x}|f > \right]^*.$$

$$= \left[\int_{-\infty}^{\infty} dx\, x g^*(x) f(x) \right]^* = \int_{-\infty}^{\infty} dx\, x g(x) f^*(x)$$

$$= < f|\hat{x}|g >$$

$$\Rightarrow \hat{x}^\dagger = \hat{x} \; : \; \text{Hermitian}$$

For the differential operator $\partial/\partial x$, from Eq. (4.4), and doing an integration by parts gives

$$< f|\frac{\partial}{\partial x}^\dagger|g > \equiv < g|\frac{\partial}{\partial x}|f >^* = \left[\int_{-\infty}^{\infty} dx < g|x >< x|\frac{\partial}{\partial x}|f > \right]^*$$

$$= \left[\int_{-\infty}^{\infty} dx\, g^*(x) \frac{\partial f(x)}{\partial x} \right]^* = -\left[\int_{-\infty}^{\infty} dx\, \frac{\partial g^*(x)}{\partial x} f(x) \right]^*$$

$$= -\int_{-\infty}^{\infty} dx\, \frac{\partial g(x)}{\partial x} f^*(x) = - < f|\frac{\partial}{\partial x}|g >$$

$$\Rightarrow \frac{\partial}{\partial x}^\dagger = -\frac{\partial}{\partial x} \; : \; \text{anti-Hermitian}$$ (4.5)

The co-ordinate operator $\hat{x}$ is Hermitian. The differential operator is anti-Hermitian, and can be made Hermitian by multiplying it by $i = \sqrt{-1}$, yielding the Hermitian operator $\partial/i\partial x$. Moreover

$$\left(\frac{\partial^2}{\partial x^2} \right)^\dagger = \frac{\partial^2}{\partial x^2} \; : \; \text{Hermitian}$$ (4.6)

An important point for future reference is that in all the matrix elements, for example $< x|\partial/\partial x|f >$, the operators acts on the **left**, that is on the **dual space**. This point occurs in a number of derivations.

The fact the operators and the state space are two independent structures in quantum mechanics can be seen from the properties of the differential operator $\partial/\partial x$. This operator has very different properties acting on a state space that consists of all functions x, with x

---

[3] The reason for studying Hermitian conjugation is because one needs to know the space that an operator acts on, namely whether it acts on $\mathcal{V}$ or on its dual $\mathcal{V}_{\text{dual}}$. For non-Hermitian operators, and these are the ones that occur in finance, the difference is important.

taking values on the real line $\Re$, from the case of the state space consisting of functions of x with a fixed periodicity.

The Hamiltonian operator, denoted by H, evolves the system in time, and hence is the most important operator in option pricing. In finance, the operator H in general is non-Hermitian.

There are special states, called eigenstates, that are of particular importance for all operators. For the co-ordinate operator, Eq. (4.4) can be re-written as

$$\hat{x}|x> = x|x> \tag{4.7}$$

The equation above shows that the state vector $|x>$, under the action of the co-ordinate operator $\hat{x}$, has the special property that it is only multilpied by a real number x. The state vector $|x>$ is called an eigenstate of the co-ordinate operator $\hat{x}$ with real eigenvalue x since $\hat{x}$ is Hermitian. The eigenvalue equation for a **non-Hermitian Hamiltonian** H is given by a generalization of Eqs. (4.59) and (4.7), since for non-Hermitian Hamiltonians the eigenvalues E are **complex**.[4] The equations for the eigenvalues and eigenfunctions are given by

$$H|\psi_E> = E|\psi_E> \quad : \ E \ \text{complex}$$
$$<\tilde{\psi}_E|H = E <\tilde{\psi}_E| \Rightarrow H^\dagger|\tilde{\psi}_E> = E^*|\tilde{\psi}_E>$$
$$<\tilde{\psi}_E|\psi_{E'}> = \frac{1}{\mu(E)}\delta(E-E')$$

where $\mu(E)$ is the density of states for eigenvalue E defined by

$$\mu(E) = \text{trace } \delta(H-E) \equiv \int_{-\infty}^{+\infty} dx <x|\delta(H-E)|x> \tag{4.8}$$

From above equations, since E is complex, for a given E, the left and right eigenfunctions are not dual to one another; in other words, $<\tilde{\psi}_E| \neq <\psi_E|$. To obtain the completeness equation, a **subset** of the eigenenergies, and their corresponding eigenfunctions $|\psi_E>$, have to be selected so that the collection yields a complete basis for the state space; denote this subset by $\mathcal{D}$. The completeness equation is then

$$\int_{\mathcal{D}} dE\mu(E)|\psi_E><\tilde{\psi}_E| = \mathcal{I} \tag{4.9}$$

More explicitly, the completeness equation yields

$$\int_{\mathcal{D}} dE\mu(E)\psi_E(x)\tilde{\psi}_E(x') = <x|\mathcal{I}|x'> = \delta(x-x') \tag{4.10}$$

[4] A solution of a barrier option is given in Appendix 4.13 using a non-Hermitian Hamiltonian

In general the Hamiltonian is an operator $H = H(x, \partial/\partial x)$. The eigenfunction equation, for example, is then written as

$$< x|H\left(x, \frac{\partial}{\partial x}\right)|\psi_E >= H\left(x, \frac{\partial}{\partial x}\right)\psi_E(x) = E\psi_E(x)$$

As mentioned earlier, the Hamiltonian operator, like all other operators, acts on the basis state to the **left**. If, for example, one evaluates $<\psi|H(x, \partial/\partial x)|x>$, one is in effect computing $<x|H^\dagger(x, \partial/\partial x)|\psi>^*$.

4.4 Black–Scholes and Merton–Garman Hamiltonians

The Black–Scholes derivation is reinterpreted in the formalism of quantum mechanics. The time-evolution equation for the Black–Scholes and the Merton–Garman equations is analyzed so as to obtain the underlying Hamiltonians that drive the option prices.

The Black–Scholes equation (3.18) for option price with constant volatility is given by

$$\frac{\partial C}{\partial t} = -\frac{1}{2}\sigma^2 S^2 \frac{\partial^2 C}{\partial S^2} - rS\frac{\partial C}{\partial S} + rC \tag{4.11}$$

Consider the change of variable

$$S = e^x \; ; \quad -\infty \leq x \leq \infty$$

This yields the Black–Scholes–Schrodinger equation

$$\frac{\partial C}{\partial t} = H_{BS}C \tag{4.12}$$

with the Black–Scholes Hamiltonian given by

$$H_{BS} = -\frac{\sigma^2}{2}\frac{\partial^2}{\partial x^2} + \left(\frac{1}{2}\sigma^2 - r\right)\frac{\partial}{\partial x} + r \tag{4.13}$$

Viewed as a quantum mechanical system, the Black–Scholes equation has one degree of freedom, namely x, with volatility being the analog of the inverse of mass, the drift term a (velocity-dependent) potential, and with the price of the option C being the analog of the Schrodinger state function.

Eqs.(4.5) and (4.6) yield

$$H_{BS}^\dagger = -\frac{\sigma^2}{2}\frac{\partial^2}{\partial x^2} - \left(\frac{1}{2}\sigma^2 - r\right)\frac{\partial}{\partial x} + r \neq H_{BS} \tag{4.14}$$

Hence, the Black–Scholes Hamiltonian is non-Hermitian due to the drift term.

In Dirac's notation, the Black–Scholes equation is written as

$$< x|H_{BS}|C >= H_{BS}\left(\frac{\partial}{\partial x}\right) < x|C >= H_{BS}\left(\frac{\partial}{\partial x}\right)C(x) \qquad (4.15)$$

Recall from Eq. (3.34) that the Merton–Garman equation for the price of an option on an equity with stochastic volatility is

$$\frac{\partial C}{\partial t} + rS\frac{\partial C}{\partial S} + (\lambda + \mu V)\frac{\partial C}{\partial V} + \frac{1}{2}VS^2\frac{\partial^2 C}{\partial S^2} + \rho\xi V^{1/2+\alpha}S\frac{\partial^2 C}{\partial S\partial V} + \xi^2 V^{2\alpha}\frac{\partial^2 C}{\partial V^2}$$
$$= rC$$

Since both S and V are positive-valued random variables, define variables x and y by

$$S = e^x, \quad -\infty < x < \infty$$
$$\sigma^2 = V = e^y, \quad -\infty < y < \infty$$

In terms of these variables, the Merton–Garman equation is [5,70]

$$\frac{\partial C}{\partial t} + \left(r - \frac{e^y}{2}\right)\frac{\partial C}{\partial x} + \left(\lambda e^{-y} + \mu - \frac{\xi^2}{2}e^{2y(\alpha-1)}\right)\frac{\partial C}{\partial y} + \frac{e^y}{2}\frac{\partial^2 C}{\partial x^2}$$
$$+ \rho\xi e^{y(\alpha-1/2)}\frac{\partial^2 C}{\partial x\partial y} + \xi^2 e^{2y(\alpha-1)}\frac{\partial^2 C}{\partial y^2} = rC \quad (4.16)$$

The above equation can be re-written as the Merton–Garman–Schrodinger equation given by

$$\frac{\partial C}{\partial t} = H_{MG}C \qquad (4.17)$$

and Eq. (4.16) yields the Merton–Garman Hamiltonian

$$H_{MG} = -\frac{e^y}{2}\frac{\partial^2}{\partial x^2} - \left(r - \frac{e^y}{2}\right)\frac{\partial}{\partial x} - \left(\lambda e^{-y} + \mu - \frac{\xi^2}{2}e^{2y(\alpha-1)}\right)\frac{\partial}{\partial y}$$
$$- \rho\xi e^{y(\alpha-1/2)}\frac{\partial^2}{\partial x\partial y} - \frac{\xi^2 e^{2y(\alpha-1)}}{2}\frac{\partial^2}{\partial y^2} + r \qquad (4.18)$$

The Merton–Garman Hamiltonian is a system with **two** degrees of freedom, and is a formidable one by any standard. The only way of solving it for general α seems to be numerical. The special case of $\alpha = 1/2$ can be solved exactly using techniques of partial differential equations [45], and $\alpha = 1$ will be seen to be soluble using path-integral methods [5].

4.5 Pricing kernel for options

The (risk-free) evolution equation for the option pricing is analyzed, so as to extract – from the pricing formula – the conditional probability that expresses the random evolution of the security in question. The conditional probability is the **pricing kernel** as it carries all the information required to price any path-independent option.

Consider for the sake of generality the random evolution of a stock price having stochastic volatility. Suppose a path-independent option matures at time T with $g(x, y)$ being the payoff function. The price of the option at time $t < T$ needs to be determined.

Let $p(x, y, T - t; x', y')$ be the (risk-neutral) conditional probability that, given security price x and volatility y at time t, it will have a value of x' and volatility y' at time T. The final value condition at $t = T$ is given by Dirac delta functions, namely

$$p(x, y, 0; x', y') = \delta(x' - x)\delta(y' - y) \tag{4.19}$$

The derivative price is given, for $t \leq T$, by the Feynman–Kac formula ($\tau = T - t$) [31]

$$C(\tau; x, y) = \int_{-\infty}^{+\infty} dx' dy' \, p(x, y, \tau; x', y') g(x'; y') \tag{4.20}$$

The expression $p(x, y, \tau; x', y')$ is the **pricing kernel** since it is the kernel of the transformation that evolves the payoff function $g(x', y')$ backwards in time to its current value at time t, and yields the price of the option $C(\tau; x, y)$.

It follows from Eq. (4.19) that, as required, $C(\tau; x, y)$ given in Eq. (4.26) satisfies the final value condition

$$C(0, x, y) = g(x, y) \tag{4.21}$$

If the payoff function has a special form, the pricing kernel can be further simplified. For example, in the case of a stock price with stochastic volatility the payoff function depends only on the stock price, and is independent of final volatility; that is $g(x', y') = g(x')$. Final volatility can be consequently integrated out, and yields

$$C(\tau; x, y) = \int_{-\infty}^{+\infty} dx' p_{MG}(x, y, \tau; x') g(x') \tag{4.22}$$

where the Merton–Garman pricing kernel is given by

$$p_{MG}(x, y, \tau; x') = \int_{-\infty}^{+\infty} dy' p(x, y, \tau; x', y') \tag{4.23}$$

For the simpler Black–Scholes case, only the stock price is evolving, and the pricing kernel yields the price of the options as

$$C(\tau; x) = \int_{-\infty}^{+\infty} dx' p_{BS}(x, \tau; x') g(x')$$
(4.24)

Digital options

The pricing kernel is the price of the difference of two digital options. The payoff function $D(x)$ has a non-zero value only if the stock price has a final value around a narrow range of the stock price, say, the value x_0. More precisely

$$D(x - x_0) = \lim_{\epsilon \to 0} \begin{cases} \frac{1}{\epsilon} & -\epsilon/2 \le (x - x_0) \le \epsilon/2 \\ 0, & \text{Otherwise} \end{cases}$$

$$= \delta(x - x_0)$$

Hence, for the Black–Scholes case

$$C(t; x) = \int_{-\infty}^{+\infty} dx' p_{BS}(x, \tau; x') \delta(x' - x_0)$$

$$= p_{BS}(x, \tau; x_0)$$
(4.26)

In other words, the pricing kernel itself can be thought of as the price of a combination of digital options.

4.6 Eigenfunction solution of the pricing kernel

The pricing kernel is determined by the Hamiltonian. Only the Black–Scholes case will be analyzed, as its extension to many variables is straightforward. From Eq. (4.12)

$$\frac{\partial C}{\partial t} = HC$$
(4.27)

with a formal solution given by

$$C(t, x) = e^{tH} C(0, x)$$
(4.28)

where $C(0, x)$ is the initial condition.

Explicitly putting in the dependence of $C(t, x)$ on the time variable, Eq. (4.27) is, in Dirac's notation

$$< x | \frac{\partial}{\partial t} | C, t > = < x | H | C, t >$$

$$\frac{\partial}{\partial t} | C, t > = H | C, t >$$

$$| C, t >= = e^{tH} | C, 0 >$$

with the final value condition given in Eq. (4.21)

$$|C, T> = e^{TH}|C, 0> = |g>$$
$$\Rightarrow |C, 0> = e^{-TH}|g>$$

Hence

$$|C, t> = e^{-(T-t)H}|g>$$

Remaining time $\tau = T - t$ runs backwards, that is when $\tau = 0$, real time $t = T$ and when $\tau = T$, real time $t = 0$. Hence

$$C(t, x) = <x|C, t> \tag{4.29}$$
$$= <x|e^{-\tau H}|g>$$

and using completeness equation (4.2)

$$C(t, x) = \int_{-\infty}^{\infty} dx' <x|e^{-\tau H}|x'> g(x')$$

yields a formal solution for the pricing kernel in terms of the Hamiltonian given by

$$p(x, \tau; x') = <x|e^{-\tau H}|x'> \tag{4.30}$$

The option price C seems to be unstable in Eq. (4.28), being represented by a growing exponential. However, expressed in terms of remaining time $\tau = T - t$ the option price is given by a decaying exponential as in Eq. (4.30). This is because the boundary condition for C is given at final time T, and Eq. (4.28) is converted to a decaying exponential in terms of remaining time τ.

It can be see from Eq. (4.30) that the pricing kernel is the matrix element of the differential operator $e^{-\tau H}$. The role of the Hamiltonian in option pricing is not to evolve the system forward in time, as is the case in quantum mechanics, but rather to **discount** the future payoff function by evolving it **backwards** in time.

The completeness equation for the eigenfunctions of the Hamiltonian, given in Eqs. (4.9) and (4.10), yields a formal and explicit expression for the pricing kernel since

$$p(x, \tau; x') = <x|e^{-\tau H}|x'>$$
$$= <x|e^{-\tau H} \int_{D} dE\mu(E)|\psi_E> < \tilde{\psi}_E|x'>$$
$$= \int_{D} dE\mu(E)e^{-\tau E}\psi_E(x)\tilde{\psi}_E(x') \tag{4.31}$$

The expression above is useful if one can evaluate all the eigenfunctions of H, and for studying the formal properties of the pricing kernel.

4.6.1 Black–Scholes pricing kernel

To evaluate the price of the European call option with constant volatility, the Feynman–Kac formula yields

$$C(t, x) = \int_{-\infty}^{\infty} dx' < x|e^{-\tau H_{BS}}|x' > g(x')$$

where recall from Eq. (4.13)

$$H_{BS} = -\frac{\sigma^2}{2} \frac{\partial^2}{\partial x^2} + \left(\frac{1}{2}\sigma^2 - r\right)\frac{\partial}{\partial x} + r$$

The Hamiltonian can be used to compute the pricing kernel

$$p_{BS}(x, \tau|x') = < x|e^{-\tau H_{BS}}|x' > \; ; \; \tau = T - t \tag{4.32}$$

Hamiltonian derivation of the Black–Scholes pricing kernel

The first step is to find the eigenfunctions of H_{BS}. This can be done efficiently by going to the 'momentum' basis in which H_{BS} is diagonal. The Fourier transform of the $|x >$ basis to momentum space, from Eq. (A.11), is given by

$$< x|x' >= \delta(x - x') = \int_{-\infty}^{\infty} \frac{dp}{2\pi} e^{ip(x-x')}$$
$$= \int_{-\infty}^{\infty} \frac{dp}{2\pi} < x|p >< p|x' >$$

that yields, for momentum space basis $|p >$ the completeness equation

$$\int_{-\infty}^{\infty} \frac{dp}{2\pi} |p >< p| = \mathcal{J} \tag{4.34}$$

with the scalar product

$$< x|p >= e^{ipx} \; ; \; < p|x >= e^{-ipx}. \tag{4.35}$$

From the definition of the Hamiltonian given in Eqs. (4.13) and (4.15)

$$< x|H_{BS}|p > \equiv H_{BS} < x|p >= H_{BS}e^{ipx}$$
$$= \left\{\frac{1}{2}\sigma^2 p^2 + i\left(\frac{1}{2}\sigma^2 - r\right)p + r\right\} e^{ipx} \tag{4.36}$$

For reference Eq. (4.14) yields[5]

$$< p|H_{BS}|x > = < x|H_{BS}^\dagger|p >^* = [H_{BS}^\dagger e^{ipx}]^*$$
$$= \left\{ \frac{1}{2}\sigma^2 p^2 + i \left(\frac{1}{2}\sigma^2 - r \right) p + r \right\} e^{-ipx}$$

It can be seen from Eq. (4.36) that functions e^{ipx} are eigenfunctions of H_{BS}, labelled by the 'momentum' index p. Eq. (4.34) shows that the eigenfunctions of H_{BS} are complete. Hence

$$p_{BS}(x, \tau; x') = < x|e^{-\tau H_{BS}}|x' > \tag{4.37}$$
$$= \int_{-\infty}^{\infty} \frac{dp}{2\pi} < x|e^{-\tau H_{BS}}|p >< p|x' >$$
$$= e^{-r\tau} \int_{-\infty}^{\infty} \frac{dp}{2\pi} e^{-\frac{1}{2}\tau\sigma^2 p^2} e^{ip(x-x'+\tau(r-\sigma^2/2))} \tag{4.38}$$

Performing the Gaussian integration in Eq. (4.38) above gives the pricing kernel for the Black–Scholes equation

$$p_{BS}(x, \tau; x') \equiv p_{BS}(x, \tau; x'; \sigma) = < x|e^{-\tau H_{BS}}|x' >$$
$$= e^{-r\tau} \frac{1}{\sqrt{2\pi\tau\sigma^2}} e^{-\frac{1}{2\tau\sigma^2}\{x-x'+\tau(r-\sigma^2/2)\}^2} \tag{4.39}$$

A derivation of the Black–Scholes pricing kernel is given in [7] using the method of Laplace transforms.[6]

Equation (4.39) states that x' has a normal distribution with mean equal to $\log(S(t)) + (r - \sigma^2/2)\tau$ and variance of $\sigma^2\tau$, as is expected for the Black–Scholes case with constant volatility.

In general for a more complicated (nonlinear) Hamiltonian, such as the one given in Eq. (4.18) for stochastic volatility, it is usually not possible to exactly diagonalize H, and consequently to exactly evaluate the matrix elements of $e^{-\tau H}$. The Feynman path integral is an efficient theoretical tool for analytic and numerical studies of such nonlinear Hamiltonians.

---

[5] One might be tempted to consider evaluating the matrix element $< p|H_{BS}|x >$ by directly differentiating on $|x >$; but $< p|\partial/\partial x|x > \neq \partial/\partial x < p|x >$ and hence this would give an incorrect answer. The operators $\partial/\partial x$ and H_{BS} are defined by their action on the dual co-ordinate basis $< x|$ and not on the basis $|x >$; for a Hermitian Hamiltonian this distinction is irrelevant since both procedures give the same answer – and hence this issue is ignored in quantum mechanics – but this is not so for the non-Hermitian case. In fact, it is precisely the non-Hermitian drift term that comes out with the wrong sign if one acts on the basis $|x >$ with H_{BS}.

[6] Recall $x' = \log(S(T))$, $x = \log(S(t))$ and $\tau = T - t$, and the earlier derivation of the pricing kernel given in Eq. (3.21) has been obtained using the martingale condition.

4.7 Hamiltonian formulation of the martingale condition

Consider an option on a security $S = e^x$ that matures at time T and has a payoff function given by $g(x)$. As discussed in Eq. (4.26), the risk-free evolution of the security is given by the Hamiltonian H, with the value of the option at time $t < T$ being given by

$$C(t, x) = \int_{-\infty}^{\infty} dx' < x|e^{-(T-t)H}|x' > g(x') \qquad (4.40)$$

The martingale condition for the risk-free evolution of the security is that the price of the security at some future time, say t_*, when discounted by a martingale measure, is equal, on average, to the price of the security at earlier time t. The equation for the martingale condition, from Eq. (A.40), states that

$$S(t) = E\left[e^{-(t_*-t)r} S(t_*)|S(t)\right] \qquad (4.41)$$

and is explicitly expressed in Eq. (3.20). Clearly, the martingale condition is an instantaneous condition since it is valid for any t_*.

From Eq. (4.40), if $g(x) = S(x)$, then evolving this back in time must yield S as required by the martingale condition given in Eq. (4.41). Hence

$$S(x) = \int_{-\infty}^{\infty} dx' < x|e^{-(t_*-t)H}|x' > S(x')$$

Or, in Dirac's notation

$$< x|S > = \int_{-\infty}^{\infty} dx' < x|e^{-(t_*-t)H}|x' >< x'|S > \qquad (4.42)$$

Using the completeness equation for a single security given by

$$\mathcal{I} = \int_{-\infty}^{\infty} dx'|x' >< x'|$$

yields from Eq. (4.42), the (eigenstate) equation

$$|S >= e^{-(t_*-t)H}|S >$$

Since time t_* is arbitrary, the instantaneous expression of the martingale condition is given by [3]

$$H|S >= 0 \qquad (4.43)$$

Hence it can be seen that the security $S = e^x$ is a very special **eigenstate** of H, namely having a zero energy eigenvalue; the equity is an element of the state space that is not normalizable. The equity having zero eigenenergy means, under a martingale evolution driven by H, the underlying security does not change.

One can easily verify that both the Black–Scholes and Merton–Garman Hamiltonians, given in Eqs. (4.13) and (4.18) respectively, satisfy the martingale condition given by Eq. (4.43). For the case of the stock price with stochastic volatility, all the volatility dependent terms in the Merton–Garman Hamiltonian appear only through terms containing $\partial/\partial y$, and these terms automatically annihilate $S(t)$.

The result given in Eq. (4.43) shows that the existence of a martingale measure is equivalent to a risk-free Hamiltonian that annihilates the underlying security S. The existence of a risk-neutral Hamiltonian in turn implies that all the derivatives of the underlying security that are priced with this Hamiltonian are free from arbitrage.

4.8 Potentials in option pricing

From Eq. (4.60) a typical quantum mechanical Hamiltonian is written as

$$H = -\frac{1}{2m}\frac{\partial^2}{\partial x^2} + V(x)$$

In contrast, all the terms in the Black–Scholes and Merton–Garman Hamiltonians depend only on derivatives of x, y; there seems to be no analog of the potential $V(x)$ that depends on the stock price x.

Can one, in principle, include a potential term in the option-pricing Hamiltonian? The answer is yes: one can in fact include a potential term in the Black–Scholes formalism, and the potential can be used to represent a certain class of path-dependent options, as has been considered by Linetsky [67]. In particular, some of the barrier options can be expressed as a problem with a potential.

Path-dependent options, such as the barrier options, are all evolved by the Black–Scholes Hamiltonian H_{BS}; the entire effect of barrier options (discussed in Section 4.9), and of some path-dependent options can be effectively realized by a potential $V(x)$ that is **added** to the Black–Scholes Hamiltonian H_{BS}, and yields an effective Hamiltonian given by

$$H_{\text{eff}} = H_{BS} + V$$

The Lagrangian given in Eq. (5.21) for a class of path-dependent options yields

$$H_{\text{path-dependent option}} = H_{BS} + igf(x)$$

where the function $f(x)$ encodes the path-dependent option. In particular, from Eq. (5.22) for the (path-dependent) Asian option, its non-Hermitian Hamiltonian

is given by

$$H_{\text{Asian option}} = H_{BS} + ige^x$$

and has been discussed in [7].

The Black–Scholes Hamiltonian can be generalized so as to include a security-dependent potential $V(x)$. Recall from Eq. (4.13) that the Black–Scholes Hamiltonian is given by

$$H_{BS} = -\frac{\sigma^2}{2}\frac{\partial^2}{\partial x^2} + \left(\frac{1}{2}\sigma^2 - r\right)\frac{\partial}{\partial x} + r$$

From the derivation of the martingale condition, it can be seen that the form of the Hamiltonian in option pricing is constrained by the requirement of annihilating the stock price $S(t)$, so as to fulfil the martingale condition given in Eq. (4.43). A straightforward generalization of the Black–Scholes Hamiltonian that fulfils the martingale condition is

$$H_V = -\frac{\sigma^2}{2}\frac{\partial^2}{\partial x^2} + \left(\frac{1}{2}\sigma^2 - V(x)\right)\frac{\partial}{\partial x} + V(x) \qquad (4.44)$$

where the potential $V(x)$ is an arbitrary function of x.

One can easily verify that this Hamiltonian H_V annihilates the security $S = e^x$. A security evolving with this Hamiltonian will yield a risk-free measure and hence can be used for pricing options. The interpretation of the potential $V(x)$ is that the security is discounted with a security-dependent discounting factor $\exp\{-\int_t^T V(x(t'))dt'\}$ [7]; for example, for a European option maturing at time T, this discounting factor would give the value of the option at time $t < T$ as

$$E_{[t,T]}\left[e^{-\int_t^T V(x(t'))dt'}(e^x - K)_+\right] \qquad (4.45)$$

with an evolution driven by the Hamiltonian $-(\sigma^2/2)\partial^2/\partial x^2 + [(1/2)\sigma^2 - V(x)]\partial/\partial x$. The discounting is chosen to be equal to the drift term in the Hamiltonian so that the martingale condition can be fulfilled.

The interpretation of the potential $V(x)$ needs to be understood from the point of view of finance. The usual discounting of a security using the spot interest rate r is determined by the argument of no arbitrage involving fixed deposits in the money market account. Whether the discounting by $V(x)$ can be realized by the market, and is consistent with the principles of finance, needs to be studied further.

The non-Hermiticity of H_V is of a particularly simple nature, and it can be shown that, for arbitrary V, H_V is equivalent via a similarity transformation to a

Hermitian Hamiltonian H_{eff}[7] given by [7]

$$H_V = e^s H_{\text{eff}} e^{-s} \qquad (4.46)$$

where

$$H_{\text{eff}} = -\frac{\sigma^2}{2}\frac{\partial^2}{\partial x^2} + \frac{1}{2}\frac{\partial V}{\partial x} + \frac{1}{2\sigma^2}\left(V + \frac{1}{2}\sigma^2\right)^2 \quad ; \quad s = \frac{1}{2}x - \frac{1}{\sigma^2}\int_0^x dy\, V(y)$$

H_{eff} is Hermitian and hence its eigenfunctions form a complete basis; from this it follows that the Hamiltonian H_V can also be diagonalized using the eigenfunctions of H_{eff}. In particular

$$H_{\text{eff}}|\phi_n> = E_n|\phi_n> \quad \Rightarrow \quad H_V|\psi_n> = E_n|\psi_n>$$

where

$$|\psi_n> = e^s|\phi_n> \quad ; \quad <\tilde{\psi}_n| = e^{-s} <\phi_n| \neq <\psi_n|$$

The Black–Scholes Hamiltonian H_{BS} has $V(x) = r$ and hence

$$H_{BS} = e^s H_{\text{eff}} e^{-s} = e^{\alpha x}\left[-\frac{\sigma^2}{2}\frac{\partial^2}{\partial x^2} + \gamma\right]e^{-\alpha x} \qquad (4.47)$$

where

$$\gamma = \frac{1}{2\sigma^2}\left(r + \frac{1}{2}\sigma^2\right)^2 \quad ; \quad \alpha = \frac{1}{\sigma^2}\left(\frac{1}{2}\sigma^2 - r\right) \qquad (4.48)$$

The effective Black–Scholes Hamiltonian will be used to solve the double barrier problem.

4.9 Hamiltonian and barrier options

To illustrate the workings of the Hamiltonian $H_{BS} + V$ for the case of path-dependent options, barrier options are analyzed. For both the cases studied, namely the down-and-out barrier option and the double-knock-out barrier option, the fundamental idea is that since the option becomes worthless the moment the stock price equals the barrier value, the **boundary conditions** are imposed on the **eigenfunctions** – that they must vanish outside the barrier. Potentials are introduced for imposing the appropriate boundary conditions on the eigenfunctions.

---

[7] For more complex Hamiltonians such as the Merton–Garman H_{MG} the equivalent Hermitian Hamiltonian H_{eff} is far from obvious.

4.9.1 Down-and-out barrier option

Consider the down-and-out barrier European call option, which is a European option with the additional constraint that the stock price $S(t)$ must always be greater than a preset barrier e^B. If the stock price equals or drops below the barrier e^B, the option becomes worthless.

The price of the barrier option is determined by the set of paths taken by $S(t)$ such that for all points on the path $S(t) > e^B$. To implement this constraint, for $S = e^x$ introduce the barrier potential $V(x)$

$$V(x) = \begin{cases} \infty, & x \le B \\ r, & x > B \end{cases}$$

and is shown in Figure 4.1.

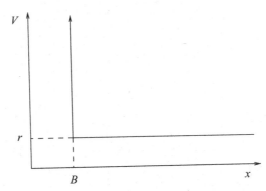

Figure 4.1 The potential $V(x)$ for the down-and-out barrier option

The **Hamiltonian** for the down-and-out barrier option is given by

$$H_{DO} \equiv H_{BS} + V(x)$$
$$= -\frac{\sigma^2}{2} \frac{\partial^2}{\partial x^2} + \left(\frac{1}{2}\sigma^2 - r \right) \frac{\partial}{\partial x} + V(x) \tag{4.49}$$

The potential ensures that only those paths that survive the barrier contribute to the pricing kernel. In effect the potential imposes the boundary condition on all the eigenfunctions $\psi_E(x)$ of H_{BS} that they vanish outside the barrier, namely that $\psi_E(x) = 0$; $x \le B$.

The derivation of the pricing kernel is given in Appendix 4.13 using techniques that directly address the non-Hermitian property of the Black–Scholes Hamiltonian. Let the price of the stock at (remaining time) τ be given by $x > B$, and at $\tau = 0$ be x'. The pricing kernel for the barrier option is then given by Eq. (4.65) as

$$p_{DO}(x, x'; \tau) = \ <x|e^{-\tau H_{DO}}|x'>$$

$$= \begin{cases} p_{BS}(x, \tau; x') - \left(\frac{e^x}{e^B}\right)^{2\alpha} p_{BS}(2B - x, \tau; x'), & x, x' > B \\ 0, & x > B; x' < B \end{cases}$$

(4.50)

For $x, x' > B$ the pricing kernel $p_{DO}(x, x'; \tau)$ is always positive, as indeed it must be, and is set to be zero in the range $x > B, x' < B$.

The pricing kernel for an **up-and-out barrier**, shown in Figure 4.2, for which the option is valid only if $x, x' < B$, has a pricing kernel identical to $p_{DO}(x, x'; \tau)$ for the range of $x, x' < B$ (for which it has positive values), and is zero for $x < B, x' > B$.

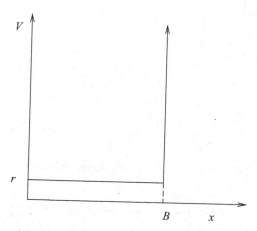

Figure 4.2 Potential barrier for up-and-out barrier option

4.9.2 Double-knock-out barrier option

A double barrier option confines the value of the stock to lie between two barriers, denoted by e^a and e^b as in shown in Figure 4.3, with the value of the option being zero if the stock price takes a value outside the barrier.

The double-knock-out barrier's Hamiltonian is

$$\hat{H}_{DB} = \hat{H}_{BS} + V(x) = e^s[H_{eff} + V(x)]e^{-s}$$

(4.51)

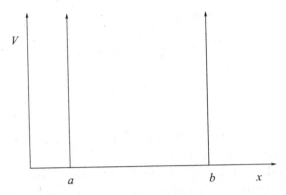

Figure 4.3 Potential barrier for double-knock-out barrier option

The Black–Scholes Hamiltonian, from Eq. (4.47), is given by

$$\hat{H}_{BS} = -\frac{\sigma^2}{2}\frac{\partial}{\partial x^2} + \left(\frac{\sigma^2}{2} - r\right)\frac{\partial}{\partial x} + r$$

$$= e^s H_{eff} e^{-s}$$

$$= e^{\alpha x}\left[-\frac{\sigma^2}{2}\frac{\partial^2}{\partial x^2} + \gamma\right]e^{-\alpha x} \qquad (4.52)$$

and the potential $V(x)$ is

$$V(x) = \begin{cases} \infty & x \le a \\ 0 & a < x < b \\ \infty & x \ge b \end{cases} \qquad (4.53)$$

The potential is accounted for by choosing eigenfunctions that vanish on both of the boundaries. This is the well-known problem of a particle in an infinitely deep quantum well. The Hamiltonian H_{eff} has (normalized) eigenfunctions $|\phi_n >$, **vanishing** at both boundaries, with eigenenergies $E_n + \gamma$ given by

$$\phi_n(x) = < x|\phi_n >= \sqrt{\frac{2}{b-a}}\sin[p_n(x-a)]$$

$$p_n = \frac{n\pi}{b-a}\ ;\ E_n = \frac{\sigma^2}{2}(p_n)^2\ ;\ n = 1, 2, \ldots \infty$$

Since energy eigenvalues for the double barrier option are discrete, the completeness equations, (4.9) and (4.10), yield, as shown in Appendix 4.14

$$\sum_{n=1}^{+\infty} |\phi_n >< \phi_n| = \mathcal{J} \ \Rightarrow\ \sum_{n=1}^{+\infty} < x|\phi_n >< \phi_n|x' >= \delta(x - x')$$

Hence the pricing kernel is given by

$$< x|e^{-\tau H_{DB}}|x' > = < x|e^{s}e^{-\tau[H_{\text{eff}}+V]}e^{-s}|x' >$$

$$= e^{-\tau\gamma}e^{\alpha(x-x')}\sum_{n=1}^{\infty}e^{-\tau E_n}\phi_n(x)\phi_n(x') \qquad (4.54)$$

The explicit answer for the pricing kernel and the price of a barrier European call option is worked out in Appendix 4.14.

The pricing of other barrier options such as the down-and-in, up-and-out, the soft barrier options, exotic options and so on can be defined by appropriate choices of $V(x)$.

4.10 Summary

The formalism of quantum mechanics was reviewed, and the problem of option pricing was seen to have a natural representation in the language of quantum mechanics. In particular, the Hamiltonian driving the evolution of the option price for the case of constant and stochastic volatility was obtained. It was shown that the pricing kernel, which in turn is determined by the Hamiltonian, contains all the information required to price a path-independent option.

The martingale condition was given a Hamiltonian formulation, and the Black–Scholes Hamiltonian was generalized by including a potential in a manner that automatically satisfies the martingale condition. It was shown how a class of options can be represented by a adding a security-dependent potential to the Black–Scholes Hamiltonian. It was further demonstrated how various barrier options can be modelled by imposing the requisite boundary conditions on the eigenfunctions of the Black–Scholes Hamiltonian, and the pricing kernels for a number of barrier options was thus obtained.

The Hamiltonian and the completeness equation are two essential ingredients in making the transition from the partial differential equation formulation of the option pricing to its path-integral representation; this transition is discussed in detail in the next chapter.

4.11 Appendix: Two-state quantum system (qubit)

For simplicity, consider a quantum system that has only two possible states, say an electron with its spin pointing up or down along some fixed axis. Since there are only two possible configurations for the system, one can choose the following two-dimensional vectors as the **basis states** of the system, namely

$$|up> = \begin{bmatrix} 1 \\ 0 \end{bmatrix}; \quad |down> = \begin{bmatrix} 0 \\ 1 \end{bmatrix}$$

$$<up| = \begin{bmatrix} 1 & 0 \end{bmatrix}; \quad <down| = \begin{bmatrix} 0 & 1 \end{bmatrix}$$

$$<up|up> = 1 = <down|down>$$

$$<up|down> = 0 = <down|up>$$

A key feature of the basis states is that they are **complete**, that is, they span the entire state space $\mathcal{V}$. In general, the **completeness equation** is given by taking the **tensor product** of the basis state with its dual state and summing over all the basis states. For the two-state system this yields the following

$$|up><up| + |down><down| = \begin{bmatrix} 1 \\ 0 \end{bmatrix}\begin{bmatrix} 1 & 0 \end{bmatrix} + \begin{bmatrix} 0 \\ 1 \end{bmatrix}\begin{bmatrix} 0 & 1 \end{bmatrix}$$

$$= \begin{bmatrix} 1 & 0 \\ 0 & 1 \end{bmatrix} = \mathcal{I} \tag{4.55}$$

Due to the completeness equation any two-dimensional vector can be represented by a linear combination of the $|up>$ and $|down>$ vectors. That is, an arbitrary state of the electron's spin is given by

$$|\psi> = a|up> + b|down> \quad ; \quad a, b : \text{complex numbers}$$

The interpretation of the state vector $|\psi>$ is that the electron's spin has the probability $|a|^2$ of being in the $|up>$ state and probability $|b|^2$ of being in the $|down>$ state. State $|\psi>$ is also known as a qubit in the field of quantum computation.

Since total probability must equal unity

$$<\psi|\psi> = |a|^2 + |b|^2 = 1 \tag{4.56}$$

From Eq. (4.56), since a, b are complex numbers, it can be seen that the state space $\mathcal{V}$ of the qubit is isomorphic to a three-dimensional sphere $\mathcal{S}^3$; there is however a redundancy in this description since the qubit can be re-scaled by a constant phase without there being any change in its description; in other words states linked by a global phase, namely $|\psi> \rightarrow e^{i\phi}|\psi>$, are equivalent. The phase forms a space isomorphic to a circle $\mathcal{S}^1$. Hence, one needs to 'divide out' $\mathcal{S}^3$ by $\mathcal{S}^1$ to form equivalence classes of qubits, which constitute the state space of the qubit [101].

The state space of a qubit is $\mathcal{V} \equiv \mathcal{S}^3/\mathcal{S}^1 = \mathcal{S}^2$; this space is called the Bloch sphere, and is equal to the two-dimensional sphere $\mathcal{S}^2$. To prove this result, one needs to construct $\mathcal{S}^3$ by a Hopf fibration, using the mathematics of fibre bundles, by fibrating the base manifold $\mathcal{S}^2$ with fibres given by $\mathcal{S}^1$.

In summary, every point of the state space $\mathcal{V} = \mathcal{S}^2$ corresponds to a unique two-dimensional (state) vector of the two-state system.

The two-state Hermitian Hamiltonian for a typical system is given by a 2×2 Hermitian matrix

$$H = \begin{bmatrix} \alpha & \beta \\ \beta^* & \gamma \end{bmatrix} = H^\dagger \; ; \; \alpha, \gamma : \text{real} \; ; \; \beta : \text{complex} \qquad (4.57)$$

In particular, the matrix elements of H are hence given by

$$< \text{up}|H|\text{up} >= \alpha \; ; \quad < \text{down}|H|\text{down} >= \gamma$$
$$< \text{up}|H|\text{down} >= \beta \; ; \quad < \text{down}|H|\text{up} >= \beta^*$$

The (normalized) eigenstates of the two-state Hamiltonian are given by

$$|\psi_+ > = N_+ \begin{bmatrix} 1 \\ \frac{\gamma - E_+}{\beta^*} \end{bmatrix} \; ; \; |\psi_- >= N_- \begin{bmatrix} \frac{\gamma - E_-}{\beta^*} \\ 1 \end{bmatrix}$$

$$E_\pm = \frac{1}{2}(\alpha + \gamma) \pm \sqrt{\frac{1}{4}(\alpha - \gamma)^2 + |\beta|^2} : \text{real}$$

$$N_\pm = \sqrt{\frac{|\beta|^2}{(\gamma - E_\pm)^2 + |\beta|^2}}$$

$$< \psi_\pm|\psi_\pm > = 1 \; ; \quad < \psi_+|\psi_- >= 0$$

The completeness equation, analogous to Eq. (4.55), is given by

$$|\psi_+ >< \psi_+| + |\psi_- >< \psi_-| = \begin{bmatrix} 1 & 0 \\ 0 & 1 \end{bmatrix} = \mathcal{J} \qquad (4.58)$$

4.12 Appendix: Hamiltonian in quantum mechanics

The most important operator in quantum mechanics, and in option pricing – since it evolves the system in time – is the **Hamiltonian**, the energy operator, denoted by H. The matrix elements of H are complex numbers; more precisely, its matrix elements between an arbitrary vector $|g >$ and a dual vector $< f|$, given by $<f|H|g>$, is a complex number. In particular, for a Hermitian Hamiltonian H_R

$$< f|H_R|g >^* \equiv < g|H_R^\dagger|f >$$
$$= < g|H_R|f > \; : \; \text{Hermitian}$$

The time evolution of the state function $|\psi(t) >$ is given by the Schrodinger equation

$$\frac{\partial}{\partial t}|\psi(t) >= -\frac{i}{\hbar}H_R|\psi(t) > \; ; \; |\psi(0) > \; : \; \text{specified}$$

where $i = \sqrt{-1}$, $\hbar$ is Planck's constant, and $|\psi(t)>$ is the Schrodinger wavefunction or state function. It is seen from above that the Hamiltonian is the infinitesimal generator of time translations. The Schrodinger equation is an initial value problem, with the value of the state function $|\psi(t)>$ specified at $t = 0$, and its future behaviour being determined by the evolution equation.

In quantum mechanics, all physical systems are described by Hermitian Hamiltonians H_R. There are special quantum states, called energy **eigenstates**, with real energy **eigenvalues**, that form a **complete set of states**,[8] and are given by

$$H_R|\psi_E> = E|\psi_E> \quad ; \quad E^* = E$$
$$<\psi_E|H_R^\dagger = <\psi_E|H_R = <\psi_E|E$$
$$\int_E dE\mu(E)|\psi_E><\psi_E| = \mathcal{J} \tag{4.59}$$

where the density of states is defined in Eq. (4.8). From the above equations, since the eigenenergy E is real, the right and left eigenstates of H_R are dual to each other; this is not the case for Hamiltonians in finance that as a rule are non-Hermitian.

The time evolution for an energy eigenstate is particularly simple, and is given by

$$|\psi_E(t)> = e^{-iEt/\hbar}|\psi_E>$$

The Hamiltonian for a quantum particle with mass m, moving in a potential $V(x)$, is given by

$$H_R\left(x, \frac{\partial}{\partial x}\right) = -\frac{\hbar^2}{2m}\frac{\partial^2}{\partial x^2} + V(x) = H_R^\dagger \tag{4.60}$$

The derivative term in H_R is a kinetic term since it constrains any change in positions x of the particle; for this reason the Hermitian operator $\hbar\partial/i\partial x$ is referred to as the 'velocity' of the particle; the term $V(x)$ is the potential as it affects the energy of the particle as a function of its position.

4.13 Appendix: Down-and-out barrier option's pricing kernel

The pricing kernel is solved by directly analyzing the non-Hermitian Hamiltonian and illustrates the new features that arise in a non-Hermitian Hamiltonian. From

---

[8] One of the most important property of a Hermitian operator is that its eigenfunctions form a complete basis.

(4.49) the Hamiltonian for the down and out options is given by

$$H_{DO} \equiv H_{BS} + V(x)$$

$$= \frac{\sigma^2}{2} \frac{\partial^2}{\partial x^2} + \left(\frac{1}{2}\sigma^2 - r\right) \frac{\partial}{\partial x} + V(x)$$

The eigenfunctions of H_{DO}, namely $\psi_E(x)$ satisfy

$$x > B :$$
$$H_{DO}|\psi_E >= E|\psi >$$
$$< \tilde{\psi}_E|H_{DO} = E < \tilde{\psi}_E|$$
$$x \le B :$$
$$|\psi_E >= 0 =< \tilde{\psi}_E|$$

Note $\tilde{\psi}_E(x) \ne \psi_E^*(x)$ since $H_{DO} \ne H_{DO}^\dagger$. To meet the requirement of the boundary condition, one constructs eigenfunctions similar to the Black–Scholes case, except that these must satisfy the condition $\psi_E(B) = 0$.

Define the quantities

$$i\lambda_\pm = \alpha \pm ip$$

$$p = \sqrt{\frac{2E}{\sigma^2} - \beta}$$

$$\alpha = \frac{\sigma^2/2 - r}{\sigma^2} \; ; \; \beta = \frac{(\sigma^2/2 + r)^2}{\sigma^4}$$

The eigenfunctions for the down-and-out barrier option are given by

$$x > B$$
$$< x|\psi_E > = e^{i\lambda_+(x-B)} - e^{i\lambda_-(x-B)} \tag{4.61}$$
$$= 2i e^{\alpha(x-B)} \sin\left[p(x - B)\right]$$
$$< \tilde{\psi}_E|x > = e^{-i\lambda_+(x-B)} - e^{-i\lambda_-(x-B)} \tag{4.62}$$
$$= -2i e^{-\alpha(x-B)} \sin\left[p(x - B)\right]$$
$$< \tilde{\psi}_E|\psi_{E'} > = \left[2\pi\sigma^2 \sqrt{2E/\sigma^2 - \beta}\right] \delta(E - E')$$
$$x \le B$$
$$< x|\psi_E > = 0 =< \tilde{\psi}_E|x >$$

The physical interpretation of the wavefunction $\psi_E(x)$, as shown in Figure 4.4, is that a travelling wave $e^{i\lambda_+(x-B)}$ comes in moving from right to the left, hits the the barrier at $x = B$ and is reflected, causing a phase shift from λ_+ to λ_-, and propagates back towards the right. A similar interpretation, with the phases and signs switched, can be given for $\tilde{\psi}_E(x)$.

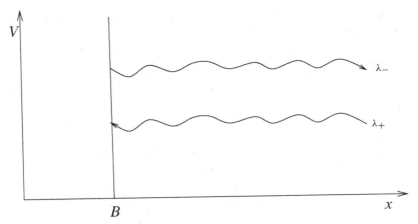

Figure 4.4 Reflection of $\psi_E(x)$ at the barrier

There is no time dependence in the problem, and one should think of the incident and reflected waves as a stationary problem with a steady flux of particles being incident and reflected at the barrier.

The completeness equation is given by

$$\int_{\sigma^2\beta/2}^{\infty} \frac{dE}{2\pi\sigma^2\sqrt{\frac{2E}{\sigma^2}-\beta}} < x|\psi_E >< \tilde{\psi}_E|x' >= \delta(x-x') \qquad (4.63)$$

Although the eigenvalues E can take any complex value, in writing the completeness equation the domain $\mathcal{D}$ of the eigenvalues E has been taken to be only the real eigenvalues, with $\sigma^2\beta/2 \le E \le +\infty$.

To prove the completeness equation one has

$$dp = dE \frac{1}{\sigma^2\sqrt{\frac{2E}{\sigma^2}-\beta}} \; ; \; 0 \le p \le \infty$$

Hence, from above and Eqs. (4.61) and (4.62), the left-hand side of Eq. (4.63) is given by

$$e^{\alpha(x-x')} \int_0^{\infty} \frac{dp}{2\pi} \left[e^{ip(x-x')} + e^{-ip(x-x')} - e^{ip(x+x'-2B)} - e^{-ip(x+x'-2B)} \right]$$

$$= e^{\alpha(x-x')} \int_{-\infty}^{\infty} \frac{dp}{2\pi} \left[e^{ip(x-x')} - e^{ip(x+x'-2B)} \right]$$

$$= \delta(x-x') + e^{\alpha(x-x')}\delta(x+x'-2B)$$

$$= \delta(x-x') \text{ since } x, x' > B \qquad (4.64)$$

The eigenfunctions can be used to evaluate the pricing kernel since they satisfy the completeness equation. Working with the p variable, for remaining time

$\tau = T - t$ and $E = \sigma^2(p^2 + \beta)/2$, performing the Gaussian integration and after some simplifications

$$
\begin{aligned}
p_{DO}(x\tau; x') &= < x|e^{-\tau H_{DO}}|x' > \\
&= e^{-\frac{\tau\beta\sigma^2}{2} + \alpha(x-x')} \int_0^\infty \frac{dp}{2\pi} e^{-\frac{1}{2}\tau\sigma^2 p^2} \\
&\quad \times \left[e^{ip(x-x')} + e^{-ip(x-x')} - e^{ip(x+x'-2B)} - e^{-ip(x+x'-2B)} \right] \\
&= p_{BS}(x, \tau; x') - \frac{1}{\sqrt{2\pi\tau\sigma^2}} e^{-\frac{\tau\beta\sigma^2}{2} - \alpha(x-x')} e^{-\frac{1}{2\tau\sigma^2}(x+x'-2B)^2} \\
&= p_{BS}(x, \tau; x') - \left(\frac{e^x}{e^B} \right)^{2\alpha} p_{BS}(2B - x, \tau; x') \ ; \ x, x' > B
\end{aligned}
$$

$$(4.65)$$

where, recall from Eq. (4.39) that

$$
\begin{aligned}
p_{BS}(x, \tau; x') &= < x|e^{-\tau H_{BS}}|x' > \\
&= e^{-r(T-t)} \frac{1}{\sqrt{2\pi\tau\sigma^2}} e^{-\frac{1}{2\tau\sigma^2}\{x-x'+\tau(r-\sigma^2/2)\}^2}
\end{aligned}
$$

The result Eq. (4.65) for the pricing kernel of the down-and-out option has been obtained in [105] using a method of images.

The result for the pricing kernel for the down-and-out option is what one intuitively expects; its price should be less than the corresponding European call option since its value is non-zero over a more restricted set of paths for the security. The potential $V(x)$ implements the constraint on the allowed paths for the barrier option by **restricting** the eigenstates of H_{DO} to be zero outside the barrier.

Similar to the reflection interpretation of the eigenfunctions, one can also give a similar interpretation to the pricing kernel. Every path that is eliminated from the allowed paths for the pricing kernel has to **hit** the barrier, and hence can be put into a one-to-one correspondence with a path that originates in the forbidden domain of $x < B$. The way one does this, as shown in Figure 4.5, is to associate with every path that originates at $x > B$, and hits the barrier for the first time at some point, a '**mirror image**' that reaches x', but which originates from forbidden region of $2B - x < B$, and crosses the barrier at the same point in time as the eliminated path. These mirror-image paths completely account for all the forbidden paths, including paths with multiple crossings of the barrier.

The pricing kernel for the barrier option is then given by the unrestricted Black–Scholes pricing kernel, but now with a **subtraction** performed for all the paths that are eliminated. The pre-factor of $(e^x/e^B)^{2\alpha}$ in the subtraction comes from the drift of the stock price e^x.

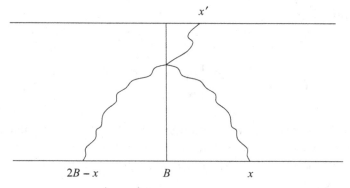

Figure 4.5 Pricing kernel from mirror image

4.14 Appendix: Double-knock-out barrier option's pricing kernel

A double barrier option is an option whose value is non-zero only if the price of the underlying instrument lies within the lower and upper barriers, which are denoted by e^a and e^b respectively [7].

The price, at time t, of a double-knock-out barrier European call option expiring at time T and with strike price K – provided it has not already been knocked out – is given by

$$e^{-r(T-t)} E\left[(e^{x(T)} - K)_+ \mathbf{1}_{a<x(t')<b,\ t<t'<T} \right] \tag{4.66}$$

where $\mathbf{1}$ stands for the indicator function that is non-zero only if the path is within the barrier. The probability distribution of $x(T)$ needs to be obtained for only those paths that do not go outside the barriers.

The restriction on the paths is equivalent to infinite potential barriers since they effectively prohibit the paths from entering the forbidden region (outside the barriers). Hence, the problem can be solved by using the Black–Scholes Hamiltonian with the barrier as a boundary condition on the eigenfunctions of the Hamiltonian.

In the Schrodinger formulation, the problem is to find the pricing kernel for a system with the Hamiltonian

$$\hat{H} = \hat{H}_{BS} + V(x) \tag{4.67}$$

where the Black–Scholes Hamiltonian is given by

$$\hat{H}_{BS} = -\frac{\sigma^2}{2} \frac{\partial}{\partial x^2} + \left(\frac{\sigma^2}{2} - r \right) \frac{\partial}{\partial x} + r \tag{4.68}$$

and the potential $V(x)$ is given by

$$V(x) = \begin{cases} \infty & x \le a \\ 0 & a < x < b \\ \infty & x \ge b \end{cases} \qquad (4.69)$$

The problem has been shown in Section 4.9 to be identical to that of a quantum mechanical particle of mass $1/\sigma^2$ (in units where $\hbar = 1$) in an infinite potential well. For the allowed momenta $p_n = n\pi/(b-a)$, the eigenfunctions are orthonormal since

$$\langle \phi_n \mid \phi_{n'} \rangle = \frac{2}{b-a} \int_a^b \sin p_n(x-a) \sin p_{n'}(x-a) dx = \delta_{n-n'} \qquad (4.70)$$

and form a complete basis since

$$\sum_{n=1}^{\infty} \langle x \mid \phi_n \rangle \langle \phi_n \mid x' \rangle = \frac{2}{b-a} \sum_{n=1}^{\infty} \sin p_n(x-a) \sin p_n(x'-a)$$

$$= \frac{1}{2(b-a)} \sum_{n=-\infty}^{\infty} \left(\exp \frac{in\pi}{b-a}(x-x') - \exp \frac{in\pi}{b-a}(x+x'-2a) \right) \qquad (4.71)$$

$$= \frac{\pi}{b-a} \left(\delta\left(\frac{\pi(x-x')}{b-a} \right) - \delta\left(\frac{\pi(x+x'-2a)}{b-a} \right) \right)$$

$$= \delta(x-x') \text{ given that } a < x, x' < b$$

The pricing kernel is hence, from Eq. (4.54), given by

$$< x|e^{-\tau H}|x' > = e^{-\gamma\tau+\alpha(x-x')} \sum_{n=1}^{\infty} e^{-\tau E_n} < x|\phi_n >< \phi_n|x' >$$

$$= \frac{1}{2(b-a)} e^{-\gamma\tau+\alpha(x-x')} \sum_{n=-\infty}^{\infty} e^{-\frac{\tau\sigma^2 p_n^2}{2}} \left(e^{ip_n(x-x')} - e^{ip_n(x+x'-2a)} \right) \qquad (4.72)$$

In the equation above remaining time τ appears inside the summation in a manner unsuitable for studying the limit of $\tau \to 0$. The inversion $\tau \to 1/\tau$ can be done using the Poisson summation formula

$$\delta(y-n) = \sum_{n=-\infty}^{\infty} e^{2\pi iny} \qquad (4.73)$$

and yields

$$\langle x|e^{-\tau H}|x'\rangle$$

$$= \frac{1}{2(b-a)} \exp\left(-\gamma\tau + \alpha(x-x')\right)$$

$$\times \sum_{n=-\infty}^{\infty} \int_{-\infty}^{+\infty} dy\,\delta(y-n)\exp\left(-\frac{y^2\pi^2\tau\sigma^2}{2(b-a)^2}\right)$$

$$\times \left(\exp\frac{iy\pi(x-x')}{b-a} - \exp\frac{iy\pi(x+x'-2a)}{b-a}\right)$$

$$= \sqrt{\frac{1}{2\pi\tau\sigma^2}} \exp\left(-\gamma\tau + \alpha(x-x')\right)$$

$$\times \sum_{n=-\infty}^{\infty}\left(\exp-\frac{(x-x'+2n(b-a))^2}{2\tau\sigma^2} - \exp-\frac{(x+x'-2a-2n(b-a))^2}{2\tau\sigma^2}\right)$$

where recall from Eq. (4.48) that

$$\gamma = \frac{(r+\sigma^2/2)^2}{2\sigma^2} \quad ; \quad \alpha = \frac{\sigma^2/2 - r}{\sigma^2}$$

The pricing kernel (apart from the drift terms) is given by an infinite sum of Gaussian distributions. The result can be checked by verifying that it reduces to the two solved cases, namely that of a single barrier and the Black–Scholes case of no barrier.

The limits $b \to \infty$ with a finite a, and $a \to -\infty$ with b finite are taken. In the former case, only the $n = 0$ term contributes and in the latter, only the $n = 0$ and $n = 1$ terms contribute. It is easy to see that, in both cases, the result reduces to the solution for the single-knock-out barrier propagator. When both limits are simultaneously active, only the first term in the $n = 0$ term exists and it is easily seen that it gives rise to the Black–Scholes result.

The price of a double barrier European call option can now be evaluated using the pricing kernel from (4.72). The result is seen to be [7]

$$C(S, K; \tau) = \sum_{n=-\infty}^{\infty}\left(e^{-2n\alpha(b-a)}\left(e^{2n(b-a)}SN(d_{n1}) - Ke^{-r\tau}N(d_{n2})\right)\right.$$

$$\left. - S^{2\alpha}e^{-2\alpha(n(b-a)-a)}\left(e^{2n(b-a)}\frac{e^{2a}}{S}N(d_{n3}) - Ke^{-r\tau}N(d_{n4})\right)\right)$$

$$(4.74)$$

where

$$d_{n1} = \frac{\ln\left(\frac{S}{K}\right) + 2n(b-a) + \tau\left(r + \frac{\sigma^2}{2}\right)}{\sigma\sqrt{\tau}}$$

$$d_{n2} = \frac{\ln\left(\frac{S}{K}\right) + 2n(b-a) + \tau\left(r - \frac{\sigma^2}{2}\right)}{\sigma\sqrt{\tau}} = d_{n1} - \sigma\sqrt{\tau}$$

$$d_{n3} = \frac{\ln\left(\frac{e^{2a}}{SK}\right) + 2n(b-a) + \tau\left(r + \frac{\sigma^2}{2}\right)}{\sigma\sqrt{\tau}}$$

$$d_{n4} = \frac{\ln\left(\frac{e^{2a}}{SK}\right) + 2n(b-a) + \tau\left(r - \frac{\sigma^2}{2}\right)}{\sigma\sqrt{\tau}} = d_{n3} - \sigma\sqrt{\tau} \qquad (4.75)$$

and agrees with the result given in [41].

4.15 Appendix: Schrodinger and Black–Scholes equations

The following are some of the general properties of Black–Scholes and Schrodinger equation.

- In quantum mechanics, the position of a quantum particle is a random variable. In finance, the price of a security is a random variable.
- The price of the option $|C>$ is analogous to the state function $|\psi(t)>$; however, unlike $|\psi(t)>$, the option price $|C>$ is directly observable and does not need a probabilistic interpretation. Hence, unlike the condition $<\psi|\psi>=1$ required by the probabilistic interpretation in quantum mechanics, the value of $<C|C>$ is arbitrary.
- The Schrodinger equation requires a complex state function $|\psi(t)>$, whereas the Black–Scholes equation is a real partial differential equation that always yields a real valued expression for the option price C. One can think of the Black–Scholes equation as the Schrodinger equation for imaginary time.
- All Hamiltonians in quantum mechanics are Hermitian as this ensures that all the eigenvalues are real. The Black–Scholes and other Hamiltonians determining the option price are not Hermitian, and this leads to eigenvalues that are complex.
- Complex eigenvalues of Hamiltonians in finance lead to a more complicated analysis than one encountered in quantum mechanics; in particular, there is no well-defined procedure applicable to all Hamiltonians for choosing the set of eigenfunctions that yield the completeness equation. The special cases where a similarity transformation leads to an equivalent Hermitian Hamiltonian yields a natural choice for the set of complete eigenfunctions.

- The Schrodinger equation is an initial value problem, whereas the Black–Scholes equation is a final value problem since the price of the option on maturity – the payoff function – is specified in advance.
- The Schrodinger equation is time reversible due to the fact that the Hamiltonian is Hermitian and its time-evolution is given by $e^{-itH/\hbar}$. In contrast, the Black–Scholes process is time-irreversible due to its Hamiltonian being non-Hermitian, and also because the pricing kernel is determined by the time-irreversible semi-group $e^{-\tau H}$.

5

Path integrals and stock options

Only a few problems in finance are directly tractable using the Hamiltonian, since in general nonlinear systems such as the Merton–Garman Hamiltonian are too difficult to solve analytically. One is hence led to recasting the problem in the formalism of path integration; this allows for a different approach and point of view to attack otherwise intractable problems, and in some cases leads to simplifications that are easy to implement analytically and numerically.

The main focus of this chapter is to develop the **Feynman Path Integral** [30, 33, 62, 90, 106] for the pricing of options as it has proven to be a powerful tool for analytical and computational studies of random systems. This entails a shift away from studying the Hamiltonian operator to the study of the **Lagrangian**. Similar to the role played by the Hamiltonian in the previous chapter, the Lagrangian L is the fundamental mathematical structure in the path-integral formulation of quantum mechanics [30, 33, 90].

This chapter is an introduction to the growing field of the applications of path integrals to option pricing [5, 7, 25, 67, 79, 82, 91].

5.1 Lagrangian and action for the pricing kernel

Recall the pricing kernel is sufficient for solving the problem of pricing a class of options. For a single security with remaining time $\tau = T - t$, Eq. (4.30) gives

$$p(x, \tau; x') = \langle x \mid e^{-\tau H} \mid x' \rangle$$

Feynman's approach to quantum mechanics rests on the fact that even though it may not be possible to evaluate the matrix elements of $e^{-\tau H}$, what one can evaluate exactly are the matrix elements of $e^{-\epsilon H}$, where $\epsilon \to 0$. For a Hamiltonian of the form $H = T + V$, where T, the kinetic term, is a differential operator, and V is a potential term, in general $TV - VT \equiv [T, V] \neq 0$. Since $e^{-\tau H} \neq e^{-\tau T} e^{-\tau V}$, to compute $e^{-\tau H}$ is usually quite intractable.[1] However, for $\tau = \epsilon$ one obtains to leading order that $e^{-\epsilon H} \simeq e^{-\epsilon T} e^{-\epsilon V}$, and this is the fundamental reason for going over to the path-integral formulation.

To determine the pricing kernel, discretize time so that there are N time steps to maturity, with each time step equal to $\epsilon = \tau/N$. The continuous-time label of the variables $x(t)$ is discretized to x_i, where $t = i\epsilon$ and $0 \le i \le N$. The matrix element $<x|e^{-\tau H}|x'>$ can then be written as an N-fold product of $e^{-\epsilon H}$ in the following manner

$$p(x, \tau; x') = \lim_{N \to \infty} <x|\left[e^{-\epsilon H}\right]^N|x'>$$
$$= \lim_{N \to \infty} <x|e^{-\epsilon H} \cdots e^{-\epsilon H}|x'> \qquad (5.1)$$

The limit of $N \to \infty$ will be eventually taken in most calculations.

Recall the completeness equation for $|x>$ is given by Eq. (4.3)

$$\mathcal{J} = \int_{-\infty}^{\infty} dx|x><x| \qquad (5.2)$$

Inserting the completeness equation $(N-1)$ times in Eq. (5.1) yields

$$p(x; \tau|x') = \left(\prod_{i=1}^{N-1} \int dx_i\right) \prod_{i=1}^{N} <x_i|e^{-\epsilon H}|x_{i-1}> \qquad (5.3)$$

with boundary conditions

$$x_N = x, \quad x_0 = x' \qquad (5.4)$$

The **Lagrangian** for the system is defined by Feynman's formula

$$<x_i|e^{-\epsilon H}|x_{i-1}> \equiv \mathcal{N}_i(\epsilon)e^{\epsilon L(x_i;x_{i-1};\epsilon)} \qquad (5.5)$$

and from Eqs. (5.3) and (5.5)

$$p(x; \tau, x') \equiv \int DX e^S \qquad (5.6)$$

where the **action** is given by

$$S = \epsilon \sum_{i=1}^{N} L(x_i; x_{i-1}; \epsilon) \qquad (5.7)$$

---

[1] In general, for non-commuting operators A and B such that $[A, B] \ne 0$, the following infinite expansion is given by the Campbell–Baker formula $\exp(A)\exp(B) = \exp(A + B + [A, B]/2 + \ldots)$ [39].

and the path-integration measure is given by[2]

$$\int DX = \mathcal{N}_N(\epsilon) \prod_{i=1}^{N-1} \int \mathcal{N}_i(\epsilon) dx_i$$

Eq. (5.6) is the discrete-time **Feynman path integral**.

In many cases, the normalization $\mathcal{N}_i$ is independent of the integration variable x_i; in such cases the normalization constant $\mathcal{N}_i$ can be ignored, and the limit of $N \to \infty$ then exists. For boundary conditions $x(\tau) = x$; $x(0) = x'$, one obtains the continuous-time Feynman path integral

$$p(x; \tau, x') = \prod_{t=0}^{\tau} \int_{-\infty}^{+\infty} dx(t) e^S$$

$$\Rightarrow \langle x \mid e^{-\tau H} \mid x' \rangle \equiv \int DX e^S \qquad (5.8)$$

with boundary conditions given by Eq. (5.4), and with

$$S = \int L\left(x, \frac{dx}{dt}\right) dt$$

One can see from equation (5.7) that at a given instant of time $t = i\epsilon$, the system has only one degree of freedom given by x_i; furthermore from Eq. (5.8) the path integration entails performing infinitely many integrations, one integration over x_t for each instant of time $t \in [0, \tau]$. The path integral is known as the Wiener integral [24, 104] in probability theory.

For the case where there are many independent variables, for example a stock price having stochastic volatility, the path integral can be extended in a straightforward manner.

Path-integral quantum mechanics is briefly reviewed in Appendix 5.9.

5.2 Black–Scholes Lagrangian

Consider the case of risk-neutral evolution of a stock price given by the Black–Scholes equation [5]. In this case there is only one independent variable, namely the stock price $S = e^x$, which yields the following

$$\langle x_i \mid e^{-\epsilon H_{BS}} \mid x_{i-1} \rangle = N_{BS}(\epsilon) e^{\epsilon L_{BS}}$$

where $N_{BS}(\epsilon)$ is a normalization constant. The function L_{BS} is the Black–Scholes Lagrangian. The pricing kernel for the Black–Scholes process given in Eq. (4.39)

---

[2] Note there there are N normalization factors and only $N - 1$ number of x_i integrations.

is exact, and setting $\tau = \epsilon$ yields

$$\langle x_i \mid e^{-\epsilon H_{BS}} \mid x_{i-1} \rangle = \frac{e^{-r\epsilon}}{\sqrt{2\pi\epsilon\sigma^2}} e^{-\frac{1}{2\epsilon\sigma^2}\{x_i - x_{i-1} + \epsilon(r - \sigma^2/2)\}^2}$$

$$= \frac{e^{-r\epsilon}}{\sqrt{2\pi\epsilon\sigma^2}} e^{-\frac{\epsilon}{2\sigma^2}\{\frac{\delta x_i}{\epsilon} + (r - \sigma^2/2)\}^2}$$

$$= N_{BS}(\epsilon) e^{\epsilon L_{BS}}$$

where $\delta x_i = x_i - x_{i-1}$. Hence

$$L_{BS}(i) = -\frac{1}{2\sigma^2}\left(\frac{\delta x_i}{\epsilon} + r - \frac{\sigma^2}{2}\right)^2 - r \tag{5.9}$$

$$S_{BS} = \epsilon \sum_{i=1}^{N} L_{BS}(i) \tag{5.10}$$

$$N_{BS}(\epsilon) = \frac{1}{\sqrt{2\pi\epsilon\sigma^2}}$$

Keeping in mind that the index i in x_i labels remaining time with $N\epsilon = \tau = T - t$, where t is real time, the boundary conditions are given by

$$x_N = x, \quad x_0 = x' \tag{5.11}$$

The completeness equation yields for the integration measure

$$\int_{BS} DX = \left[\frac{1}{2\pi\epsilon\sigma^2}\right]^{N/2} \prod_{i=1}^{N-1} \int_{-\infty}^{+\infty} dx_i \tag{5.12}$$

The Black–Scholes pricing kernel is given by the path integral

$$p_{BS}(x, \tau; x') = \int_{BS} DX\, e^{S_{BS}} \tag{5.13}$$

5.2.1 Black–Scholes path integral

The path integration for the Black–Scholes pricing kernel is explicitly performed. From Eq. (5.10) the action is given by

$$S_{BS} = -\frac{1}{2\epsilon\sigma^2} \sum_{i=1}^{N} \left(x_i - x_{i-1} + \epsilon(r - \frac{\sigma^2}{2})\right)^2 - r\epsilon N$$

Define the following change of variables

$$\zeta_i = x_i - x_{i-1} + \epsilon \left(r - \frac{\sigma^2}{2} \right) \tag{5.14}$$

$$d\zeta_i = dx_i \; ; \; i = 1, \ldots, N$$

There are only $N - 1$ original x_i integration variables, whereas N-independent integration variables have been defined for the ζ_i's. Hence, one needs to introduce a constraint on the ζ_i variables, and this will be done by implementing the boundary conditions given in Eq. (5.11). The boundary condition at x_0 is automatically fulfilled, as can be seen in Eq. (5.70), and to implement the boundary condition at x_N impose the following constraint on the ζ_i variables

$$x_N = x_0 - N\epsilon \left(r - \frac{\sigma^2}{2} \right) + \sum_{i=1}^{N} \zeta_i$$

$$\Rightarrow m - \sum_{i=1}^{N} \zeta_i = 0 \; ; \; m \equiv x - x' + \tau \left(r - \frac{\sigma^2}{2} \right) \tag{5.15}$$

Implementing the boundary condition above as a delta-function inside the $\int D\zeta$ path integral, yields from Eq. (5.13)

$$p_{BS}(x, \tau; x') = \langle x \mid e^{-\tau H_{BS}} \mid x' \rangle = \int_{BS} DX e^{S_{BS}}$$

$$= e^{-r\tau} \left[\frac{1}{2\pi\epsilon\sigma^2} \right]^{N/2} \prod_{i=1}^{N} \int_{-\infty}^{+\infty} d\zeta_i e^{-\frac{1}{2\epsilon\sigma^2} \sum_{i=1}^{N} \zeta_i^2} \delta \left(m - \sum_{i=1}^{N} \zeta_i \right)$$

Using the integral representation of the delta function given in Eq. (A.11) factorizes the ζ_i integrations into uncoupled Gaussian integrations, and hence

$$p_{BS}(x, \tau; x') = e^{-r\tau} \left[\frac{1}{2\pi\epsilon\sigma^2} \right]^{N/2} \times$$

$$\int_{-\infty}^{+\infty} \frac{dp}{2\pi} \prod_{i=1}^{N} \int_{-\infty}^{+\infty} d\zeta_i e^{-\frac{1}{2\epsilon\sigma^2} \sum_{i=1}^{N} \zeta_i^2} e^{ip[m - \sum_{i=1}^{N} \zeta_i]}$$

$$= e^{-r\tau} \left[\frac{1}{2\pi\epsilon\sigma^2} \right]^{N/2} \int_{-\infty}^{+\infty} \frac{dp}{2\pi} e^{ipm} \prod_{i=1}^{N} \int_{-\infty}^{+\infty} d\zeta_i e^{-\frac{1}{2\epsilon\sigma^2} \zeta_i^2 + ip\zeta_i}$$

$$= e^{-r\tau} \int_{-\infty}^{+\infty} \frac{dp}{2\pi} e^{ipm} e^{-\frac{N\epsilon\sigma^2}{2} p^2}$$

$$= e^{-r\tau} \frac{1}{\sqrt{2\pi\tau\sigma^2}} e^{-\frac{1}{2\tau\sigma^2} \{x - x' + \tau(r - \sigma^2/2)\}^2} \tag{5.16}$$

where $N\epsilon = \tau$ has been used to obtain the expected Black–Scholes pricing kernel given in Eq. (4.39).

Black–Scholes velocity correlation functions

The velocity correlators are evaluated using the discretized Black–Scholes Lagrangian defined on the time lattice $t = n\epsilon$; since ϵ is expressed in terms of remaining time $\tau = T - t = N\epsilon$, all time derivatives with respect to t pick up a negative sign when expressed in terms of remaining time $\tau = T - t$.

The expectation value of some function $\mathcal{B}$ of the stock price $S = e^x$ is given by

$$< \mathcal{B} > = \frac{1}{Z_{BS}} \int_{BS} DX e^{S_{BS}} \mathcal{B} \; ; \; Z_{BS} = \int_{BS} DX e^{S_{BS}} \qquad (5.17)$$

where, since $< 1 > = 1$, division by Z_{BS} is necessary to obtain a probability measure $e^{S_{BS}}/Z_{BS}$.

Consider a path integral more general than the one evaluated in Eq. (5.16), in that the path integral is performed in the presence of an external current j_i that is coupled to the ζ_i, and yields the moment-generating partition function

$$Z_{BS}[j] = \int_{BS} DX e^{S_{BS}} e^{\sum_{i=1}^{N} j_i \zeta_i}$$

$$= e^{-r\tau} \left[\frac{1}{2\pi\epsilon\sigma^2} \right]^{N/2} \prod_{i=1}^{N} \int_{-\infty}^{+\infty} d\zeta_i e^{-\frac{1}{2\epsilon\sigma^2} \sum_{i=1}^{N} \zeta_i^2} \delta\left(m - \sum_{i=1}^{N} \zeta_i \right) e^{\sum_{i=1}^{N} j_i \zeta_i}$$

Similar to the result obtained in Eq. (5.16), Gaussian integrations yield ($\mathcal{N}$ is a normalization constant)

$$Z_{BS}[j] = \mathcal{N} e^{F[j]}$$

$$F[j] = \frac{1}{2}\epsilon\sigma^2 \sum_{i=1}^{N} j_i^2 - \frac{m^2}{2\epsilon\sigma^2 N} + \frac{m}{N} \sum_{i=1}^{N} j_i - \frac{\epsilon}{2N}\sigma^2 \left(\sum_{i=1}^{N} j_i \right)^2$$

where $m = x - x' + \tau(r - \sigma^2/2)$ is defined in Eq. (5.15). The generating function yields, keeping in mind that time derivatives have an extra negative when taken with respect to remaining time, the following

$$\frac{1}{\epsilon} < \zeta_n > = \frac{1}{\epsilon} \frac{\partial F[j]}{\partial j_n}\bigg|_{j=0} = \frac{m}{\epsilon N} \qquad (5.18)$$

$$\Rightarrow < \frac{dx}{dt} > = -\frac{x - x'}{\tau} = \frac{x' - x}{T - t}$$

The reason that the average velocity does not depend on σ is because in the limit of $\sigma \to 0$, the random velocity reduces to a deterministic value given by the classical velocity $(x - x')/(T - t)$. Furthermore, for $t = n\epsilon, t' = i\epsilon$

$$\frac{1}{\epsilon^2} < \zeta_i \zeta_n > = \frac{1}{\epsilon^2} \frac{\partial^2 F[j]}{\partial j_i \partial j_n}\Big|_{j=0} = \frac{1}{\epsilon} \sigma^2 \delta_{n-i} - \frac{\sigma^2}{\epsilon N} + \left(\frac{m}{\epsilon N}\right)^2 \qquad (5.19)$$

$$\Rightarrow < \frac{dx(t)}{dt} \frac{dx(t')}{dt} > = \sigma^2 \delta(t - t') - \frac{\sigma^2}{T - t} + \left(\frac{x' - x}{T - t}\right)^2$$

A measure of the **fluctuations** of the velocity that depart from its average value is given by

$$< \left[\frac{dx(t)}{dt} - < \frac{dx(t)}{dt} >\right]\left[\frac{dx(t')}{dt} - < \frac{dx(t)}{dt} >\right] > = \sigma^2 \delta(t - t') - \frac{\sigma^2}{T - t}$$

The result shows that all departures of the velocity from its deterministic value are due to $\sigma \neq 0$, as expected.

Continuum limit

The $\int_{BS} DX$ path integral for the Black–Scholes case, from Eq. (5.12), has an integration measure that is essentially the measure for the flat space R^N; the limit of $N \to \infty$ for DX can consequently be taken and a well-defined continuous-time path integral is obtained. From Eqs. (5.10) and (5.12), taking the continuum limit of $\epsilon \to 0$, yields the following

$$S_{BS} = \int_0^\tau dt L_{BS} = -\frac{1}{2\sigma^2} \int_0^\tau dt \left(\frac{dx}{dt} + r - \frac{1}{2}\sigma^2\right)^2$$

with boundary conditions $x(0) = x'$ and $x(\tau) = x$.

It can be seen from above that the Black–Scholes case of constant volatility corresponds to the evolution of a free quantum mechanical particle with mass given by $1/\sigma^2$. Up to an irrelevant constant, the continuous-time path integral for the Black–Scholes pricing kernel is the following

$$p_{BS}(x, \tau | x') = < x|e^{-\tau H_{BS}}|x' > = \int_{BS} DX e^{S_{BS}} \qquad (5.20)$$

$$\int_{BS} DX = \prod_{t=0}^\tau \int_{-\infty}^{+\infty} dx(t)$$

Similar to the quantum mechanical case, Eq. (5.20) for the Black–Scholes pricing kernel is a continuous functional integration over all the stock price variables $x(t)$, and can be graphically represented – as shown in Figure (5.1) of Appendix 5.9 – by a summation over all possible values that the stock price takes in its evolution from x to x'.

5.3 Path integrals for path-dependent options

Path-dependent options come in many varieties. The two main categories are options that are typically like barrier options that restrict the values of the stock price to take values within a pre-specified range for the option to have a non-zero value. The other category is options with payoff functions that depend on the all the values of the stock price from its initial value till the maturity of the option.

Consider for concreteness the pricing kernel of the double-knock-out barrier option considered in Appendix 4.14; the value of the stock price leads to the restriction that $a \leq x(t') \leq b$; $t \leq t' \leq T$, where t is current time and T is the time for the maturity of the option. The path-integration is over the domain DB which similar to Eq. (5.12), is given by

$$\int_{DB} DX = \left[\frac{1}{2\pi\epsilon\sigma^2}\right]^{N/2} \prod_{i=1}^{N-1} \int_a^b dx_i$$

and, similar to the Black–Scholes case, the pricing kernel for the double-knock-out option is given by

$$p_{DB}(x, \tau; x') = \int_{DB} DX e^{S_{BS}}$$

Note the important fact that the action for the double-knock-out option is given by the Black–Scholes action S_{BS}, and only the domain of integration DB for the path-integral changes to account for the barrier option. The path integral for the barrier option is evaluated more efficiently using the Hamiltonian, and the pricing kernel $p_{DB}(x, \tau; x')$ is given in Eq. (4.54).

The payoff function for a large class of path-dependent options is given by ($\tau = T - t$)

$$g[x, K] = \max\left(K, \frac{1}{\tau}\int_0^\tau ds f(x(s))\right)_+$$

For the Asian option $f(x(s)) = e^{x(s)}$. Matacz [72] has treated a number of such path-dependent options using a path-integral representation.

The price of the option at present time t is given by the Black–Scholes path integral, namely that

$$C(x_0; \tau; K) = \int_{BS} DX e^{S_{BS}} g(x, K)$$
$$= \int_{-\infty}^{+\infty} d\xi g(\xi, K) \left[\int_{BS} DX \delta\left[\xi - \frac{1}{\tau}\int_0^\tau ds f(x(s))\right] e^{S_{BS}}\right]$$

$$\text{mixed boundary conditions}: x(\tau) = x_0 \ ; \ \frac{dx(0)}{dt} = 0$$

Using the Fourier representation of the delta function given in Eq. (A.11) yields

$$C(x_0, \tau; K) = \int_{-\infty}^{+\infty} \frac{d\xi \, dp}{2\pi} e^{ip\xi} Z(p) g(\xi, K)$$

$$\Rightarrow Z(p) = \int_{BS} DX e^{S_{\text{eff}}(p)}$$

where $S_{\text{eff}}(p) = S_{BS} - i\frac{p}{\tau} \int_0^\tau ds f(x(s))$ (5.21)

In particular for the Asian option

$$L_{\text{Asian option}} = L_{BS} - i\frac{p}{\tau} e^x$$ (5.22)

5.4 Action for option-pricing Hamiltonian

Recall that the Hamiltonian with an arbitrary potential $V(x)$ is given by

$$H_V = -\frac{\sigma^2}{2} \frac{\partial^2}{\partial x^2} + \left(\frac{1}{2}\sigma^2 - V(x)\right) \frac{\partial}{\partial x} + V(x)$$ (5.23)

It can be shown, similar to the analysis for the Black–Scholes case, that

$$p_V(x, \tau | x') = < x | e^{-\tau H_V} | x' >$$
$$= \int_{BS} DX e^{S_V}$$

where

$$S_V = -\int_0^\tau dt \left[\frac{1}{2\sigma^2} \left(\frac{dx}{dt} - V(x) + \frac{1}{2}\sigma^2\right)^2 + V(x) \right]$$ (5.24)

5.5 Path integral for the simple harmonic oscillator

The path integral for the simple harmonic oscillator is one of the simplest – as well one of the most useful – models in quantum mechanics. The simple harmonic oscillator is exactly soluble, and is a special case of infinite-dimensional Gaussian integration.

Consider a quantum particle, with random position given by variable x, moving in a potential $V(x) = \frac{1}{2}\omega^2 x^2$; this corresponds to a restoring force on a particle that is proportional to its displacement from the origin at $x = 0$, namely Hooke's law. Let the particle have mass m and spring constant ω, and be subject to an external force j; the particle's Lagrangian and action, from initial time and position t_i, x_i

to final time and position t_f, x_f, is given by

$$S = \int_{t_i}^{t_f} dt\, L\left[x, \frac{dx}{dt}\right]$$

$$L = -\frac{1}{2}m\left(\frac{dx}{dt}\right)^2 - \frac{1}{2}m\omega^2 x^2 + jx \tag{5.25}$$

The transition amplitude is given by

$$p(x; t_i; x_f, t_f) = <x_f|e^{-(t_f-t_i)H}|x_i> = \int_{BS} DX e^S$$

Consider the case when the initial and final positions x_i, x_f are random. Then

$$Z(t_i, t_f; j) = \int_{-\infty}^{+\infty} dx_i dx_f\, p(x_i, t_i; x_f, t_f) \tag{5.26}$$

The integration over the initial and final positions that results in $Z(t_i, t_f; j)$ has a simple expression in terms of the boundary conditions imposed on the path-integration measure $\int DX$. Instead of the initial and final positions being fixed, the paths $x(t)$ now have

$$\frac{dx(t_i)}{dt} = 0 = \frac{dx(t_f)}{dt} \quad : \text{Neumann B.C.'s.} \tag{5.27}$$

The Neumann boundary conditions allow one to do an integration by parts of the action given in Eq. (5.25) and yields the following action

$$S = -\frac{1}{2}m \int_{t_i}^{t_f} dt\, x(t)\left[-\frac{d^2}{dt^2} + \omega^2\right]x(t) + \int_{t_i}^{t_f} dt\, j(t)x(t) \tag{5.28}$$

The generating functional[3] is given by the path integral

$$Z(t_i, t_f; j) = \int_N DX e^S \tag{5.29}$$

where the subscript N denotes the Neumann boundary condition on the path integral. The formal solution for Z can be read off from Eqs. (A.19) and (A.20). Defining

$$m\left[-\frac{d^2}{dt^2} + \omega^2\right]D(t, t'; t_i, t_f) = \delta(t - t') \quad : \text{Neumann B.C.'s}$$

---

[3] The term-generating functional is used instead as a generating function as in Eq. (A.16) to indicate that one is considering a system with infinitely many variables.

gives from Eq. (A.20) that

$$Z(t_i, t_f; j) = e^{\frac{1}{2} \int_{t_i}^{t_f} dt dt' j(t) D(t,t';t_i,t_f) j(t')}$$ (5.30)

The function $D(t, t'; t_i, t_f)$ is the **propagator** for the simple harmonic oscillator.

The simple harmonic path integral: Fourier expansion

The path integral $\int DX e^S$ is performed over all paths (functions) $x(t)$ that satisfy the Neumann boundary conditions given in Eq. (5.27). All such functions can be expanded in a Fourier cosine series as follows

$$x(t) = a_0 + \sum_{n=1}^{\infty} a_n \cos\left[n\pi \frac{(t - t_i)}{\tau} \right] ; \quad \tau \equiv t_f - t_i$$

$$\int_N DX = \mathcal{N} \prod_{n=0}^{\infty} \int_{-\infty}^{+\infty} da_n \quad : \text{infinite multiple integral}$$

($\mathcal{N}$ is a normalization constant). The orthogonality equations

$$\int_{t_i}^{t_f} dt \cos\left[n\pi \frac{(t - t_i)}{\tau} \right] \cos\left[m\pi \frac{(t - t_i)}{\tau} \right] = \int_{t_i}^{t_f} dt \sin\left[n\pi \frac{(t - t_i)}{\tau} \right] \sin\left[m\pi \frac{(t - t_i)}{\tau} \right]$$

$$= \frac{\tau}{2} \delta_{m-n} ; \quad m, n \geq 1$$ (5.31)

yields for the action given in Eq. (5.28) the following

$$S = -\frac{1}{2} m\omega^2 \tau \left\{ a_0^2 + \frac{1}{2} \sum_{n=1}^{\infty} \left[1 + \left(\frac{n\pi}{\omega\tau} \right)^2 \right] a_n^2 \right\}$$

$$+ \int_{t_i}^{t_f} dt j(t) \left\{ a_0 + \sum_{i=1}^{\infty} a_n \cos\left[n\pi \frac{(t - t_i)}{\tau} \right] \right\}$$

$$= -\frac{1}{2} \sum_{n=0}^{\infty} \kappa_n a_n^2 + \sum_{n=0}^{\infty} j_n a_n$$ (5.32)

where

$$\kappa_0 = m\omega^2 \tau ; \quad \kappa_n = \frac{1}{2} m\omega^2 \tau \left[1 + \left(\frac{n\pi}{\omega\tau} \right)^2 \right]; \quad n \geq 1$$

$$j_n = \int_{t_i}^{t_f} dt j(t) \cos\left[n\pi \frac{(t - t_i)}{\tau} \right] ; \quad n = 0, 1, \ldots \infty$$

All the Gaussian integrations over the variables a_n have decoupled in the action S given in Eq. (5.32). The path integral has been reduced to an infinite product of single Gaussian integrations, each of which can be performed using Eq. (A.16). Hence, from Eqs. (5.29) and (5.32)

$$
\begin{aligned}
Z(t_i, t_f; j) &= \int DX e^S \\
&= \mathcal{N} \prod_{n=0}^{\infty} \left[\int_{-\infty}^{+\infty} da_n e^{-\frac{1}{2}\kappa_n a_n^2 + j_n a_n} \right] \\
&= e^{\frac{1}{2} \sum_{n=0}^{\infty} j_n \frac{1}{\kappa_n} j_n}
\end{aligned}
\tag{5.33}
$$

Using Eq. (5.30) to factor out the $j(t)$'s from above equation yields

$$
D(t, t'; t_i, t_f) = \frac{1}{m\omega^2 \tau} \left\{ 1 + 2 \sum_{n=1}^{\infty} \cos\left[n\pi \frac{(t - t_i)}{\tau} \right] \left[\frac{1}{1 + \left(\frac{n\pi}{\omega\tau}\right)^2} \right] \cos\left[n\pi \frac{(t' - t_i)}{\tau} \right] \right\}
\tag{5.34}
$$

Let $\theta = t - t_i > 0$ and $\theta' = t' - t_i > 0$; then

$$
2 \sum_{n=1}^{\infty} \frac{\cos(n\pi\theta/\tau)\cos(n\pi\theta'/\tau)}{1 + \left(\frac{n\pi}{\omega\tau}\right)^2} = \left(\frac{\omega\tau}{\pi}\right)^2 \sum_{n=1}^{\infty} \frac{\cos(n\pi(\theta + \theta')/\tau) + \cos(n\pi(\theta - \theta')/\tau)}{(\frac{\omega\tau}{\pi})^2 + n^2}
\tag{5.35}
$$

The summation over integer n is performed using the identity[4]

$$
\sum_{n=1}^{\infty} \frac{\cos(n\theta)}{a^2 + n^2} = \frac{\pi}{2a} \frac{\cosh(\pi - |\theta|)a}{\sinh \pi a} - \frac{1}{2a^2}
\tag{5.36}
$$

and yields the result

$$
D(t, t'; t_i, t_f) = \frac{\cosh \omega \left\{ \tau - |\theta - \theta'| \right\} + \cosh \omega \left\{ \tau - (\theta + \theta') \right\}}{2m\omega \sinh \omega\tau}
\tag{5.37}
$$

Hence, from Eq. (5.37) and since $\tau = t_f - t_i$, the propagator is given by

$$
D(t, t'; t_i, t_f) = \frac{\cosh \omega \left\{ (t_f - t_i) - |t - t'| \right\} + \cosh \omega \left\{ (t_f - t_i) - (t + t' - 2t_i) \right\}}{2m\omega \sinh \omega(t_f - t_i)}
\tag{5.38}
$$

Given the importance of the propagator for field theory of the forward interest rates, and for field theory in general, the derivation of the propagator is studied in some detail in Appendix A.7 for various boundary conditions. The propagator

---

[4] The formula given in Eq. (5.36) is valid for any complex number a, and will be applied in later discussions for a case where a is indeed a complex number. The branch of the square root of a^2 that is taken on the right-hand side need not be specified since the right-hand side is a function of a^2.

with Neumann boundary conditions is evaluated using eigenfunctions discussed in Appendix A.7.1, and a Greens function derivation given in Appendix A.7.2.

The result given in Eq. (5.38) for the propagator $D(t, t'; t_i, t_f)$ – from a seemingly artificial example of quantum mechanics – is of fundamental importance in the analysis of the quantum field theory model of the forward rates discussed in Chapters 7–10.

The case of infinite time for Eq. (5.38) is obtained by taking the limit of $t_i \rightarrow -\infty$, $t_f \rightarrow +\infty$, and yields

$$D(t, t') = \frac{1}{2m\omega} e^{-\omega|t-t'|} \tag{5.39}$$

and which is derived in Eq. (A.3.3) using a different method.

5.6 Lagrangian for stock price with stochastic volatility

So far, in the path-integral solution of the Black–Scholes pricing kernel, and of the simple harmonic oscillator, only linear theories have been treated, which can be solved exactly using Gaussian integration. The problem of solving nonlinear systems with actions S that have cubic or higher powers in the random variables is notoriously difficult, and only a handful of such systems can be exactly solved.

For this reason, the case of a stock price with stochastic volatility [5] is analyzed in some detail as it can be exactly solved using path integration. This system consists of two degrees of freedom (two random variables at each instant of time), namely the random stock price $S(t)$ and its volatility $V(t)$. This example is important in its own right, and is also sufficiently complex to serve as an exemplar for demonstrating the mathematical derivations that arise in the path-integral approach to option pricing.

Recall that the Merton–Garman Hamiltonian in Eq. (3.34) is valid for a process with an arbitrary value for the parameter α. Analysis of data from option prices [8] shows that it is quite insensitive to the precise value of α, and in fact any value of $\frac{1}{2} \leq \alpha \leq 1$ fits data equally well.

Only the special case of $\alpha = 1$ will be considered as this considerably simplifies all the calculations, and also yields an efficient numerical algorithm [8, 70]. Consider the case of $\lambda = 0$ in the stochastic equation driving volatility as given in Eq. (3.27); the fundamental stochastic equation for stock price $S(t)$ and its volatility V takes the form

$$\frac{dV}{dt} = \mu V + \xi V Q; \quad \frac{dS}{dt} = \phi S + \sigma S R \tag{5.40}$$

The Merton–Garman Hamiltonian for $\alpha = 1$, $\lambda = 0$ is given by Eq. (3.34) as

$$H_{MG} = -\frac{e^y}{2}\frac{\partial^2}{\partial x^2} + \left(\frac{1}{2}e^y - r\right)\frac{\partial}{\partial x} - \xi\rho e^{y/2}\frac{\partial^2}{\partial x\partial y} - \frac{\xi^2}{2}\frac{\partial^2}{\partial y^2} + \left(\frac{1}{2}\xi^2 - \mu\right)\frac{\partial}{\partial y}$$

$$(5.41)$$

Derivation of the Merton–Garman Lagrangian

To obtain the Merton–Garman Lagrangian from its Hamiltonian H_{MG} (henceforth, in this derivation, the subscript on the Hamiltonian is dropped for simplicity) one needs to use the momentum basis introduced in Section 4.6.1; from Eq. (4.34)

$$< x, y|e^{-\epsilon H}|x', y' > = \int_{-\infty}^{\infty}\frac{dp_y}{2\pi}\frac{dp_y}{2\pi} < x, y|e^{-\epsilon H}|p_x, p_y >< p_x, p_y|x', y' >$$

$$= \int_{-\infty}^{\infty}\frac{dp_x}{2\pi}\frac{dp_y}{2\pi}e^{ip_x(x-x')}e^{ip_y(y-y')}e^{-\epsilon H(x,y,p_x,p_y)} \quad (5.42)$$

and from Eq. (5.41) the matrix elements of the Hamiltonian are given by

$$H(x, y, p_x, p_y) = \frac{e^y}{2}p_x^2 + \left(\frac{1}{2}e^y - r\right)ip_x + \xi\rho e^{y/2}p_x p_y + \frac{\xi^2}{2}p_y^2 + \left(\frac{1}{2}\xi^2 - \mu\right)ip_y$$

Re-writing Eq. (5.42) in matrix notation yields

$$< x, y|e^{-\epsilon H}|x', y' >$$

$$= \int_{-\infty}^{\infty}\frac{dp_x}{2\pi}\frac{dp_y}{2\pi}\exp\left\{-\frac{\epsilon}{2}\begin{bmatrix} p_x & p_y \end{bmatrix}\mathcal{M}\begin{bmatrix} p_x \\ p_y \end{bmatrix} + i\begin{bmatrix} A & B \end{bmatrix}\begin{bmatrix} p_x \\ p_y \end{bmatrix}\right\}$$

$$(5.43)$$

where

$$A = x - x' + \epsilon r - \frac{\epsilon}{2}e^y$$

$$B = y - y' + \epsilon\mu - \frac{\epsilon}{2}\xi^2$$

and

$$\mathcal{M} = \begin{bmatrix} e^y & \xi\rho e^{y/2} \\ \xi\rho e^{y/2} & \xi^2 \end{bmatrix}$$

To perform the Gaussian integrations over p_y and p_y one needs

$$\det\mathcal{M} = \xi^2 e^y(1 - \rho^2)$$

and

$$\mathcal{M}^{-1} = \frac{1}{\xi^2(1 - \rho^2)}\begin{bmatrix} \xi^2 e^{-y} & -\xi\rho e^{-y/2} \\ -\xi\rho e^{-y/2} & 1 \end{bmatrix}$$

Performing the two-dimensional Gaussian integrations in Eq. (5.43) using Eq. (A.18), and extending Eq. (5.5) to the case of stock price with stochastic volatility, one obtains

$$< x, y|e^{-\epsilon H}|x', y' > = \frac{1}{2\pi\epsilon\sqrt{\det\mathcal{M}}}e^{\epsilon L} \tag{5.44}$$

where

$$L_{MG} = -\frac{1}{2\epsilon^2(1-\rho^2)}\left(e^{-y}A^2 + \frac{1}{\xi^2}B^2 - 2\frac{\rho}{\xi}e^{-y/2}AB\right) + O(\epsilon) \tag{5.45}$$

Simplifying Eq. (5.45) above, and for $\delta x = x - x'$ and $\delta y = y - y'$, the (negative definite) **Merton–Garman Lagrangian** for $\alpha = 1$ is given by

$$L_{MG} = -\frac{1}{2\xi^2}\left(\frac{\delta y}{\epsilon} + \mu - \frac{1}{2}\xi^2\right)^2$$

$$-\frac{e^{-y}}{2(1-\rho^2)}\left[\frac{\delta x}{\epsilon} + r - \frac{1}{2}e^y - \frac{\rho}{\xi}e^{y/2}\left(\frac{\delta y}{\epsilon} + \mu - \frac{1}{2}\xi^2\right)\right]^2 + O(\epsilon)$$

$$\equiv L_0 + L_X \tag{5.46}$$

For $\epsilon \to 0$, one recovers the expected result that

$$\lim_{\epsilon\to 0} < x_i, y_i|e^{-\epsilon H}|x_{i-1}, y_{i-1} > = \delta(x_i - x_{i-1})\delta(y_i - y_{i-1}) + O(e) \tag{5.47}$$

and the pre-factors to the exponential on the right-hand side of Eq. (5.44) ensure the correct limit.

The action S is

$$S = \epsilon\sum_{i=1}^{N} L(i) + O(e)$$

$$\equiv S_0 + S_X \tag{5.48}$$

S is **quadratic** in x_i and non-linear in the y_i variables, and hence in principle the path integral over the stock prices x_i can be done exactly.

Collecting results from Eqs. (5.6) and (5.48), the Merton–Garman pricing kernel is given by the following path integral

$$p(x, y, \tau; x') = \int_{-\infty}^{\infty} dy' p(x, y, \tau; x', y')$$

$$= \lim_{N\to\infty}\int DXDY e^S \tag{5.49}$$

where, for $\epsilon = \tau/N$

$$\int DX = \frac{e^{-y_N/2}}{\sqrt{2\pi\epsilon(1-\rho^2)}}\prod_{i=1}^{N-1}\int_{-\infty}^{\infty}\frac{dx_i e^{-y_i/2}}{\sqrt{2\pi\epsilon(1-\rho^2)}} \tag{5.50}$$

and

$$\int DY = \int_{-\infty}^{\infty} dy_0 \prod_{i=1}^{N-1} \int_{-\infty}^{\infty} \frac{dy_i}{\sqrt{2\pi \epsilon \xi^2}}. \tag{5.51}$$

An extra $dy_0 = dy'$ integration has been included in $\int DY$ given in Eq. (5.51) due to the integration over y' in Eq. (5.49).

Eq. (5.49) is the path integral for stochastic stock price with stochastic volatility with $\alpha = 1$; the path integral for arbitrary α in its full generality is treated in the Appendices 5.14–5.13.

To evaluate expressions such as the correlation of $S(t)$ with $V(t')$, the path integral in Eq. (5.49) has to be used.

5.7 Pricing kernel for stock price with stochastic volatility

An exact solution for the Merton–Garman pricing kernel is obtained by performing all the integrations that appear in the path integral for the pricing kernel. The solution is given in the form of an infinite series expansion that requires some residual integrations, all of which can also be carried out exactly. Hence the problem can be considered to have been formally solved.

Eq. (5.81) provides a mathematically rigorous basis for taking the $N \to \infty$ limit for the case of stochastic volatility. On exactly performing the $\int DX$ path integral, the remaining $\int DY$ path integral has a measure, namely $DY = [\frac{1}{2\pi \epsilon \xi^2}]^{N/2} \prod_{i=1}^{N} dy_i$ that is, just as in the Black–Scholes case, essentially the measure for the flat space $\mathbf{R}^N$ (and unlike the nonlinear expression in Eq. (5.50)); hence the $N \to \infty$ limit for $\int DY$ and for $S_0 + S_1$ can be taken, and results in a well-defined continuous-time path integral.

Taking the limit of $\epsilon \to 0$ yields $t = i\epsilon$, $\epsilon \sum_{i=1}^{N} e^{y_i} \to \int_0^{\tau} dt\, e^{y(t)} \equiv \tau w$, and $\delta y_i/\epsilon \to dy/dt$. Hence, from Eqs. (5.82) and (5.83)

$$S = S_0 + S_1 \tag{5.52}$$

where

$$S_0 = -\frac{1}{2\xi^2} \int_0^{\tau} dt \left(\frac{dy}{dt} + \mu - \frac{1}{2}\xi^2\right)^2$$

$$S_1 = -\frac{1}{2(1-\rho^2)w} \left\{ x - x' + r\tau - \frac{1}{2} \int_0^{\tau} dt\, e^{y(t)} \right.$$

$$\left. + \frac{2\rho}{\xi} \left(e^{y(0)/2} - e^{y/2}\right) - \frac{\rho}{\xi} \left(\mu - \frac{\xi^2}{2}\right) \int_0^{\tau} dt\, e^{y(t)/2} \right\}^2 \tag{5.53}$$

with boundary value

$$y(\tau) = y$$

and the pricing kernel is given by

$$p(x, y, \tau; x') = \int DY \frac{e^S}{\sqrt{2\pi(1 - \rho^2)\tau w}}. \tag{5.54}$$

Taking the continuum limit is possible only after the discrete $\int DX$ path integration has been performed, since the nonlinear measure given by Eq. (5.50) does not have a finite continuum limit.

For $\rho \neq 0$,[5] it can be seen from Eq. (5.54) that $p(x, y; \tau, x')$ depends on w as well as on $u = \frac{1}{\tau}\int_0^\tau e^{y(t)/2}dt$ and $e^{y(0)/2}$. Hence, from Eq. (5.54)

$$p(x, y, \tau; x') = \int_0^\infty dw \; du \; dv \; \frac{e^{S_1(u,v,w)}}{\sqrt{2\pi(1 - \rho^2)\tau w}} g(u, v, w) \tag{5.55}$$

where

$$S_1(u, v, w) =$$

$$-\frac{1}{2(1 - \rho^2)\tau w}\left[x - x' + r\tau - \frac{\tau}{2}w + \frac{2\rho}{\xi}(v - e^{y/2}) - \frac{\rho}{\xi}\left(\mu - \frac{\xi^2}{2}\right)u\right]^2$$

with

$$u = \frac{1}{\tau}\int_0^\tau e^{y(t)/2}dt \; ; \; v = e^{y(0)/2}$$

$$w = \frac{1}{\tau}\int_0^\tau e^{y(t)}dt$$

and $g(u, v, w)$ is the probability distribution for the moments given by the path integral

$$g(u, v, w)$$

$$= \int DY e^{S_0} \delta\left\{v - e^{y(0)/2}\right\} \delta\left\{u - \frac{1}{\tau}\int_0^\tau e^{y(t)/2}dt\right\} \delta\left\{w - \frac{1}{\tau}\int_0^\tau e^{y(t)}dt\right\}$$

$$\tag{5.56}$$

From Eq. (5.56) above it can be seen that $g(u, v, w)$ is the probability density for u, v, and w and $p(x, y, \tau; x')$ is the weighted average of the integrand $e^{S_1(u,v,w)}/\sqrt{2\pi(1 - \rho^2)\tau w}$ with respect to $g(u, v, w)$.

---

[5] For the case of $\rho = 0$ to recover the Black–Scholes formula, set $w = \sigma^2$, where $\sigma = e^{y/2}$ is the volatility at time $t = 0$. For the case of $\rho = 0$, as has been noted earlier in the discussion on Merton's theorem in Appendix 3.9, and also by Hull and White [49], the pricing kernel depends on stochastic volatility $y(t)$ only through the combination $w = \frac{1}{\tau}\int_0^\tau e^{y(t)}dt$.

The path integral for $g(u, v, w)$ is nonlinear and cannot be performed exactly.

The function $g(u, v, w)$ has the remarkable property that it is **independent** of ρ. Following Hull and White [49] one can expand the integrand in Eq. (5.55) in an infinite power series in u, v, w and reduce the evaluation of $p(x, y, \tau; x')$ to finding all the moments of u, v and w; in other words one needs to evaluate

$$< u^n w^m v^p > \equiv \int_0^\infty du \; dw \; u^n \; w^m \; v^p \; g(u, v, w) \tag{5.57}$$

$$= \int DY \left[\frac{1}{\tau} \int_0^\tau e^{y(t)/2} dt \right]^n \left[\frac{1}{\tau} \int_0^\tau e^{y(t)} dt \right]^m e^{py(0)/2} e^{S_0}$$

The path integral for $< u^n w^m v^p >$ can be performed exactly. Rewrite Eq. (5.57) as

$$< u^n w^m v^p > = \frac{1}{\tau^{n+m}} \int_0^\tau dt_1 \cdots dt_n dt_{n+1} \cdots dt_{n+m} Z(j, y, p) \tag{5.58}$$

where

$$Z(j, y, p) = \int DY \exp\left[\int_0^\tau dt j(t) y(t) \right] e^{S_0} \tag{5.59}$$

and from Eqs. (5.57)–(5.59)

$$j(t) = \frac{1}{2} \sum_{i=1}^n \delta(t - t_i) + \sum_{i=n+1}^{n+m} \delta(t - t_i) + \frac{p}{2}\delta(t)$$

$$\equiv \sum_{i=1}^{n+m} a_i \delta(t - t_i) + \frac{p}{2}\delta(t). \tag{5.60}$$

The path integral for $Z(j, y, p)$ is evaluated exactly in Appendix 5.12 and yields[6]

$$< u^n w^m v^p > = \sigma^p \frac{e^{y \sum_i a_i}}{\tau^{n+m}} \left[\prod_{i=1}^{n+m} \int_0^\tau dt_i \right] e^F$$

$$: \; \alpha = 1 \; \text{exact solution} \tag{5.61}$$

where from Eqs. (5.60), (5.93) and (5.94), and after some simplifications[7]

$$F = \left(\mu - \frac{1}{2}\xi^2 \right) \sum_{i=1}^{n+m} a_i t_i + \xi^2 \sum_{i,j=1}^{n+m} a_i a_j \Theta(t_i - t_j) t_j + F' \tag{5.62}$$

---

[6] For $m + n = 0$ $\prod_{i=1}^{n+m} \int_0^\tau dt_i = 1$ and $\sum_{i=1}^{n+m} = 1$.

[7] In obtaining Eq. (5.62) the identity $\int_0^\tau dt \delta(t - \tau) = \frac{1}{2}$ has been used, which follows from Eq. (A.9).

with

$$F' = \frac{1}{8}p^2\xi^2\tau + \frac{1}{2}p\tau\left(\mu - \frac{1}{2}\xi^2\right) + \frac{1}{2}p\xi^2\sum_{i=1}^{n+m}a_i t_i$$

Eq. (5.61) constitutes an **exact solution** for the $\alpha = 1$ case; all the t_i integrations can be performed exactly since the exponent is linear in all the t_i's as in Eq. (5.62). Some of the moments of stock price and its stochastic volatility are evaluated in Appendix 5.13.

5.8 Summary

The path-integral formulation of option pricing is a natural complement to the Hamiltonian-based differential calculus approach of the earlier Chapter 4. The main motivation for introducing path integrals is to enlarge the mathematical tools for studying option pricing theory.

To illustrate the general features of the path integral, three important cases were studied in some detail. Firstly, the path integral for the well-known Black–Scholes pricing kernel was solved exactly by a change of the path-integration variables. Secondly, the path integral for the simple harmonic oscillator with Neumann boundary conditions was solved using a Fourier expansion of the path-integration variables. And lastly, for the pricing kernel of stock price with stochastic volatility, a perturbation expansion of the nonlinear path integral was obtained.

The three techniques used, namely a change of variables, Fourier expansion (or more generally a 'normal mode expansion') and perturbation expansion are important methods that are useful for solving a wide class of path integrals.

All three cases were eventually reduced to performing Gaussian integrations. There are many techniques for performing path integrals that go beyond Gaussian integrations, but nevertheless Gaussian path integrals, appropriately generalized, remain one of the bedrocks of path integration.

5.9 Appendix: Path-integral quantum mechanics

Consider the position of a particle $x(t)$ as a function of time t. Let the particle be experimentally observed to be at position x_i at time t_i, allowed to evolve without any measurement till time t_f and then again observed to be at position x_f.

How does one describe the evolution of the particle in quantum mechanics?

There are **three independent** and **equivalent** descriptions of quantum mechanics.[8]

---

[8] The quantum evolution is discussed for physical Minkowski time, and to make the connection with finance one has to analytically continue physical time in the pure imaginary direction, called Euclidean time.

In the Schrodinger representation of quantum mechanics [39], the probability for the particle to be found in the interval $[x_0, x_0 + dx]$ at time $t_0 \in [t_i, t_f]$ is given by[9]

$$P(x_0 < x < x_0 + dx, t_0) = |\psi(x_0, t_0)|^2 dx$$

and is shown in Figure 5.1.

The second quantum description is provided by the Heisenberg operator equations [39] in which the position, momentum, energy and so on of the particle are considered to be non-commuting Hermitian operators acting on the Hilbert space of physical states, and operator evolution equations define the dynamics of the system. The fundamental commutation equation for quantum mechanics is given by

$$[x, p] = i\hbar \tag{5.63}$$

where Hermitian operator x represents the quantum particle's position and p is its momentum.

The third description of the quantum particle is given by the Feynman path integral [33, 106]. In this representation, the position of the particle $x(t)$ at each instant $t \in [t_i, t_f]$ is considered to be an **independent random variable**. The probability amplitude for making a transition from its initial to its final position is given by the square of the absolute value of the transition amplitude, namely $| < x_f | e^{-(i/\hbar)(t_f - t_i)H} | x_i > |^2$, where H is the Hamiltonian operator driving the evolution of the quantum particle.[10] The transition amplitude is obtained by integrating over all possible values of all the random variables $x(t)$, and yields the Feynman path integral [33, 106]

$$< x_f | e^{-\frac{i}{\hbar}(t_f - t_i)H} | x_i > = \mathcal{N} \prod_{t_i < t < t_f} \int_{-\infty}^{+\infty} dx(t) e^{S_{QM}} \tag{5.64}$$

with the boundary condition $x(t_i) = x_i$ and $x(t_f) = x_f$. The (functional) integration given in Eq. (5.64) can be thought of as summing over **all possible virtual paths** that the quantum particle takes between points x_i and x_f, and hence the term path integration.[11] One can think of the quantum particle **simultaneously** taking, in a virtual sense, all the possible paths from x_i to x_f.

---

[9] Recall the complex valued function $\psi(x, t)$ is the Schrodinger wavefunction, which obeys the Schrodinger equation.

[10] The transition amplitude in quantum mechanics is analogous to the pricing kernel in option pricing.

[11] A path in space is mathematically speaking a function of time; hence, path integration can also be thought of as integrating the exponential of the action, namely e^S, over all possible functions that start at $x(t_i) = x_i$ and end at $x(t_f) = x_f$. The mathematical term of functional integration is also used for path integration.

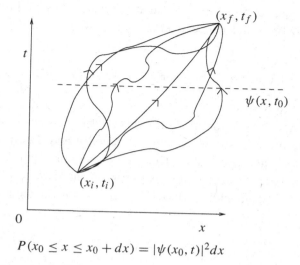

$$P(x_0 \le x \le x_0 + dx) = |\psi(x_0, t)|^2 dx$$

Figure 5.1 Quantum particle simultaneously taking all possible virtual paths from (x_i, t_i) to (x_f, t_f)

The quantum virtual paths are figuratively shown in Figure 5.1, together with the interpretation of the Schrodinger wavefunction as determining the probability for the particle to be found near some point x.

The action S is a functional of the paths, and can be constructed from the Hamiltonian H. A general form for the action, corresponding to the Hamiltonian given in Eq. (4.60), is

$$S_{QM} = \frac{i}{\hbar} \int_{t_i}^{t_f} dt \left[\frac{1}{2} m \left(\frac{dx}{dt} \right)^2 + V(x) \right]$$

where m is the mass of the particle and $V(x)$ is the potential it is moving in.

The formalism of quantum mechanics is based on conventional mathematics of partial-differential equations and functional analysis. The infinite-dimensional integration measure given by $\prod_{t_i < t < t_f} \int_{-\infty}^{+\infty} dx(t)$ can be given a rigorous, measure theoretic, definition as the integration over all continuous, but nowhere differentiable, paths running between points x_i and x_f.

In Feynman's formulation of quantum mechanics, one computes the **matrix elements** of operators using functional integration, and hence the structure of the Hilbert space of states, the non-commuting operator algebra acting on this space and so on, are present in an implicit manner in the path integral.

Only the quantum mechanical evolution of a single particle has been discussed, and it is not too difficult to see that the formalism extends without much change to that of N particles.

5.10 Appendix: Heisenberg's uncertainty principle in finance

Almost synonymous with quantum mechanics is the Heisenberg uncertainty principle that follows from the Heisenberg commutation equation. It is shown how Heisenberg's results emerge in finance.

The random evolution of the stock price $S(t)$ implies that if one knows the value of the stock price, then one has no information regarding its velocity; in terms of $x = \ln S$, the random evolution states that once x is observed, one cannot simultaneously observe the value of dx/dt. In the formalism of operators and state space, if one represents the logarithm of the value of a stock price and its velocity by **operators** on state space (hat is to emphasize the operator nature of the symbols) $\hat{x}$ and $d\hat{x}/dt \equiv \hat{\dot{x}}$, then the random evolution of stock price implies, similar to Eq. (5.63), the following commutation equation

$$[\hat{x}(t), \hat{\dot{x}}(t)] \equiv \hat{x}(t)\hat{\dot{x}}(t) - \hat{\dot{x}}(t)\hat{x}(t) \neq 0 \qquad (5.65)$$

In the path-integral formulation the commutation equation is represented by taking the expectation value of both sides of Eq. (5.65); if one naively takes the expectation value, the left-hand side of Eq. (5.65) apparently seems to be zero. However, following Feynman [33], one needs to first discretize time $t = n\epsilon$ to take the expectation value. The expectation value of Eq. (5.65) is evaluated using the option-pricing Lagrangian; hence ϵ is expressed in terms of remaining time $\tau = T - t = N\epsilon$, and consequently all time derivatives pick up a negative sign when expressed in terms of remaining time. The commutation equation is interpreted to mean that the operator on the **left**-hand side is at *later* real time, and therefore for remaining time the operator on the left-hand side is at an earlier remaining time. Hence one obtains the following representation

$$< [\hat{x}(t), \hat{\dot{x}}(t)] > \; \rightarrow \; - < x_n \left(\frac{x_{n+1} - x_n}{\epsilon} \right) - \left(\frac{x_n - x_{n-1}}{\epsilon} \right) x_n > \; \equiv \; < \mathcal{C} > \qquad (5.66)$$

The Black–Scholes Lagrangian together with a potential is given from Eq. (5.10) as

$$S_V = -\frac{1}{2\epsilon\sigma^2} \sum_{i=1}^{N} \left(x_i - x_{i-1} + \epsilon \left(r - \frac{\sigma^2}{2} \right) \right)^2 - r\epsilon N - \epsilon \sum_{i=1}^{N-1} V(x_i) \qquad (5.67)$$

To evaluate $< \mathcal{C} >$, note that e^{S_V} is zero for $x_i = \pm\infty$ and yields the identity [33]

$$0 = \int_{BS} DX \frac{\partial}{\partial x_n} \left(x_n e^{S_V} \right)$$

$$\Rightarrow 1 = -\frac{1}{Z_V} \int_{BS} DX x_n \frac{\partial S_V}{\partial x_n} e^{S_V}; \quad \left(Z_V = \int_{BS} DX e^{S_V} \right)$$

$$\equiv - < x_n \frac{\partial S_V}{\partial x_n} > \qquad (5.68)$$

From Eq. (5.67)

$$\frac{\partial S_V}{\partial x_n} = -\frac{1}{\epsilon \sigma^2}\left(2x_n - x_{n-1} - x_{n+1}\right) - \epsilon \frac{\partial V(x_n)}{\partial x_n}$$

Hence from above and Eq. (5.68)

$$\sigma^2 = \, <C> + \epsilon \sigma^2 < x_n \frac{\partial V(x_n)}{\partial x_n} >$$

$$\rightarrow \, <C> \tag{5.69}$$

The above result shows that the potential, which in finance represents some (path-dependent) option, has only a vanishingly small $O(\epsilon)$ effect on the commutation equation, and with no contribution from the boundary conditions on S_V.

The result for the commutation equation can also be obtained from the earlier derivation of the velocity correlation functions of the Black–Scholes Lagrangian given in Section 5.2.1. Eq. (5.14) yields the following change of variables

$$\zeta_i = x_i - x_{i-1} + \epsilon\left(r - \frac{1}{2}\sigma^2\right)$$

$$x_n = x_0 - n\epsilon\left(r - \frac{\sigma^2}{2}\right) + \sum_{i=1}^{n} \zeta_i \tag{5.70}$$

$$\Rightarrow C = -\left[x_0 - n\epsilon\left(r - \frac{\sigma^2}{2}\right) + \sum_{i=1}^{n} \zeta_i\right]\left[\zeta_{n+1} - \zeta_n\right]$$

Eqs. (5.18) and (5.19) yield

$$<C> = -\frac{1}{\epsilon}\sum_{i=1}^{N} < \zeta_i(\zeta_{n+1} - \zeta_n) > = \sigma^2 \tag{5.71}$$

and hence yields the commutation equation[12]

$$[\hat{x}(t), \hat{\dot{x}}(t)] = \sigma^2 \; : \; \text{Heisenberg's commutation equation} \tag{5.72}$$

One realization of the commutation equation is given by

$$\Rightarrow \hat{x} = x; \quad \hat{\dot{x}} = -\sigma^2 \frac{\partial}{\partial x}$$

where $\hat{x}$ is a Hermitian operator and $\hat{\dot{x}}$ is anti-Hermitian.

---

[12] In quantum mechanics, the operator $\hat{x}$ represents the position of a quantum particle with mass m, and with the Heisenberg commutation equation given by $[\hat{x}(t), \hat{\dot{x}}(t)]_{QM} = i\hbar/m$; in finance one is working with Euclidean time and hence there is no factor of i, and $\sigma^2 = \hbar/m$.

The commutation equation is independent of the boundary conditions on the path integral, and of any potential that the Lagrangian may have. The commutation equation is also independent of the drift terms in the Black–Scholes Lagrangian, and hence does not depend of the martingale condition. The commutation equations holds for any Lagrangian that has a kinetic term containing the square of the first-order time derivative; this is the reason the Black–Scholes Lagrangian is sufficient to derive a result that is of great generality.

Define the uncertainty of x by $\Delta x = \sqrt{<f|x^2|f> - <f|x|f>^2}$ for any state vector $|f>$, and similarly for $\dot{x}$. For any two positive normed state vectors the relationship $<f|f><g|g> \geq |<f|g>|^2$ follows from the Cauchy–Schwartz inequality. Applying this inequality to Eq. (5.72),[13] it can be shown that [39]

$$\Delta x \Delta \dot{x} \geq \frac{\sigma^2}{2} \quad : \text{ Heisenberg's uncertainty principle} \qquad (5.73)$$

5.11 Appendix: Path integration over stock price

The path integration over the $x(t)$ variables in Eq. (5.49) can be done exactly since recall the action S is quadratic in the $x(t)$ variables.

From Eq. (5.49)

$$p(x, y; \tau, x') = \int DY e^{S_0} \left[\int DX e^{S_X} \right] \equiv \int DY e^{S_0} Q \qquad (5.74)$$

$$\Rightarrow Q = \int DX e^{S_X} \qquad (5.75)$$

where from Eq. (5.46) the x-dependent term of the Lagrangian $L_{MG} = L_0 + L_X$ is given by

$$L_X(i) = -\frac{e^{-y_i}}{2(1-\rho^2)} \left[\frac{\delta x_i}{\epsilon} + r - \frac{1}{2}e^{y_i} - \frac{\rho}{\xi}e^{y_i/2}\left(\frac{\delta y_i}{\epsilon} + \mu - \frac{1}{2}\xi^2\right) \right]^2 + O(\epsilon) \qquad (5.76)$$

Let

$$c_i = r - \frac{1}{2}e^{y_i} - \frac{\rho}{\xi}e^{y_i/2}\left(\frac{\delta y_i}{\epsilon} + \mu - \frac{1}{2}\xi^2\right)$$

Then

$$S_X = -\frac{1}{2\epsilon(1-\rho^2)} \sum_{i=1}^{N} e^{-y_i}(x_i - x_{i-1} + \epsilon c_i)^2 \qquad (5.77)$$

---

[13] The simplest way to proceed is to multiply Eq. (5.72) by i and define the Hermitian operator $p = i\hat{x}$, thus making Eq. (5.72) equivalent to Eq. (5.63); one can then follow the standard proof given in [39].

with boundary values given by

$$x_0 = x', \quad x_N = x$$

Make a change of variables

$$x_i = z_i - \epsilon \sum_{j=1}^{i} c_j, \qquad dx_i = dz_i, \quad i = 1, 2, \ldots, N-1$$

with boundary values

$$z_0 = x'; \quad z_N = x + \epsilon \sum_{j=1}^{N} c_j$$

Hence, from Eq. (5.77)

$$S_X \equiv S_Z = -\frac{1}{2\epsilon(1-\rho^2)} \sum_{i=1}^{N} e^{-y_i}(z_i - z_{i-1})^2 \tag{5.78}$$

and, from Eqs. (5.75) and (5.78)

$$Q = \frac{e^{-y_N/2}}{\sqrt{2\pi\epsilon(1-\rho^2)}} \prod_{i=1}^{N-1} \int_{-\infty}^{+\infty} \frac{dz_i e^{-y_i/2}}{\sqrt{2\pi\epsilon(1-\rho^2)}} e^{S_Z}$$

All the z_i integrations can be performed exactly; one starts from the boundary by first integrating over z_1, and then over $z_2, \ldots$ and finally over z_{N-1}. The z_1 integration yields

$$\int_{-\infty}^{+\infty} \frac{dz_1 e^{-y_1/2}}{\sqrt{2\pi\epsilon(1-\rho^2)}} \exp\left\{-\frac{1}{2\epsilon(1-\rho^2)}\left[e^{-y_2}(z_2-z_1)^2 + e^{-y_1}(z_1-z_0)^2\right]\right\}$$

$$= \frac{e^{y_2/2}}{\sqrt{e^{y_1}+e^{y_2}}} \exp\left\{-\frac{1}{2\epsilon(1-\rho^2)}\frac{1}{e^{y_1}+e^{y_2}}(z_2-z_0)^2\right\} \tag{5.79}$$

Repeating this procedure $(N-1)$ times yields

$$Q = \frac{e^{S_1}}{\sqrt{2\pi\epsilon(1-\rho^2)\sum_{i=1}^{N} e^{y_i}}}$$

where

$$S_1 = -\frac{1}{2\epsilon(1-\rho^2)\sum_{i=1}^{N} e^{y_i}}(z_N - z_0)^2$$

$$= -\frac{1}{2\epsilon(1-\rho^2)\sum_{i=1}^{N} e^{y_i}}\left\{x - x' + \epsilon \sum_{i=1}^{N} c_i\right\}^2 \tag{5.80}$$

For the case of constant volatility $\xi = 0 = \rho$ and $e^{y_i} = \sigma^2 = $ constant for all i; Eq. (5.80) then reduces to the Black–Scholes given in Eq. (4.39).

From Eqs. (5.80) and (5.74)

$$p(x, y, \tau; x') = \int DY \frac{e^{S_0 + S_1}}{\sqrt{2\pi\epsilon(1 - \rho^2)\sum_{i=1}^{N} e^{y_i}}} \tag{5.81}$$

where the first term in Eq. (5.48) gives

$$S_0 = -\frac{\epsilon}{2\xi^2} \sum_{i=1}^{N} \left(\frac{\delta y_i}{\epsilon} + \mu - \frac{\xi^2}{2}\right)^2 \tag{5.82}$$

and S_1 is the result of the DX path integration given by Eq. (5.80) as

$$S_1 = -\frac{1}{2(1 - \rho^2)\epsilon \sum_{i=1}^{N} e^{y_i}} \left(x - x' + \epsilon \sum_{i=1}^{N} \left(r - \frac{1}{2}e^{y_i}\right)\right)$$
$$-\frac{\rho}{\xi} \sum_{i=1}^{N} e^{y_i/2} \left[\delta y_i + \epsilon\left(\mu - \frac{\xi^2}{2}\right)\right]\right)^2 \tag{5.83}$$

Based on Eqs. (5.81)–(5.83), extensive numerical studies of the pricing of European call options are discussed in [8], and the algorithm used in the study is discussed in Appendix 5.16.

5.12 Appendix: Generating function for stochastic volatility

The following partition function is evaluated

$$Z(j, y, p) = \int DY \exp\left[\int_0^\tau dt\, j(t)y(t)\right] e^{py(0)/2} e^{S_0} \tag{5.84}$$

From Eq. (5.53)

$$S_0 = -\frac{1}{2\xi^2} \int_0^\tau dt \left(\frac{dy}{dx} + \mu - \frac{1}{2}\xi^2\right)^2$$

with boundary condition

$$y(\tau) = y \tag{5.85}$$

Consider $Z(j, y, y')$ that is given by the path-integral in Eq. (5.84), but with the boundary conditions

$$y(0) = y', y(\tau) = y \tag{5.86}$$

Then, from Eq. (5.84)

$$Z(j, y, p) = \int_{-\infty}^{\infty} dy' Z(j, y, y') e^{py'/2} \tag{5.87}$$

Define new path-integration variables $z(t)$ by

$$z(t) = y(t) - y' + \frac{t}{\tau}(y' - y) \tag{5.88}$$

which, due to Eq. (5.88), have boundary conditions

$$z(0) = 0 = z(\tau). \tag{5.89}$$

Hence

$$Z(j, y, y') = e^{W_0} \int DZ e^{S_z} \tag{5.90}$$

with

$$W_0 = -\frac{1}{2\tau\xi^2} \left(y' - y - \mu\tau + \frac{1}{2}\xi^2\tau^2 \right)^2 + y' \int_0^\tau dt\, j(t)$$
$$- \frac{y' - y}{\tau} \int_0^\tau dt\, j(t)t \tag{5.91}$$

and

$$S_z = \int_0^\tau dt\, j(t) z(t) - \frac{1}{2\xi^2} \int_0^\tau dt \left(\frac{dz}{dt} \right)^2 \tag{5.92}$$

To perform the path integral over $z(t)$, the boundary conditions Eq. (5.89) yield the following Fourier sine expansion

$$z(t) = \sum_{n=1}^{\infty} \sin(\pi n t/\tau) z_n$$

From Eqs. (5.89) and (5.92)

$$S_z = -\frac{\pi^2}{4\tau\xi^2} \sum_{n=1}^{\infty} n^2 z_n^2 + \sum_{n=1}^{\infty} \left[\int_0^\tau dt\, j(t) \sin(\pi n t/\tau) \right] z_n$$

The path integration over $z(t)$ factorizes into infinitely many Gaussian integrations over the z_ns and hence

$$\int DZ e^{S_z} = C' \int_{-\infty}^{\infty} dz_1 dz_2 dz_3 \cdots dz_\infty e^{S_z}$$
$$= C(\tau) e^W$$

where

$$W = \frac{\xi^2 \tau}{\pi^2} \int_0^\tau dt dt' \, j(t) D_0(t, t') j(t')$$

$$= \frac{\xi^2}{\tau} \int_0^\tau dt \int_0^t dt' \, j(t)(\tau - t)t' j(t')$$

since

$$D_0(t, t') = \sum_{n=1}^\infty \frac{1}{n^2} \sin(\pi nt/\tau) \sin(\pi nt'/\tau)$$

$$= \frac{\pi^2}{2\tau} \left[t' \Theta(t - t') + t \Theta(t' - t) - \frac{tt'}{\tau} \right]$$

where the step function is given by Eq. (A.7). In Eq. (A.46) the propagator, up to a normalization, has been derived using the method of Greens functions.

The normalization function $C(\tau)$ can be evaluated by first considering the discrete and finite version of the DZ path integral. The result is given [33] by

$$C(\tau) = \frac{1}{\sqrt{2\pi \xi^2 \tau}}$$

Collecting the results yields

$$Z(j, y', y) = \frac{e^{W_0 + W}}{\sqrt{2\pi \xi^2 \tau}}$$

and performing the y' Gaussian integration in Eq. (5.87) finally yields

$$Z(j, y, p) = e^F$$

with

$$F = y \int_0^\tau dt j(t) + \left(\mu - \frac{1}{2}\xi^2 \right) \int_0^\tau dt(\tau - t) j(t)$$

$$+ \xi^2 \int_0^\tau dt j(t)(\tau - t) \int_0^t dt' \, j(t') + F' \qquad (5.93)$$

where

$$F' = \frac{1}{2} yp + \frac{1}{2}\xi^2 p \int_0^\tau dt(\tau - t) j(t) + \frac{1}{2} p\tau \left(\mu - \frac{1}{2}\xi^2 \right) + \frac{1}{8} p^2 \xi^2 \tau \qquad (5.94)$$

5.13 Appendix: Moments of stock price and stochastic volatility

The expression Eq. (5.61) for $< u^n w^m v^p >$ generalizes the result of Hull and White [49] since it is an exact expression for all the moments of u, v and w as well as for all their cross correlators.

The first few moments are evaluated using Eq. (5.61).

$$< w > = \frac{e^y}{\tau} \int_0^\tau dt\, e^{\mu t}$$

$$= \frac{V}{\mu\tau}(e^{\mu\tau} - 1) \qquad (5.95)$$

where $V = e^y$

$$< w^2 > = \frac{2e^{2y}}{\tau^2} \int_0^\tau dt_1 \int_0^{t_1} dt_2\, e^{\mu(t_1+t_2)} e^{\xi^2 t_2}$$

$$= \frac{2V^2}{\tau^2} \left[\frac{e^{(2\mu+\xi^2)\tau}}{(\xi^2+\mu)(2\mu+\xi^2)} - \frac{e^{\mu\tau}}{\mu(\mu+\xi^2)} + \frac{1}{\mu(2\mu+\xi^2)} \right] \qquad (5.96)$$

$< w^3 >$ is evaluated for the case of $\mu = 0$; from Eq. (5.61)

$$< w^3 > = \frac{6e^{3y}}{\tau^3} \int_0^\tau dt_1 \int_0^{t_1} dt_2 \int_0^{t_2} dt_3\, e^{\xi^2(t_2+2t_3)}$$

$$= \frac{V^3}{3\xi^6\tau^2} \left(e^{3\xi^2\tau} - 9e^{\xi^2\tau} + 8 + 6\xi^2\tau \right) \qquad (5.97)$$

Eqs. (5.95), (5.96) and (5.97) agree exactly with the results stated (without derivation) in Hull and White [49].

It is reassuring to see that two very different formalisms agree exactly and this increases ones confidence in the path-integral approach.

A few more moments are computed (recall $\sigma = e^{y/2}$).

$$< u > = \frac{e^{y/2}}{\tau} \int_0^\tau dt\, e^{(\mu/2-\xi^2/8)t} = \frac{2\sigma}{\tau(\mu - \xi^2/4)} \left[e^{(\mu/2-\xi^2/8)\tau} - 1 \right]$$

Furthermore

$$< u^2 > = \frac{2e^y}{\tau^2} \int_0^\tau dt_1 \int_0^{t_1} dt_2\, e^{\mu(t_1+t_2)/2} e^{-\xi^2(t_1-t_2)/8}$$

$$= \frac{4V}{\tau^2} \left[\frac{e^{(\mu/2-\xi^2/8)\tau}}{(\mu^2 + \xi^2/4)(\mu - \xi^2/8)} - \frac{2e^{(\mu/2-\xi^2/4)\tau}}{(\mu + \xi^2/4)(\mu - \xi^2/2)} \right.$$

$$\left. - \frac{1}{(\mu - \xi^2/2)(\mu - \xi^2/8)} \right]$$

and, lastly, for $a_1 = \frac{1}{2}, a_2 = 1$ in Eq. (5.62)

$$< uw > = \frac{2e^{3y/2}}{\tau^2} \int_0^\tau dt_1 \int_0^{t_1} dt_2 e^{\mu(t_1/2+t_2)} e^{\xi^2(t_1-6t_2)/8}$$

$$= \frac{4\sigma V}{\tau^2} \left[\frac{e^{(3\mu/2+3\xi^2/8)\tau}}{3(\mu+\xi^2/2)(\mu+\xi^2/4)} - \frac{e^{(\mu/2-\xi^2/8)\tau}}{(\mu+\xi^2/2)(\mu^2-\xi^2/4)} + \frac{1}{3(\mu-\xi^4/16)} \right]$$

All the other moments $< u^n w^m v^p >$ can similarly be evaluated from Eq. (5.61), and which in turn yields an infinite series solution for $p(x, y, \tau; x')$ in Eq. (5.55).

5.14 Appendix: Lagrangian for arbitrary α

The value for α that is chosen by the market is not *à priori* known. Hence, one needs to repeat the calculations carried out for the case of $\alpha = 1$ for arbitrary α, and then calibrate α from the market. Recall market data show that all values of $\alpha \in [\frac{1}{2}, 1]$ are equally valid.

For stochastic volatility with arbitrary α [8, 70]

$$N(\epsilon)e^{\epsilon L} = \langle x, y \mid e^{-\epsilon H} \mid x', y' \rangle$$

$$= \int_{-\infty}^\infty \frac{dp_x}{2\pi} \int_{-\infty}^\infty \frac{dp_y}{2\pi} \langle x, y \mid e^{-\epsilon H} \mid p_x, p_y \rangle \langle p_x, p_y \mid x', y' \rangle$$

The Hamiltonian in the momentum basis is given by ($\delta x = x - x'; \delta y = y - y'$)

$$H = \frac{e^y}{2} p_x^2 + \xi\rho e^{y(\alpha-1/2)} p_x p_y + \frac{\xi^2 e^{2y(\alpha-1)}}{2} p_y^2$$

$$+ \left(\frac{e^y}{2} - r - \frac{\delta x}{\epsilon} \right) i p_x + \left(\frac{\xi^2 e^{2y(\alpha-1)}}{2} - \lambda e^{-y} - \mu - \frac{\delta y}{\epsilon} \right) i p_y$$

Hence

$$N(\epsilon)e^{\epsilon L} = \int_{-\infty}^\infty \frac{dp_x}{2\pi} \int_{-\infty}^\infty \frac{dp_y}{2\pi} \exp -\epsilon \left(\frac{e^y}{2} p_x^2 + \xi\rho e^{y(\alpha-1/2)} p_x p_y \right.$$

$$+ \frac{\xi^2 e^{2y(\alpha-1)}}{2} p_y^2 - \left(\frac{\delta x}{\epsilon} + r - \frac{e^y}{2} \right) i p_x$$

$$\left. - \left(\frac{\delta y}{\epsilon} + \lambda e^{-y} + \mu - \frac{\xi^2 e^{2y(\alpha-1)}}{2} \right) i p_y \right)$$

Performing the Gaussian integrations over p_x, p_y yields

$$N(\epsilon) = \frac{e^{y(1/2-\alpha)}}{2\pi\epsilon\xi\sqrt{1-\rho^2}}$$

with

$$L = -\frac{e^{-y}}{2(1-\rho^2)}\left(\frac{\delta x}{\epsilon} + r - \frac{e^y}{2} - \frac{\rho e^{y(3/2-\alpha)}}{\xi}\left(\frac{\delta y}{\epsilon} + \lambda e^{-y} + \mu - \frac{\xi^2 e^{2y(\alpha-1)}}{2}\right)\right)^2$$
$$-\frac{e^{2y(1-\alpha)}}{2\xi^2}\left(\frac{\delta y}{\epsilon} + \lambda e^{-y} + \mu - \frac{\xi^2 e^{2y(\alpha-1)}}{2}\right)^2$$

The above Lagrangian is exact in the limit $N \to \infty$.

5.15 Appendix: Path integration over stock price for arbitrary α

Recall the action is defined as $S = \int L dt$. The discretized version of the action is given by $S = \epsilon \sum_{i=1}^{N} L_i + O(\epsilon)$ where L_i is the Lagrangian at time step i. Hence, the pricing kernel can be written in terms of the action as

$$\langle x, y \mid e^{-\hat{H}\tau} \mid x'\rangle = \int_{-\infty}^{\infty} dy' \langle x, y \mid e^{-\hat{H}\tau} \mid x', y'\rangle$$
$$= \lim_{N \to \infty} \int DX DY e^S$$

where

$$DX = \frac{e^{-y_N/2}}{\sqrt{2\pi\epsilon(1-\rho^2)}} \prod_{i=1}^{N-1} \int_{-\infty}^{\infty} \frac{dx_i e^{-y_i/2}}{\sqrt{2\pi\epsilon(1-\rho^2)}}$$

$$DY = \int_{-\infty}^{\infty} dy_0 \left(\prod_{i=1}^{N-1} \int_{-\infty}^{\infty} \frac{dy_i e^{y_i(1-\alpha)}}{\sqrt{2\pi\epsilon\xi}}\right)$$

(again $x_0 = x'$, $x_N = x$, $y_0 = y'$ and $y_N = y$). The action is quadratic in x. This enables one, similar to the case of $\alpha = 1$, to integrate over the stock price. As in Eq. (5.75) define

$$Q = \int DX e^{S_x} = \frac{e^{-y_N/2}}{\sqrt{2\pi\epsilon(1-\rho^2)}} \prod_{i=1}^{N-1} \int_{-\infty}^{\infty} \frac{dx_i e^{-y_i/2}}{\sqrt{2\pi\epsilon(1-\rho^2)}} e^{S_x}$$

which is the integral of the action over the stock price. Q can be evaluated by re-tracing the steps taken for $\alpha = 1$. The x-dependent term in the Lagrangian is

$$L_X(i) = -\frac{e^{-y_i}}{2(1-\rho^2)}\left(\frac{\delta x_i}{\epsilon} + r - \frac{e^{y_i}}{2} - \frac{\rho e^{y_i(3/2-\alpha)}}{\xi}\right.$$
$$\left. \times \left(\frac{\delta y_i}{\epsilon} + \lambda e^{-y_i} + \mu - \frac{\xi^2 e^{2y_i(\alpha-1)}}{2}\right)\right)^2$$

Let

$$c_i = r - \frac{e^{y_i}}{2} - \frac{\rho e^{y_i(3/2-\alpha)}}{\xi}\left(\frac{\delta y_i}{\epsilon} + \lambda e^{-y_i} + \mu - \frac{\xi^2 e^{2y_i(\alpha-1)}}{2}\right)$$

Hence

$$S_x = -\frac{1}{2\epsilon(1-\rho^2)}\sum_{i=1}^{N} e^{-y_i}(x_i - x_{i-1} + \epsilon c_i)^2$$

which is identical to Eq. (5.77) obtained earlier for the case of $\alpha = 1$. Integrating over all the variables z_i yields

$$Q = \frac{e^{S_1}}{\sqrt{2\pi\epsilon(1-\rho^2)\sum_{i=1}^{N} e^{y_i}}}$$

where, from Eq. (5.80)

$$
\begin{aligned}
S_1 &= -\frac{1}{2\epsilon(1-\rho^2)\sum_{i=1}^{N} e^{y_i}}(z_N - z_0)^2 \\
&= -\frac{1}{2\epsilon(1-\rho^2)\sum_{i=1}^{N} e^{y_i}}\left(x - x' + \epsilon\sum_{i=1}^{N} c_i\right)^2 \\
&= -\frac{1}{2\epsilon(1-\rho^2)\sum_{i=1}^{N} e^{y_i}}\left(x - x' + \epsilon\sum_{i=1}^{N}\left(r - \frac{e^{y_i}}{2} - \frac{\rho e^{y_i(3/2-\alpha)}}{\xi}\right.\right. \\
&\qquad \left.\left.\times\left(\frac{\delta y_i}{\epsilon} + \lambda e^{-y_i} + \mu - \frac{\xi^2 e^{2y_i(\alpha-1)}}{2}\right)\right)\right)^2
\end{aligned}
$$

(5.98)

On taking the limit of $N \to \infty$[14]

$$
\begin{aligned}
S_1 = -\frac{1}{2(1-\rho^2)\omega}&\left(x - x' + r\tau - \frac{\omega}{2}\right. \\
&\left. - \frac{\rho(e^{y(\tau)(3/2-\alpha)} - e^{y(0)(3/2-\alpha)})}{(3/2-\alpha)\xi} - \frac{\rho\lambda}{\xi}\theta - \frac{\rho\mu}{\xi}\eta + \frac{\rho\xi}{2}\zeta\right)^2
\end{aligned}
$$

(5.99)

---

[14] For $\alpha \neq 3/2$ the term $e^{y(0)(3/2-\alpha)}$ arises from the fact that $\int_0^\tau dt e^{y(3/2-\alpha)}(dy/dt) = \int_{y(0)}^{y(\tau)} dy e^{y(3/2-\alpha)}$ and it can be easily seen that when $\alpha = 3/2$, that term in the action is replaced by $(\rho/\xi)(y(\tau) - y(0))$.

Similar to the $\alpha = 1$ case, one has the following

$$\omega = \int_0^\tau e^y dt = \int_0^\tau V dt \qquad (5.100)$$

$$\theta = \int_0^\tau e^{y(1/2-\alpha)} dt = \int_0^\tau V^{1/2-\alpha} dt \qquad (5.101)$$

$$\eta = \int_0^\tau e^{y(3/2-\alpha)} dt = \int_0^\tau V^{3/2-\alpha} dt \qquad (5.102)$$

$$\zeta = \int_0^\tau e^{y(\alpha-1/2)} dt = \int_0^\tau V^{\alpha-1/2} dt \qquad (5.103)$$

when V follows the random process (3.27).

For the case of $\alpha = 1$, $\lambda = 0$, the quantity θ is decoupled from the problem; furthermore $\omega \to w$, with $\eta = \zeta \to u$, and $f(\omega, \theta, \eta, \zeta, v) \to g(u, v, w)$.

Similar to the case of $\alpha = 1$, if one can find the joint probability density functions for ω, θ, η, ζ and $v = e^{y(0)(3/2-\alpha)}$ one can obtain an analytic solution for the problem that will be given by

$$\langle x, y \mid e^{-\hat{H}\tau} \mid x' \rangle = \int_0^\infty d\omega d\theta d\eta d\zeta dv \frac{e^{S_1(\omega, \theta, \eta, \zeta, v)}}{\sqrt{2\pi\epsilon(1-\rho^2)\omega}} f(\omega, \theta, \eta, \zeta, v)$$

$$(5.104)$$

where f is the joint probability distribution function, similar to $g(u, v, w)$ given in Eq. (5.56). The equation above is the analog of the expression obtained in Eq. (5.55) for the case of $\alpha = 1$. However, unlike the $\alpha = 1$ case, for the general case of α it has not been possible to solve for the joint probability distribution function $f(\omega, \theta, \eta, \zeta, v)$.

The discrete solution finally gives

$$\langle x, y \mid e^{-\tau H} \mid x' \rangle = \int DY \frac{e^{S_0 + S_1}}{\sqrt{2\pi\epsilon(1-\rho^2) \sum_{i=1}^N e^{y_i}}} \qquad (5.105)$$

where S_1 is given in Eq. (5.99) and

$$S_0 = -\frac{\epsilon}{2\xi^2} \sum_{i=1}^N e^{2y_i(1-\alpha)} \left(\frac{\delta y_i}{\epsilon} + \lambda e^{-y_i} + \mu - \frac{\xi^2 e^{2y_i(\alpha-1)}}{2} \right)^2 \qquad (5.106)$$

$$\int DY = \int dy_0 \left(\prod_{i=1}^{N-1} \int \frac{dy_i e^{y_i(1-\alpha)}}{\sqrt{2\pi\epsilon\xi}} \right) \qquad (5.107)$$

For $\alpha \neq 1$ case, the result of performing the functional integration over the stock price $x(t)$ yields a nontrivial measure term for the remaining $y(t)$ integrations; this is the reason that the $N \to \infty$ limit cannot be taken.

Fortunately, it is possible to derive a Monte Carlo algorithm to calculate option prices with the volatility performing the stochastic process (3.27) with $-1 \leq \rho \leq 1$, and having almost the same efficiency as the straightforward solution when $\rho = 0$. However, the method has a disadvantage in that it cannot handle lumpy dividends (for a continuous dividend yield q, one can simply replace r by $r - q$).

5.16 Appendix: Monte Carlo algorithm for stochastic volatility

The path-integral formulation of option pricing is well known in finance, and Monte Carlo methods useful for solving these path-integrals numerically are widely used [55]. Familiarity with these techniques is assumed, and the numerical advantages that a path-integral approach may provide in some problems is discussed in this appendix. Only the Monte Carlo algorithm is discussed, and no numerical results based on this algorithm will be presented. The numerical results are extensively discussed in [8, 70].

The reasons for discussing the Monte Carlo algorithm for the case of stochastic volatility are (a) the problem is sufficiently complex to illustrate the general characteristics of a numerical evaluation of the path integral and (b) due to the exact integration over the stock price path integral, the unique advantages of the path integral can be illustrated.

The main result of this appendix is that numerically solving the volatility path integral using Monte Carlo methods is a few hundred times faster than solving it using the defining stochastic differential equations.

The algorithm for the path-independent European call option only is considered, as the large number of parameters make the study of other path-dependent options more computationally intensive.

The Monte Carlo algorithm that is discussed is valid for any system driven by a Langevin stochastic differential [44, 92], be it a stock price or the spot rate or a Treasury Bond. The algorithm is quite generic in nature, and could be applied for numerically studying a wide variety of stochastic processes.

The Monte Carlo algorithm solves for the option prices when the stock price and the volatility are undergoing the following stochastic processes, given by Eqs. (3.28) and (3.29), as

$$dS = (\phi S + \sigma S W)dt \tag{5.108}$$

$$dV = (\lambda + \mu V + \xi V^{\alpha} Q)dt \tag{5.109}$$

where ϕ, λ, μ and ξ are constants, $V = \sigma^2$ and W and Q are white noise processes whose correlation is given by $-1 < \rho < 1$. As required by the principle of risk-neutral valuation, assume that the expected growth rate of all securities is given by the risk-free interest rate r, and hence this replaces ϕ in Eq. (5.108).

From Eqs. (5.105) and (5.98) the pricing kernel is given by the volatility path integral

$$p(x, y, \tau; x') = \int dy_0 \left(\prod_{i=1}^{N-1} \int \frac{dy_i \, e^{y_i(1-\alpha)}}{\sqrt{2\pi \epsilon \xi^2}} \right) \frac{e^{S_0+S_1}}{\sqrt{2\pi \epsilon (1 - \rho^2) \sum_{i=1}^{N} e^{y_i}}} \qquad (5.110)$$

For this problem, choose the following probability density function

$$p(Y) = \left(\prod_{i=1}^{N-1} \frac{e^{y_i(1-\alpha)}}{\sqrt{2\pi \epsilon \xi}} \right) e^{S_0} \qquad (5.111)$$

where Y is the set of variables y_i (and is hence $N - 1$ dimensional) and S_0 is given in (5.106). Hence for S_1 given in (5.98) and

$$g(Y) = \frac{e^{S_1}}{\sqrt{2\pi \epsilon (1 - \rho^2) \sum_{i=1}^{N} e^{y_i}}} \qquad (5.112)$$

the integral that is being performed is

$$\int DY \left(\prod_{i=1}^{N-1} e^{y_i(1-\alpha)} \right) \frac{e^{S_0+S_1}}{\sqrt{2\pi \epsilon (1 - \rho^2) \sum_{i=1}^{N} e^{y_i}}} = \int DY p(Y) g(Y)$$

with

$$DY = dy_0 \prod_{i=1}^{N-1} \frac{dy_i}{\sqrt{2\pi \epsilon \xi}}$$

One needs to produce configurations Y with the probability distribution $p(Y)$. While $p(Y)$ looks rather complicated, it has a simple interpretation. Since only the path integration over the stock price was performed, the volatility is, as expected, the probability distribution for a discretized random walk performed by y. To see this, using Ito's lemma for the stochastic differential equation

$$dV = (\lambda + \mu V)dt + \xi V^\alpha Q dt \; ; \; V = e^y$$

yields the following process for y

$$dy = \left(\lambda e^{-y} + \mu - \frac{\xi^2 e^{2y(\alpha-1)}}{2} \right) dt + \xi e^{y(\alpha-1)} Q dt \qquad (5.113)$$

Discretize the process using Euler's method to obtain

$$\delta y_i = \left(\lambda e^{-y_i} + \mu - \frac{\xi^2 e^{2y_i(\alpha-1)}}{2} \right) \epsilon + \xi e^{y_i(\alpha-1)} Z \sqrt{\epsilon}$$

where $\delta y_i = y_i - y_{i-1}$, ϵ is the time step and $Z = N(0,1)$ is the standard normal random variable. Since $\tau = T - t$ is the time variable, the time step is actually $-\epsilon$. Hence, δy_i is a normal random variable with mean $(-\lambda e^{-y_i} - \mu + \xi^2 e^{2y_i(\alpha-1)}/2)\epsilon$ and variance $\xi^2 e^{2y_i(\alpha-1)}\epsilon$, with its probability density function given by

$$f_i = \frac{e^{y_i(1-\alpha)}}{\sqrt{2\pi\epsilon}\xi} \exp\left(-\frac{\epsilon e^{2y_i(1-\alpha)}}{2\xi^2} \left(\frac{\delta y_i}{\epsilon} + \lambda e^{-y_i} + \mu - \frac{\xi^2 e^{2y_i(\alpha-1)}}{2} \right)^2 \right)$$

Hence, the joint probability density function for the discretized process is given by

$$f = \prod_{i=1}^{N-1} f_i = \left(\prod_{i=1}^{N-1} \frac{e^{y_i(1-\alpha)}}{\sqrt{2\pi\epsilon}\xi} \right) e^{S_0}$$

which is the same as (5.111).

In the simulation Euler's discretization is used for derivatives as this is sufficiently accurate for generating the volatility paths.

The algorithm to find a Monte Carlo estimate of the pricing kernel $p = \langle x, y \mid e^{-\hat{H}\tau} \mid x' \rangle$ is as follows [70]

1. $p := 0$ (Initialization)
2. For $i := 1$ to N
3. Generate a path Y for y using (5.113)
4. $p := p + g(Y)/N$ (where $g(Y)$ is defined in (5.112))
5. End For

The paths must be generated backwards starting from y_N which is the initial value of $\ln V$ to obtain all the y_i ending at y_0. This can be done by reversing the drift terms since the equations are time symmetric. The end point y_0 is allowed to be arbitrary, and this procedure in effect automatically ends up performing the $\int dy_0$ integration. This will have to be repeated for all the points x' that will be integrated over. During implementation it is found to be more advantageous to generate the paths only once, storing the following terms ($t = N\epsilon$)

$$t_1 = \sum_{i=1}^{N} e^{y_i}$$

$$t_2 = \sum_{i=1}^{N} e^{y_i(3/2-\alpha)} \left(\frac{\delta y_i}{\epsilon} + \lambda e^{-y_i} + \mu - \frac{\xi^2 e^{2y_i(\alpha-1)}}{2} \right)$$

which are sufficient to determine S_1 once x' is given, namely

$$S_1 = -\frac{1}{2\epsilon(1-\rho^2)t_1} \left(x - x' + (r-q)t - \epsilon \left(\frac{t_1}{2} + \frac{\rho}{\xi}t_2 \right) \right)^2 \qquad (5.114)$$

That S_1 can be computed using this limited information is fortunate as storing all the paths explicitly would require a very large memory (10MB for 10,000 configurations as compared with 160kB when storing only the essential combinations of terms). The alternative of generating paths for each value of x' from the coupled Langevin equation given in Eq. (5.108) is inefficient due to the very large run time required.

The pricing kernel must finally be multiplied by the payoff function $g(x')$, and integrated over the variable x', to obtain the option price. The accuracy of this numerical quadrature depends on the spacing h between successive values of x'. This means that one has to find the pricing kernel for several values of x' to obtain reasonable accuracy, which is computationally very expensive. In [8, 70] the pricing kernel was determined using the above Monte Carlo method for only about 100 equally spaced values of x' over the range of the quadrature and cubic splines were used to interpolate it at the other quadrature points. This produces excellent results as the pricing kernel is seen to be an extremely smooth function of x'.

Hence, the algorithm to generate the option price is of the following form.

1. For $i := 1$ to N
2. Generate a path Y for y using (5.113)
3. Store t_1 and t_2 for the path
4. End For
5. For $x' := $ beginning of range to end of range
6. Find the Monte Carlo estimate for the pricing kernel at large intervals of x' using t_1 and t_2 from the paths.
7. End For
8. For $x' := $ beginning of range to end of range

9. Find the pricing kernel at small intervals of x' using cubic spline interpolation over the values of the pricing kernel found previously and integrate over the final payoff function

10. End For

11. Return the option price

In summary, the path-integral framework allows for introducing efficient algorithms into option pricing. The simple manner in which the correlation parameter ρ appears in the volatility path integral yields an efficiency that is a few hundred times faster than the standard Langevin-based algorithms [8, 70].

5.17 Appendix: Merton's theorem for stochastic volatility

The case $\rho = 0$ is considered, and the result that was stated earlier in Appendix 3.9 is derived. For $\rho = 0$ the Hamiltonian and Lagrangian are given by

$$H = -\left(r - \frac{e^y}{2}\right)\frac{\partial}{\partial x} - \left(\lambda e^{-y} + \mu - \frac{\xi^2 e^{2y(\alpha-1)}}{2}\right)\frac{\partial}{\partial y} - \frac{e^y}{2}\frac{\partial^2}{\partial x^2} - \frac{\xi^2 e^{2y(\alpha-1)}}{2}\frac{\partial^2}{\partial y^2}$$

$$\hat{L} = -\frac{e^{-y}}{2}\left(\frac{\delta x}{\epsilon} + r - \frac{e^y}{2}\right)^2 - \frac{e^{2y(1-\alpha)}}{2\xi^2}\left(\frac{\delta y}{\epsilon} + \lambda e^{-y} + \mu - \frac{\xi^2 e^{2y(\alpha-1)}}{2}\right)^2$$

Integrating out the stock price yields, from Eq. (5.98)

$$S = S_0 + S_1$$

$$S_1 = -\frac{1}{2\epsilon \sum_{i=1}^N e^{y_i}}\left(x - x' + \epsilon \sum_{i=1}^N \left(r - \frac{e^{y_i}}{2}\right)\right)^2$$

where S_0 is given in Eq. (5.106).

The expression for S_1 is what finally determines the option price, and is the same as that for the Black–Scholes case with $\frac{\epsilon}{\tau}\sum_{i=1}^N e^{y_i}$ replacing $\sigma^2 = e^y$. In other words, one needs to replace the constant volatility in the Black–Scholes equation by the average volatility during the time period under consideration, and average it over its probability of occurrence.

More precisely, inserting

$$1 = \int_{-\infty}^{+\infty} d\eta\, \delta\left(e^\eta - \frac{\epsilon}{\tau}\sum_{i=1}^N e^{y_i}\right)$$

into the pricing kernel given in Eq. (5.105), yields for the $p = 0$ case

$$\langle x, y \mid e^{-\tau H} \mid x' \rangle = \int DY \int_{-\infty}^{+\infty} d\eta \delta \left(e^{\eta} - \frac{\epsilon}{\tau} \sum_{i=1}^{N} e^{y_i} \right) \frac{e^{S_0 + S_1}}{\sqrt{2\pi \epsilon \sum_{i=1}^{N} e^{y_i}}}$$

$$= \int_{-\infty}^{+\infty} d\eta \, p_{BS}(x; \tau, x'; \sigma = e^{\eta/2}) P_M(\eta)$$

where the Black–Scholes pricing kernel $p_{BS}(x; \tau, x'; \sigma)$ is given by Eq. (4.39). The probability distribution for average volatility has been defined in Eq. (3.35), and an explicit expression for it is given by

$$P_M(\eta) = \int DY \delta \left(e^{\eta} - \frac{\epsilon}{\tau} \sum_{i=1}^{N} e^{y_i} \right) e^{S_0} \tag{5.115}$$

This is the content of Merton's theorem [75]. Although a specific process for the volatility has been assumed, the final result does not depend on the process as long as S_0 is independent of x, x'.

6

Stochastic interest rates' Hamiltonians and path integrals

All stochastic models of spot and forward interest rates are based on a finite number of degrees of freedom, and are precursors of the more general modelling of interest rates based on quantum field theory, which forms the subject of all the subsequent chapters.

The formalism of quantum field theory requires one to make a fairly large transition in the level of mathematical complexity. The path-integral and Hamiltonian analysis of stochastic interest rate models is undertaken to smoothen this transition, as well as for its intrinsic importance. The key ideas that will be later given a field theory generalization are introduced in stochastic models that have, at each instant, only a finite number of independent random variables.

6.1 Spot interest rate Hamiltonian and Lagrangian

The spot interest rate $r(t)$ is the interest rate for an overnight loan at time t. Spot rate models are useful for modelling the short time behaviour of the interest rates' yield curve [51], as well as in the study of the stock market [21]. Furthermore, since central bank policies intervene in determining the spot rate, jumps and discontinuities in the spot rate are particularly important, and need to be considered separately from the remaining yield curve.

We consider only the arbitrage-free, and not the empirical, martingale time evolution of the spot interest rate, as is required for pricing its derivatives. The interest spot rate models can hence be directly modelled using the Langevin equation.

Similar to a security, the spot rate is driven by a Langevin equation

$$\frac{dr}{dt} = a(r, t) + \sigma(r, t)R(t) : \quad R(t) : \text{ white noise } ; \quad t_0 \le t \le T \qquad (6.1)$$

with either the initial or final value of the spot rate specified as follows

$$\text{EITHER } r(t_0) = r_0 : \text{initial condition OR } r(T) = R : \text{final condition} \quad (6.2)$$

where in general the function a and volatility σ can be arbitrary functions of the spot rate r and of time t. The spot rate can be evolved either forward or backward in time using the Langevin equation, and hence two distinct boundary conditions are given in Eq. (6.2). Note at each instant in time the system has only one random variable $R(t)$, and hence has only one degree of freedom.

The evolution of the conditional probability that results from the Langevin equation yields the Fokker–Planck forward and backward equations, with their respective Hamiltonians.

Let $P_F(r, t; r_0)$ be the forward conditional probability that the spot rate has value r at time t, given that the value of r_0 occurred at an earlier time $t_0 < t$. The **forward Fokker–Planck Hamiltonian** H_F is given from Eq. (6.65) by[1]

$$\frac{\partial}{\partial t} P_F(r, t; r_0) = -H_F P_F(r, t; r_0)$$

$$\Rightarrow H_F = -\frac{1}{2}\frac{\partial^2}{\partial r^2}\sigma^2(r) + a(r)\frac{\partial}{\partial r} + \frac{\partial a(r)}{\partial r} \tag{6.3}$$

where H_F is the non-Hermitian forward Fokker–Planck Hamiltonian, and from Eq. (6.68)

$$P_F(r, t; r_0) = \, <r|e^{-(t-t_0)H_F}|r_0> \tag{6.4}$$

$$r(t_0) = r_0 : \quad \text{initial condition} \tag{6.5}$$

The backward conditional probability $P_B(R, t; r)$ is similarly defined as the probability that the spot interest rate will have the value of r at time t given that it has the value of R at some future time $T > t$. Eq. (6.70) yields the **backward Fokker–Planck Hamiltonian** (the time variable t has the opposite sign in the following equation compared with Eq. (6.3), indicating that time is $-t$ and hence is flowing backwards)

$$\frac{\partial}{\partial t} P_B(R, t; r) \equiv +H_B P_B(R, t; r) \tag{6.6}$$

$$H_B = -\frac{1}{2}\sigma^2(r)\frac{\partial^2}{\partial r^2} - a(r)\frac{\partial}{\partial r} \tag{6.7}$$

$$= H_F^\dagger$$

where, from Eq. (6.71)

$$P_B(r, t; r_0) = \, <r|e^{-(T-t)H_B}|R> \tag{6.8}$$

$$r(T) = R : \quad \text{final condition}$$

The forward Fokker–Planck equation is required when the initial value of the spot rate is specified. The backward Fokker–Planck Hamiltonian needs to be used

---

[1] The detailed derivations of the forward and backward Fokker–Planck Hamiltonians are given in Appendix 6.8.

for any problem involving the present value of an option on the spot interest rate, with the payoff function being specified in the future.

The generalized option pricing Hamiltonian H_V given in Eq. (4.44) can be seen to be equivalent to a backward Fokker–Planck Hamiltonian given by $-(\sigma^2/2)\partial^2/\partial x^2 + [(1/2)\sigma^2 - V(x)]\partial/\partial x$, with the important additional feature that the discounting of the payoff function needs to have a nontrivial factor containing the potential as in Eq. (4.45).

Black–Scholes Hamiltonian

Suppose that one has no knowledge of the Black–Scholes equation, and instead starts with the generalized backward Hamiltonian, given in Eq. (6.7),[2] for pricing options on a stock given by $S = e^x$. The evolution of the option price has to satisfy the martingale condition given in Eq. (4.43), and hence (the superscript label is to indicate that H_B refers to the evolution of a stock price driven by the Black–Scholes process)

$$[H_B^{bs} + r]e^x = 0 \; ; \quad \Rightarrow a(x) = r - \frac{1}{2}\sigma^2(x)$$

Hence the Black–Scholes Hamiltonian is shown to be given by

$$H_{BS} = H_B^{bs} + r$$
$$= -\frac{1}{2}\sigma^2(x)\frac{\partial^2}{\partial x^2} - \left(r - \frac{1}{2}\sigma^2(x)\right)\frac{\partial}{\partial x} + r$$

The martingale condition allows stochastic volatility to be an arbitrary function of the stock price e^x, a result that can also be obtained from the Black–Scholes analysis discussed in Section 3.5.

The reason that one does not start the Black–Scholes analysis from the backward Fokker–Planck Hamiltonian is that, in order to complete the analysis, one needs to first have a Hamiltonian formulation of the martingale condition. The martingale condition requires the introduction of the spot rate r into the Hamiltonian, a quantity that is not contained in the stochastic differential equation, but instead is something that results from the concepts of hedging and of no arbitrage.

6.1.1 Stochastic quantization

The forward Fokker–Planck Lagrangian, action and partition functions are given from Eq. (6.73) by

$$L_F = -\frac{1}{2\sigma^2}\left(\frac{dr}{dt} - a(r)\right)^2 - \frac{\partial a(r)}{\partial r}$$
$$Z_F = \int Dr\, e^{\int_{t_0}^{T} L_F dt} \tag{6.9}$$

---

[2] From Eq. (3.12), the Black–Scholes stochastic differential equation is a special case of the Langevin equation given in Eq. (6.1) with $a = \phi - \sigma^2/2$.

The backward Lagrangian can be obtained from the backward Hamiltonian H_B, but a procedure known as **stochastic quantization** [29] is employed to illustrate another approach to the path integral for stochastic systems.

In stochastic quantization the partition function is defined by a path integral over **both** the spot rate $r(t)$ and white noise $R(t)$. The fact that these two stochastic processes are connected by the Langevin equation is realized by a functional delta function constraint in the path integral. Hence the path integral gives the backward Fokker–Planck partition function as follows

$$Z_B = \int DRDr \prod_{t=t_0}^{T} \delta \left[\frac{dr}{dt} - a(r,t) - \sigma(r,t)R(t) \right] e^{-\frac{1}{2}\int_{t_0}^{T} R^2(t)dt}$$

$$= \int Dr e^{S_B} \tag{6.10}$$

where to obtain Eq. (6.10) the path integration over the white noise $R(t)$ has been performed. The backward Fokker–Planck action is hence given by

$$S_B = -\frac{1}{2} \int_{t_0}^{T} dt \frac{1}{\sigma^2(t)} \left(\frac{dr}{dt} - a(r,t) \right)^2 \tag{6.11}$$

Since the propagation for the action S_B is backwards in time as given in Eq. (6.6), it is more transparent to write the action S_B in terms of remaining time $\tau = T - t$; this results in changing the sign of the drift term $a(r,t)$ in the action. Hence

$$S_B = -\frac{1}{2} \int_{0}^{T-t_0} d\tau \frac{1}{\sigma^2(\tau)} \left(\frac{dr}{d\tau} + a(r,\tau) \right)^2 \quad ; \quad \tau = T - t \tag{6.12}$$

The forward and backward Fokker–Planck Lagrangians differ by the term $\partial a(r)/\partial r$, and which results from the manner in which the path integral $\int DR$ is carried out. For the forward Lagrangian, if one repeats the calculation expressed in Eq. (6.10), the path integration over $R(t)$ yields an extra Jacobian term that gives precisely the extra term, namely $\partial a(r)/\partial r$, which is absent in the backward Lagrangian [106].

6.2 Vasicek model's path integral

The Vasicek model [47, 51, 102] can be solved exactly, with an exact path-integral solution being given by Otto [82]. The Vasicek model provides a prototypical example on how to apply path integrals to the spot interest rate.

Consider a zero coupon bond $P(t_0, T)$. Recall from Eq. (2.6) that the time value of money yields the relation between the zero coupon bond and the spot interest rate as given by

$$P(t_0, T) = E\left[e^{-\int_{t_0}^T dt\, r(t)} | r(t_0) = r_0\right] \tag{6.13}$$

The Vasicek model for the spot rate is given by [51, 102]

$$\frac{dr}{dt} = a(b - r) + \sigma R(t)$$

$$r(t_0) = r_0 : \text{initial condition} ; \ t_0 \leq t \leq T$$

Since the present value of a Treasury Bond is obtained by discounting its future value, the spot rate is specified at the future time T, and which propagates backwards so as to have the value of r_0 at time t_0. Hence the backward Fokker–Planck action is used for evolving the spot rate $r(t)$, and from Eq. (6.73) the quadratic (Gaussian) action is given by

$$S_V = -\frac{1}{2\sigma^2} \int_{t_0}^T dt \left[\frac{dr}{dt} - a(b - r)\right]^2 \tag{6.14}$$

Using the action functional given in Eq. (6.14), one can evaluate the expectation value in Eq. (6.13) for $P(t_0, T)$; the class of functions of $r(t)$ over which the path integration is to be performed needs to be specified.

Since the spot rate evolves over the finite time interval $t_0 \leq t \leq T$, one needs to specify the boundary conditions on $r(t)$ at the two end points. At initial time $t = t_0$, the interest rate is fixed to r_0. The final value of the spot interest rate at $t = T$, namely $r(T)$, is free to take all possible values and yields the Neumann boundary condition $dr(T)/dt = 0$. Hence the probability distribution for the spot rate is given by

$$e^{S_V}/Z ; \text{ probability distribution}$$

$$Z = \int Dr\, e^{S_V} ; \quad \int Dr \equiv \int_{-\infty}^{+\infty} \prod_{t=t_0}^T dr(t)$$

$$\text{boundary conditions } r(t_0) = r_0 , \quad \frac{dr(T)}{dt} = 0 \tag{6.15}$$

The denominator Z is necessary to correctly normalize the probability distribution, and is also the reason that the overall constants in S_V can ignored as they cancel out.

The zero coupon bond is given by averaging the discount factor over the probability distribution given in Eq. (6.15), and yields

$$P(t_0, T) = \frac{1}{Z} \int Dr\, e^{S_V}\, e^{-\int_{t_0}^T r(t)dt} \tag{6.16}$$

Path-integral solution for the Treasury Bond in Vasicek's model

From Eqs. (6.16) and (6.14)

$$P(t_0, T) = \frac{1}{Z} \int Dr \, e^S \tag{6.17}$$

$$S \equiv S_V - \int_{t_0}^{T} r(t) dt$$

$$= -\frac{1}{2\sigma^2} \int_{t_0}^{T} dt \left[\frac{dr}{dt} - a(b - r) \right]^2 - \int_{t_0}^{T} r(t) dt$$

Similar to Eq. (5.14), the Vasicek path integral is solved by a change of variables. The continuum formulation is used for notational simplicity. Since the path-integral measure $\int Dr$ is invariant under the translation $r(t) \to r(t) + b$

$$S = -\frac{1}{2\sigma^2} \int_{t_0}^{T} dt \left[\frac{dr}{dt} + ar \right]^2 - \int_{t_0}^{T} [r(t) + b] dt \tag{6.18}$$

Define new variables $v(t)$ by

$$v(t) = \frac{dr}{dt} + ar \tag{6.19}$$

$$\Rightarrow r(t) = e^{-a(t - t_0)} r_0 + e^{-at} \int_{t_0}^{t} dt' e^{at'} v(t') \tag{6.20}$$

$$\int_{t_0}^{T} dt \, r(t) = B(t_0, T) r_0 + \int_{t_0}^{T} dt \, B(t, T) v(t) \tag{6.21}$$

where $B(t, T) \equiv \dfrac{1 - e^{-a(T - t)}}{a}$

The initial condition at $t = t_0$ is fulfilled by Eq. (6.20); for the final condition at $t = T$, since $dr(T)/dt = 0$, which is equivalent to $r(T)$ being arbitrary, one can see from Eq. (6.19)[3] that $v(T)$ is free to take all possible values. Hence the boundary conditions are satisfied by integrating over all variables $v(t)$ for $t_0 \le t \le T$.[4] Hence, from Eqs. (6.16), (6.18), (6.19) and (6.21)

$$P(t_0, T) = e^{-b(T - t_0) - B(t_0, T) r_0} \frac{1}{Z} \int Dv \, e^{-\frac{1}{2\sigma^2} \int_{t_0}^{T} dt [v^2(t) + 2\sigma^2 B(t, T) v(t)]}$$

$$= e^{-b(T - t_0) - B(t_0, T) r_0} e^{\frac{\sigma^2}{2} \int_{t_0}^{T} dt \, B^2(t, T)} \tag{6.22}$$

The $v(t)$ integrations that have been performed to obtain Eq. (6.22) are decoupled Gaussian integrations, with the overall normalization being cancelled by the factor of Z. The result

---

[3] If, for example, $a = 0$ in Eq. (6.19), then $v(T) = dr(T)/dt = 0$ and hence one would need to constrain the variable $v(T)$ to be zero in the $\int Dv$ path integral; the $a = 0$ case can be recovered from $a \to 0$ since the limit is uniform.

[4] The change of variables in Eq. (6.19) yields $Dv = \det(\frac{d}{dt} + a) Dr$, and the Jacobian $J = \det(\frac{d}{dt} + a)$ is cancelled in the expression for $P(t_0, T)$ by the denominator Z in Eq. (6.16).

agrees with the one obtained in [82], where a different method has been used to do the path integral.

Simplifying Eq. (6.22), and for notational simplicity defining future time $\theta = T - t_0$, yields Vasicek's result [51, 102]

$$P(t_0, T) = A(\theta)e^{-B(\theta)r_0} \tag{6.23}$$

where recall $r(t_0) = r_0$, and

$$B(\theta) = \frac{1 - e^{-a\theta}}{a} \tag{6.24}$$

$$A(\theta) = \exp\left[\left(\frac{\sigma^2}{2a^2} - b\right)(\theta - B(\theta)) - \frac{\sigma^2}{4a}B^2(\theta)\right] \tag{6.25}$$

The forward interest rates are given, from Eqs. (2.9) and (6.23), by

$$f(t_0, T) = -\frac{\partial}{\partial T}\ln P(t_0, T) \tag{6.26}$$

$$= r_0 + (b - r_0)(1 - e^{-a\theta}) - \frac{\sigma^2}{2a^2}(1 - e^{-a\theta})^2 \tag{6.27}$$

The financial interpretation of the forward rates obtained above is discussed in [18].

6.3 Heath–Jarrow–Morton (HJM) model's path integral

The industry standard HJM model [43], and has been studied extensively both analytically and empirically [34]. The HJM model is reformulated in the language of path integration [4, 25]. The reason for the path-integral re-formulation is two fold, namely to understand the HJM model in the formalism of path integration, and, secondly, to be able to generalize the model and construct a quantum field theory of the forward interest rates.

From Eq. (2.9) the collection of zero coupon bonds $P(t, T)$ yield all the forward rates $f(t, x)$ for the interval $t \leq x \leq T$. Since the spot interest rate model yields an expression for $P(t, T)$, for example as in the Vasicek model, why should there be any need to directly model the forward rates [56]?

The reasons for modelling the forward rates are the following:

- As discussed in Appendix 6.9, most spot interest rate models yield an affine expression for the zero coupon bonds. However, it is known from the market that zero coupon bonds do not follow an affine behaviour.
- The forward rates have a complicated behaviour and it is unlikely that any reasonably simple spot rate model can generate such complex forward rates.
- The debt market directly trades in the forward rates and provides an enormous amount of data on these. It is sensible to create models that take the forward rates as the primary instrument so as to match the behaviour of the market.
- Spot rate models can in principle produce the initial forward rate curve,[5] but the evolution of the spot rate models cannot maintain the condition of no arbitrage on the future evolution of the forward rates.
- The HJM model takes the initial forward rate $f(t_0, x)$ as the input to be determined from the market, instead of trying to derive it from a model of the spot rate. A distinct advantage of the HJM approach is that it, consequently, fully incorporates all the information on the forward rates that is available from the market.
- The HJM approach yields, once the discounting factor has been fixed, a unique arbitrage-free evolution of the forward rate curve.

Recall from Eq. (2.8) that the zero coupon Treasury Bond is given by

$$P(t, T) = e^{-\int_t^T dx f(t, x)} \tag{6.28}$$

The time evolution of the forward rates is modelled to behave in a stochastic manner, and is given by generalizing the stochastic differential equation for equity as discussed in Section 3.3. In the K-factor HJM model [43, 58, 86] the time evolution of the forward rates is driven by K-independent white noises $W_i(t)$, and is given by

$$\frac{\partial f}{\partial t}(t, x) = \alpha(t, x) + \sum_{i=1}^{K} \sigma_i(t, x) W_i(t) \tag{6.29}$$

where $\alpha(t, x)$ is the drift velocity term and $\sigma_i(t, x)$ are the deterministic volatilities of the forward rates.

Note that although the HJM model evolves an entire curve $f(t, x)$, at each instant of time t it is driven by K random variables given by $W_i(t)$, and hence has only K degrees of freedom. The term **stochastic interest rates models** is used for any forward or spot interest rate models that are driven by a finite number of white noises, to distinguish them from models based on quantum field theory.

---

[5] For example, Hull and White [51] have proposed an extension of the Vasicek model in which the parameters a and b in Eq. (6.26) are made time dependent, and are adjusted to fit the entire initial curve $f(t_0, T)$ taken from the market.

From Eq. (6.29)

$$f(t, x) = f(t_0, x) + \int_{t_0}^{t} dt' \alpha(t', x) + \int_{t_0}^{t} dt' \sum_{i=1}^{K} \sigma_i(t', x) W_i(t') \qquad (6.30)$$

The initial forward rate curve $f(t_0, x)$ is determined from the market, and so are the volatility functions $\sigma_i(t, x)$. Similar to the Black–Scholes analysis discussed in Section 3.5, the drift term $\alpha(t, x)$ is fixed to ensure that the forward rates have a martingale time evolution, which makes it a function of the volatilities $\sigma_i(t, x)$.

For every value of time t, the stochastic variables $W_i(t), i = 1, 2, \ldots, K$ are independent Gaussian random variables given by

$$E(W_i(t) W_j(t')) = \delta_{ij} \delta(t - t')$$

The forward rates $f(t, x)$ are driven by random variables $W_i(t)$ which give the same random 'shock' at time t to all the future forward rates $f(t, x)$, $x > t$. To bring in the maturity dependence of the random shocks on the forward rates, the volatility function $\sigma_i(t, x)$, at given time t, weighs this 'shock' differently for each x.

White noise for the HJM model

The salient properties of white noise are discussed in Appendix A.4, and are reviewed for completeness. To write the probability measure for $W_i(t)$, note that t takes values in a finite interval depending on the problem of interest; as usual, discretize $t \to m\epsilon$, with $m = 1, 2, \ldots, M$, and with $W_i(t) \to W_i(m)$. The probability measure is given by

$$\mathcal{P}[W] = \prod_{m=1}^{M} \prod_{i=1}^{K} e^{-\frac{\epsilon}{2} \sum_{i=1}^{K} W_i^2(m)} \qquad (6.31)$$

$$\int dW = \prod_{m=1}^{M} \prod_{i=1}^{K} \sqrt{\frac{\epsilon}{2\pi}} \int_{-\infty}^{+\infty} dW_i(m)$$

The limit of $\epsilon \to 0$ is taken for notational simplicity; for purposes of rigor, the continuum notation is simply a short-hand for taking the continuum limit of the discrete multiple integrals given above. Hence, for $t_1 < t < t_2$

$$\mathcal{P}[W, t_1, t_2] \to e^S \qquad (6.32)$$

$$S \equiv S[W, t_1, t_2] = -\frac{1}{2} \sum_{i=1}^{K} \int_{t_1}^{t_2} dt W_i^2(t) \qquad (6.33)$$

$$\int dW \to \int DW \qquad (6.34)$$

The action functional S_0 is ultra-local with all the variables being decoupled; generically, $\int DW$ stands for the (path) integration over all the random variables $W(t)$ which appear in

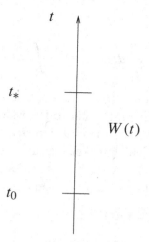

Figure 6.1 Independent $W(t) \equiv \{W_i(t)|i = 1, 2, \ldots, K\}$ random variables in the HJM model

the problem. The integration variables $W_i(t)$ are shown in Figure 6.1, where each point t in the interval $t \in [t_0, t_*]$ represents the independent random variables $W_i(t)$.

A useful formula is the generating functional for W given, from Eq. (A.20), by the path integral

$$Z[j, t_1, t_2] = \int DW e^{\sum_{i=1}^{K} \int_{t_1}^{t_2} dt j_i(t) W_i(t)} e^{S[W, t_1, t_2]}$$

$$= e^{\frac{1}{2} \sum_{i=1}^{K} \int_{t_1}^{t_2} dt j_i^2(t)} \tag{6.35}$$

6.4 Martingale condition in the HJM model

From the discussion in Section 2.5, to obtain a risk-neutral measure for the forward interest rates, one needs to impose the martingale condition on the evolution of the forward rates. All derivatives of the forward rates priced using the risk-neutral martingale measure, according to the fundamental theorem of finance [40] discussed in Appendix A.6, are free from arbitrage opportunities.

From Eq. (2.10), the martingale condition for the HJM model is given by

$$P(t_0, T) = E_{[t_0, t_*]} \left[e^{-\int_{t_0}^{t_*} r(t) dt} P(t_*, T) \right]$$

$$\Rightarrow P(t_0, T) = \int DW e^{-\int_{t_0}^{t_*} r(t) dt} P(t_*, T) e^{S[W, t_0, t_*]} \tag{6.36}$$

where the last equation has been obtained by writing out Eq. (2.10) using Eq. (6.33).

Domains of integration $\mathcal{T}$, Δ_0 and $\mathcal{R}$

The domains of integration, namely $\mathcal{T}$, Δ_0 and $\mathcal{R}$, appear in calculations for options and other derivatives in any model that one is using for the Treasury Bonds. These domains are discussed here in the specific context of fixing the drift velocity using the martingale condition.

From Eq. (6.36), for the HJM model

$$P(t_0, T) = \int DW \, e^{-X} e^{S[W, t_0, t_*]} \tag{6.37}$$

where

$$X = \int_{t_0}^{t_*} dt \, r(t) + \int_{t_*}^{T} dx f(t_*, x) \tag{6.38}$$

Recall from Eq. (6.30)

$$f(t, x) = f(t_0, x) + \int_{t_0}^{t} dt' \alpha(t', x) + \int_{t_0}^{t} dt' \sum_{i=1}^{K} \sigma_i(t', x) W_i(t')$$

Hence, from equation above, and using $r(x) = f(x, x)$, yields

$$X = \int_{t_0}^{t_*} dx \left[f(t_0, x) + \int_{t_0}^{x} dt \alpha(t, x) + \int_{t_0}^{x} dt \sum_{i=1}^{K} \sigma_i(t, x) W_i(t) \right]$$

$$+ \int_{t_*}^{T} dx \left[f(t_0, x) + \int_{t_0}^{t_*} dt \alpha(t, x) + \int_{t_0}^{t_*} dt \sum_{i=1}^{K} \sigma_i(t, x) W_i(t) \right]$$

For example

$$\int_{t_0}^{t_*} dx \int_{t_0}^{x} dt \alpha(t, x) + \int_{t_*}^{T} dx \int_{t_0}^{t_*} dt \alpha(t, x)$$

$$= \int_{t_0}^{t_*} dt \int_{t}^{t_*} dx \alpha(t, x) + \int_{t_0}^{t_*} dt \int_{t_*}^{T} dx \alpha(t, x)$$

$$= \int_{t_0}^{t_*} dt \int_{t}^{T} dx \alpha(t, x)$$

The last two equations can be written more graphically as

$$\int_{\Delta_0} \alpha(t, x) + \int_{\mathcal{R}} \alpha(t, x) = \int_{\mathcal{T}} \alpha(t, x) \tag{6.39}$$

where the domains of integration Δ_0 and $\mathcal{R}$ are shown in Figure 6.2. The integration over the domain Δ_0 arises from the discounting by the spot rate and that over domain $\mathcal{R}$ arises from the Treasury Bond $P(t_*, T)$. As shown in Figure 6.3, the two domains of the triangle and a rectangle combine to give $\mathcal{T}$

$$\mathcal{T} = \Delta_0 \oplus \mathcal{R}$$

where the *trapezoidal* domain $\mathcal{T}$ is given in Figure 6.4. Simplifying the other terms in Eq. (6.38) in a similar manner yields

$$X = \int_{t_0}^{T} dx f(t_0, x) + \int_{\mathcal{T}} \alpha(t, x) + \sum_i \int_{\mathcal{T}} \sigma_i(t, x) W_i(t) \qquad (6.40)$$

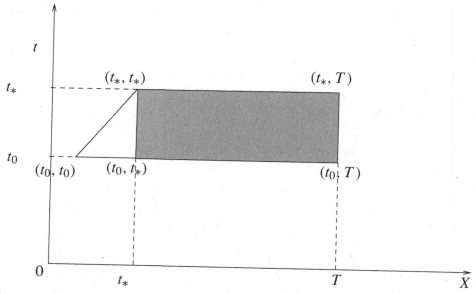

Figure 6.2 Domain $\mathcal{R}$ is shaded and domain Δ_0 is the empty triangle

Hence, for the HJM model, from Eqs. (6.36), (6.37), (6.38) and (6.40)

$$P(t_0, T) = P(t_0, T) e^{-\int_{\mathcal{T}} \alpha(t,x)} \int DW e^{-\sum_i \int_{\mathcal{T}} \sigma_i(t,x) W_i(t)} e^{S[W, t_0, t_*]} \qquad (6.41)$$

On performing the W integrations yields, from Eqs. (6.35) and (6.41)

$$e^{\int_{\mathcal{T}} \alpha(t,x)} = e^{\frac{1}{2} \int_{t_0}^{t_*} dt \sum_i [\int_t^T dx \sigma_i(t,x)]^2} \qquad (6.42)$$

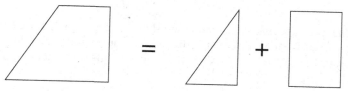

Figure 6.3 Domain $\mathcal{T}$ = domain $\Delta_0 \oplus$ domain $\mathcal{R}$

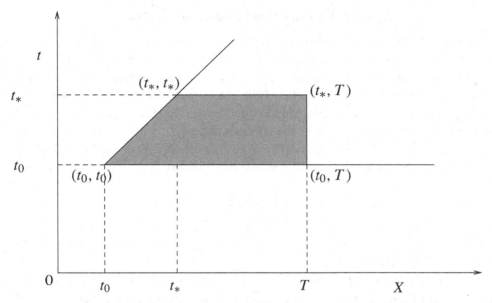

Figure 6.4 Trapezoidal domain $\mathcal{T}$ for the martingale condition

Dropping the integration over t one obtains [58]

$$\int_t^T dx \alpha(t, x) = \frac{1}{2} \sum_{i=1}^K \left[\int_t^T dx \sigma_i(t, x) \right]^2 \tag{6.43}$$

or equivalently, the drift velocity in the HJM model is given by

$$\alpha(t, x) = \sum_{i=1}^n \sigma_i(t, x) \int_t^x dy \sigma_i(t, y) \tag{6.44}$$

$$: \quad \text{condition for martingale measure}$$

As expected, the martingale condition leads to the well-known [58] no-arbitrage condition that expresses the drift velocity of the forward rates in terms of its volatility.

Consider the two-factor HJM model with volatilities given by

$$\sigma_1(t, x) = \sigma_1; \sigma_2(t, x) = \sigma_2 e^{-\lambda(x-t)} \tag{6.45}$$

The no-arbitrage condition given in Eq. (6.44) yields

$$\alpha(t, x) = \sigma_1^2(x - t) + \frac{\sigma_2^2}{\lambda} e^{-\lambda(x-t)} \left(1 - e^{-\lambda(x-t)} \right)$$

6.5 Pricing of Treasury Bond futures in the HJM model

A general discussion of forward and futures contracts was given in Section 3.1, and these concepts are now applied to the specific case of Treasury Bonds.

The future and forward contracts on a zero coupon coupon bond are instruments that are traded in the capital market. Both the forward and futures contracts are entered into at time t_0 for a zero coupon bond, maturing at time T, to be delivered to the buyer at the conclusion of the contract at time t_*, where $t_0 < t_* < T$. The forward price of a Treasury Bond $P(t, T)$ is denoted by $F(t_0, t_*, T)$, and, since there is only one cash flow, it can be shown that [52, 58]

$$F(t_0, t_*, T) = \frac{P(t_0, T)}{P(t_0, t_*)} = e^{-\int_{t_*}^{T} dx f(t_0, x)}$$

The forward price of a Treasury Bond is independent of any models for the time evolution of the Treasury Bonds.

The futures price of $P(t, T)$ is denoted by $\mathcal{F}(t_0, t_*, T)$. The difference in the forward and futures price, as was discussed in Section 3.1, is that for a forward contract there is only a single cash flow at t_*: at the expiry date of the contract. For a futures contract on the other hand there is a continuous cash flow from time t_0 to t_* such that all variations in the price of $P(t + dt, T)$ away from $P(t, T)$, for $t_0 < t < t_*$, are settled continuously between the buyer and the seller, with a final payment of $P(t_*, T)$ at time t_*. If the time evolution of $P(t, T)$ was deterministic, it is easy to see that the forward and futures price would be equal.

It can be shown that the price of the futures $\mathcal{F}$ is given by [58]

$$\mathcal{F}(t_0, t_*, T) = E_{[t_0, t_*]}[P(t_*, T)] \tag{6.46}$$

The result is model independent, and of great generality, since the expression applies to any model for the time evolution of the Treasury Bonds.

For the HJM model, from Eqs. (6.30) and (6.32)

$$\mathcal{F}(t_0, t_*, T) = \int DW e^{-\int_{t_*}^{T} dx f(t_*, x)} e^{S[W, t_0, t_*]} \tag{6.47}$$

$$= F(t_0, t_*, T) \exp \Omega_{\mathcal{F}} \tag{6.48}$$

Since there is no discounting by the spot interest rate in Eq. (6.47), the futures price is defined by an integration of the forward rates over only the *rectangular* domain $\mathcal{R}$, given in Figure 6.2, where recall from Eq. (6.39)

$$\int_{\mathcal{R}} \equiv \int_{t_0}^{t_*} dt \int_{t_*}^{T} dx$$

Hence

$$\exp \Omega_{\mathcal{F}} = e^{\Omega} e^{-\int_{\mathcal{R}} \alpha(t,x)} \tag{6.49}$$

with

$$e^{\Omega} = \int DW e^{-\sum_{i=1}^{K} \int_{\mathcal{R}} \sigma_i(t,x) W_i(t)} e^{S} \tag{6.50}$$

$$= \exp\left\{\frac{1}{2} \sum_{i=1}^{K} \int_{t_0}^{t_*} dt \left[\int_{t_*}^{T} dx \sigma_i(t,x)\right]^2\right\} \tag{6.51}$$

where Eq. (6.51) has been obtained by performing the path integration over the W variables using Eq. (6.35).

Using the martingale condition given in Eq. (6.44), and after some simplifications, one obtains from Eq. (6.49) that

$$\Omega_{\mathcal{F}}(t_0, t_*, T) = -\sum_{i=1}^{K} \int_{t_0}^{t_*} dt \int_{t}^{t_*} dx \sigma_i(t,x) \int_{t_*}^{T} dx' \sigma_i(t,x') \tag{6.52}$$

As is expected, the future and forward prices of the zero coupon bond are equal if the volatility is zero, that is, the evolution of the zero coupon bond is deterministic.

Consider the two-factor HJM model with volatilities given in Eq. (6.45). Eq. (6.52) yields

$$\Omega_{\mathcal{F}}(t_0, t_*, T) = -\sigma_1^2 (T - t_*)(t_* - t_0)^2$$
$$- \frac{\sigma_2^2}{2\lambda^3} \left(1 - e^{-\lambda(T-t_*)}\right) \left(1 - e^{-\lambda(t_*-t_0)}\right)^2$$

which is the result given in [52, 65].

6.6 Pricing of Treasury Bond option in the HJM model

Suppose one needs the price, at time t_0, of a derivative instrument of a zero coupon Treasury Bond $P(t, T)$ for a contract that expires at $t_* < T$. For concreteness consider the price of a European call option on a zero coupon bond [58, 86], namely $C(t_0, t_*, T, K)$; the option has a strike price of K and exercise time at $t_* > t_0$.

The (final) value of the option at maturity, namely at $t_0 = t_*$ is, as required by the contract, given by

$$C(t_*, t_*, T, K) = (P(t_*, T) - K)_+$$
$$\equiv \max((P(t_*, T) - K), 0)$$

For $t_0 < t_*$, the price of C is given by the expectation value of the discounted value of the payoff function, namely

$$C(t_0, t_*, T, K) = E_{[t_0, t_*]}\left[e^{-\int_{t_0}^{t_*} dt r(t)}(P(t_*, T) - K)_+\right] \quad (6.53)$$

The expectation value in Eq. (6.53) is taken by evolving the payoff function $(P(t_*, T) - K)_+$ backward from t_* to t_0, continuously discounted by stochastic spot rate $r(t) = f(t, t)$.

The payoff function is re-written in a form that is more suited to path integral calculations using the following identity given in Eq. (A.11)

$$\delta(z) = \frac{1}{2\pi}\int_{-\infty}^{+\infty} dp e^{ipz} \quad (6.54)$$

Hence, since $P(t_*, T) = \exp(-\int_{t_*}^{T} dx f(t_*, x))$, one has the following

$$(P(t_*, T) - K)_+ = \int_{-\infty}^{+\infty} dG \delta\left[G + \int_{t_*}^{T} dx f(t_*, x)\right](e^G - K)_+$$

$$= \int_{-\infty}^{+\infty} dG \frac{dp}{2\pi} e^{ip(G+\int_{t_*}^{T} dx f(t_*,x))}(e^G - K)_+ \quad (6.55)$$

Re-write Eq. (6.53) as

$$C(t_0, t_*, T, K) = \int_{-\infty}^{+\infty} dG \Psi(G, t_*, T)(e^G - K)_+ \quad (6.56)$$

where

$$\Psi(G, t_*, T) = \int_{-\infty}^{+\infty} \frac{dp}{2\pi} E_{[t_0, t_*]}\left[e^{-\int_{t_0}^{t_*} dt f(t,t)} e^{ip(G+\int_{t_*}^{T} dx f(t_*,x))}\right] \quad (6.57)$$

$$= P(t_0, t_*) \int_{-\infty}^{+\infty} \frac{dp}{2\pi} e^{\Lambda} e^{ip\Lambda_0} \quad (6.58)$$

with

$$\Lambda_0 = G + \int_{t_*}^{T} dx f(t_0, x) + \int_{\mathcal{R}} \alpha(t, x)$$

Similar to the derivation of Eq. (6.41), one has the following

$$e^{\Lambda} = e^{-\int_{\Lambda_0} \alpha(t,x)} \int DW e^{-\int_{\Lambda_0} \sigma_i(t,x)W_i(t)+ip\sum_i^K \int_{\mathcal{R}} \sigma_i(t,x)W_i(t)} e^S \quad (6.59)$$

The interplay of the sub-domains Λ_0 and $\mathcal{R}$ in Eq. (6.59) determines the price of the option.[6]

Using (6.35) to perform the integrations over W yields, after considerable simplifications[7] and using the martingale condition given in Eq. (6.43), that

$$\Lambda = -\frac{q^2}{2}p^2 \tag{6.60}$$

with

$$q^2 = \sum_{i=1}^{K} \int_{t_0}^{t_*} dt \left[\int_{t_*}^{T} dx \sigma_i(t, x) \right]^2 \tag{6.61}$$

$$= 2 \int_{\mathcal{R}} \alpha(t, x)$$

where the last equation is derived in Eq. (9.9). Eqs. (6.56)–(6.61) yield the result

$$\Psi(G, t_*, T) = P(t_0, t_*) \int_{-\infty}^{+\infty} \frac{dp}{2\pi} e^{-\frac{q^2}{2}p^2} e^{ip\Lambda_0}$$

$$= P(t_0, t_*) \sqrt{\frac{1}{2\pi q^2}} \exp{-\frac{1}{2q^2}\left(G + \int_{t_*}^{T} dx f(t_0, x) + \frac{q^2}{2}\right)^2} \tag{6.62}$$

Hence, from the equation above and (6.56) the well-known result [22, 57] is obtained that the European option on a Treasury Bond has a Black–Scholes like formula with volatility given by q^2.

For the two-factor HJM model given in Eq. (6.45) [22]

$$q^2 = \sigma_1^2 (T - t_*)^2 (t_* - t_0)$$
$$+ \frac{\sigma_2^2}{2\lambda^3} \left(1 - e^{-\lambda(T-t_*)}\right)^2 \left(1 - e^{-2\lambda(t_*-t_0)}\right) \tag{6.63}$$

6.7 Summary

The martingale evolution of the spot interest rate was modelled using the stochastic Langevin equation, and the Fokker–Planck Hamiltonians that determine the

---

[6] It will be seen in Section 7.7 that a change of numeraire greatly simplifies the calculation. In the field theory calculation for the option price done in Section 9.2 only the domain $\mathcal{R}$ appears, with no reference being made to the domain Λ_0.

[7] The identity

$$\int_{t_0}^{t_*} dt \left[\int_{t_*}^{T} dx \alpha(t, x) - \sum_{i=1}^{K} \int_{t}^{t_*} dx \sigma_i(t, x) \int_{t_*}^{T} dy \sigma_i(t, y) \right] = \frac{1}{2}q^2$$

has been used to obtain Eq. (6.60).

evolution of its conditional probabilities were obtained. The asymmetry in the forward and backward evolution of the Langevin equation was reflected in the backward and forward Fokker–Planck Hamiltonians. Since the evolution of the spot rate is assumed to satisfy the martingale condition, the analysis is simpler than the derivation of the Black–Scholes Hamiltonian.

The action and path integral for spot interest rate models were obtained using both the Fokker–Planck Hamiltonian and the procedure of stochastic quantization. The path integral for the Vasicek model was exactly solved using a change of variables slightly more complicated than the one needed to solve the Black–Scholes path integral. The various affine models for forward interest rates, and the generalization to non-affine models, were briefly discussed.

The reasons for modelling the forward interest rates as the primary and fundamental instrument of the debt market, instead of the spot rate, were enumerated.

The industry-standard HJM model of forward interest rates was re-formulated in terms of path integration, and which was then used to compute various quantities of the model. The path-integral formulation of the HJM model is important in its own right, and provides a new perspective on the model. The martingale measure and Treasury Bond option were shown to be calculable in a straightforward manner using path integration.

The main motivation for re-deriving the well-known results of the HJM model in Section 6.6 was, firstly, to understand the path integral formulation of interest rate derivatives, and, secondly, to prepare the framework for generalizing these quantities to the case of quantum field theory.

The remaining chapters of this book are all focussed on modelling the forward rates using quantum field theory, and it will be seen with hindsight that many of the features of the path integral that first appear in the context of the HJM model are simplified expressions of much more complex and involved derivations. Hence, in this sense, the HJM model is a useful preparation for the material that is covered in the subsequent chapters.

6.8 Appendix: Spot interest rate Fokker–Planck Hamiltonian

The spot rate $r(t)$ is the interest rate for an overnight loan at time t. The spot rate is driven by a Langevin equation given in Eq. (6.1) with boundary condition given in Eq. (6.2), and hence

$$\frac{dr}{dt} = a(r, t) + \sigma(r, t)R(t) \ : \ R(t) \text{ white noise} \quad t_0 \leq t \leq T$$

EITHER $r(t_0) = r_0$: initial condition OR $r(T) = R$: final condition where in principle the function a and volatility σ are arbitrary functions of the spot rate r and of time t.

The derivations employ the following properties of white noise $R(t)$

$$<R(t)> = 0 \quad ; \quad <R^2(t)> = \frac{1}{\epsilon}; \ t = n\epsilon$$

Let $P(r, t; r_0)$ be the conditional probability that the spot rate has value r at time t, given that the value of r_0 occurred at an earlier time $t_0 < t$. The forward conditional probability requires the propagation of the spot rate **forward** in time.

The conditional probability at time $t + \epsilon$ is given by the following. Discretizing the spot rate equation yields, in simplified notation

$$r(t + \epsilon) = r(t) + \epsilon[a + \sigma R(t)]$$
$$\Rightarrow r = r' + \epsilon[a(r') + \sigma(r')R(t)] \tag{6.64}$$

The forward conditional probability $P(r, t; r_0)$ is given by evolving the spot rate into the future using the Langevin equation, and averaging over the white noise. Hence, Taylor expanding the argument of the δ-function yields

$$P(r, t + \epsilon; r_0) \equiv \int dr' < \delta(r - r' - \epsilon\{a(r') + \sigma(r')R(t)\}) > P(r', t; r_0)$$
$$\simeq \int dr' < \delta(r - r') + \epsilon\{a(r') + \sigma(r')R(t)\}\frac{\partial}{\partial r'}\delta(r - r')$$
$$+ \frac{1}{2}\epsilon^2\{a(r') + \sigma(r')R(t)\}^2\frac{\partial^2}{\partial r'^2}\delta(r - r') > P(r', t; r_0)$$

Since $< R^2(t) >= 1/\epsilon$ the term of $O(\epsilon^2)$ in the equation above, which would be zero for ordinary functions, now yields a non-zero contribution. This is the manner in which the results of Ito calculus, discussed in Section 3.4, appear in derivations based on the Langevin equation. Hence

$$P(r, t + \epsilon; r_0) = P(r, t; r_0) + \int dr' \left[\epsilon a(r')\frac{\partial}{\partial r'}\delta(r - r') \right.$$
$$\left. + \frac{1}{2}\epsilon^2\sigma^2(r') \times \frac{1}{\epsilon} \times \frac{\partial^2}{\partial r'^2}\delta(r - r') \right] P(r', t; r_0)$$
$$= P(r, t; r_0) - \epsilon \left[\frac{\partial}{\partial r}a(r) - \frac{1}{2}\frac{\partial^2}{\partial r^2}\sigma^2(r) \right] P(r, t; r_0)$$

Taking the limit of $\epsilon \to 0$ yields the **forward Fokker–Planck equation** for the conditional probability

$$\frac{\partial}{\partial t}P(r,t;r_0) = \left[\frac{1}{2}\frac{\partial^2}{\partial r^2}\sigma^2(r) - \frac{\partial}{\partial r}a\right]P(r,t;r_0) \tag{6.65}$$

$$\equiv -H_F P(r,t;r_0) \tag{6.66}$$

$$\Rightarrow H_F = -\frac{1}{2}\frac{\partial^2}{\partial r^2}\sigma^2(r) + \frac{\partial}{\partial r}a$$

$$= -\frac{1}{2}\frac{\partial^2}{\partial r^2}\sigma^2(r) + a(r)\frac{\partial}{\partial r} + \frac{\partial a(r)}{\partial r} \tag{6.67}$$

where H_F is the non-Hermitian forward Fokker–Planck Hamiltonian.

It can be recognized that the conditional probability is nothing but the pricing kernel for the spot interest rate, and in fact

$$P_F(r,t;r_0) = <r|e^{-(t-t_0)H_F}|r_0> \tag{6.68}$$

$$r(t_0) = r_0 : \text{initial condition}$$

The **backward Fokker–Planck equation** is required in the pricing of options, since the payoff function is propagated **backwards** in time. Hence one needs to propagate the final value of the interest at time T, namely R, to its value at an earlier time t with the value of r, and time consequently flows backwards.

The backward conditional probability $P_B(R,t;r')$ is defined as the probability that the spot rate will have the value of r' at time t given that the value of R has occurred at some future time $T > t$. Hence, for $r = r(t+\epsilon)$ and $r' = r(t)$, Eq. (6.64) yields

$$P_B(R,t;r') = \int dr < \delta[r - r' - \epsilon\{a(r') - \sigma(r')R(t)\}] > P(R,t+\epsilon;r) \tag{6.69}$$

The coefficients $a(r'), \sigma(r')$ have the argument of the spot rate of the earlier time t, and which cannot be changed to r, with an error that is of $O(\epsilon)$, due to the singular nature of white noise. This is the fundamental reason why the forward and backward Fokker–Planck Hamiltonians are different. Similar to the derivation for the forward case, the **backward Fokker–Planck Hamiltonian** is given by

$$\frac{\partial}{\partial t}P_B(R,t;r) \equiv +H_B P_B(R,t;r)$$

$$H_B = -\frac{1}{2}\sigma^2(r)\frac{\partial^2}{\partial r^2} - a(r)\frac{\partial}{\partial r} \tag{6.70}$$

$$= H_F^{\dagger}$$

In the defining Eq. (6.70) for $P_B(R,t;r)$ the evolution equation has the **opposite** sign compared with Eq. (6.66) for the forward evolution. Since for the backward case the time variable is $-t$ and hence running backward, the backward evolution equation has a negative sign compared with the forward equation. The backward conditional probability $P_B(R, t; r)$ is hence driven by the operator e^{tH_B}.

To write out $P_B(R, t; r)$ as a matrix element, as has been written in Eq. (6.68) for the forward case, one needs to discuss the Dirac notation for representing backward time evolution. The convention is that the ket vector always represents |starting state > and the bra (dual) vector always represents < ending state|. In the case of backward time evolution, one starts from the final value $|R>$, and using H_B evolves this value backwards in time (hence time is given by $-t$) to the initial value of $<r|$. The final value is expressed by the final condition that at time $t = T$ one must have $r(T) = R$. Collecting these facts gives

$$P_B(R, t; r) = <r|e^{-(T-t)H_B}|R> \tag{6.71}$$
$$r(T) = R : \text{final condition}$$

The forward Fokker–Planck Lagrangian

Recall from Eq. (5.5) that the relation between the Lagrangian and the Hamiltonian is given by

$$\langle x \mid e^{-\epsilon H} \mid x'\rangle \equiv \mathcal{N}(\epsilon)e^{\epsilon L(x;x';\epsilon)} \tag{6.72}$$

Hence for the forward Fokker–Planck Hamiltonian, using the completeness equation in the momentum basis given in Eq. (4.34), the Lagrangian is given by ($a' \equiv \partial a(r)/\partial r$)

$$
\begin{aligned}
e^{\epsilon L_F[r,\tilde{r}]} &= <r|e^{-\epsilon H_F}|\tilde{r}> \\
&= \int \frac{dp}{2\pi} <r|e^{-\epsilon H_F}|p><p|\tilde{r}> \\
&= \int \frac{dp}{2\pi} e^{-\epsilon\left(\frac{\sigma^2}{2}p^2 + iap + a'\right)} e^{ip(r-\tilde{r})} \\
&= \frac{1}{\sqrt{2\pi\epsilon\sigma^2}} e^{-\frac{\epsilon}{2\sigma^2}[(r-\tilde{r}-\epsilon a)^2 - \epsilon a']}
\end{aligned}
$$

Taking the limit of $\epsilon \to 0$ gives $(r - \tilde{r})/\epsilon \to dr/dt$ and yields the forward Fokker-Planck Lagrangian

$$L_F = -\frac{1}{2\sigma^2}\left(\frac{dr}{dt} - a(r)\right)^2 - \frac{\partial a(r)}{\partial r} \tag{6.73}$$

6.9 Appendix: Affine spot interest rate models

The result obtained in Eq. (6.23) is generic to a wide class of spot rate models – called **affine models** – all of which have an exponential dependence of the Treasury Bond on the spot rate that is linear; in other words [46,47,51]

$$P(t, T) = \alpha(t, T)e^{-\beta(t,T)r} \quad : \text{ affine models}$$

Given below are some of the affine spot rate models [51]

- Vasicek model

$$\frac{dr}{dt} = a(b - r) + \sigma R(t)$$

- The Rendelman and Barter model

$$\frac{dr}{dt} = \mu r + \sigma r R(t)$$

- The Ho and Lee model

$$\frac{dr}{dt} = \theta(t) + \sigma R(t)$$

- The Cox–Ingersoll–Ross (CIR) model

$$\frac{dr}{dt} = a(b - r) + \sigma \sqrt{r} R(t)$$

- The Hull and White model

$$\frac{dr}{dt} = \theta(t) - ar + \sigma R(t)$$

All these models have been extensively discussed in the literature, as well as tested empirically [26], and yield path integrals that can be used for analytical and numerical studies.

One can also consider a multi-factor affine model in which the spot interest rate is driven by N white noises, and is given by

$$\frac{dr}{dt} = a(r, t) + \sum_{n=1}^{N} \sigma_n(t) R_n(t)$$

Although the multi-factor model may look more powerful, its major disadvantage is that the volatility functions implied by the models can be very different in the future than from what is determined from existing market data, and hence cannot be used effectively in pricing spot interest rate derivatives and so on.

6.10 Appendix: Black–Karasinski spot rate model

One of the most important **non-affine** spot interest rate models is the Black–Karasinski model given by a lognormal distribution for the spot rate. Consider the general process

$$r(t) = r_0 e^{\phi(t)} > 0$$
$$\frac{d\phi}{dt} = \alpha(\phi, t) + \sigma(\phi, t) R(t) \tag{6.74}$$

Black–Karasinski : $\alpha(\phi; t) = \theta(t) - a(t)\phi(t)$; $\sigma(\phi; t) = \sigma(t)$

where the salient property of this model is that the spot rate $r(t)$ is always positive. In the market, interest rates are always positive, and hence this is a major advantage of the Black–Karasinski (and of the CIR) model.

As discussed in Section 6.2 for the Vasicek model, the backward Lagrangian for the generalized Black–Karasinski model is required for determining the Treasury Bond, and is given by

$$S_B^{bk} = -\frac{1}{2} \int_{t_0}^{T} dt \frac{1}{\sigma^2(\phi; t)} \left[r_0 \frac{d\phi}{dt} - \alpha(\phi) \right]^2 \tag{6.75}$$

Consider the zero coupon bond given by

$$P(t_0, T) = E\left[e^{-r_0 \int_{t_0}^{T} dt\, e^{\phi(t)}} |\phi(t_0) = 0 \right] \tag{6.76}$$

Similar to the analysis of the Vasicek model, the zero coupon bond is given by averaging the discount factor over the probability distribution, and yields

$$P(t_0, T) = \frac{1}{Z} \int D\phi\, e^{S_B^{bk}} e^{-r_0 \int_{t_0}^{T} e^{\phi(t)} dt} \tag{6.77}$$

$$Z = \int D\phi\, e^{S_B^{bk}} \;;\; \int D\phi \equiv \int_{-\infty}^{+\infty} \prod_{t=t_0}^{T} d\phi(t) \tag{6.78}$$

boundary conditions $\phi(t_0) = 0$; $\dfrac{d\phi(T)}{dt} = 0$

The denominator cancels all the terms independent of r_0. The resulting zero coupon bond is known from numerical simulations to be non-affine.

Recall from Eq. (6.74) that the Black–Karasinski model is specified by

$$r(t) = r_0 e^{\phi(t)}$$
$$d\phi = (\theta(t) - a(t)\phi)dt + \sigma(t)dz$$

where dz is a Wiener process.

The simulation for the spot rate can be performed by generating random configurations starting from the Langevin equation using the Euler approximation for time derivatives. For simplicity, θ and a are chosen to be zero. The variable ϕ is updated according to

$$\phi(t + \epsilon) = \phi(t) + Z\sqrt{\epsilon}$$

where $Z = N(0, 1)$ is the standard normal random variable.

In order to investigate the non-affine nature of the Black–Karasinski model, the price of a ten-year zero coupon bond with different initial interest rates is numerically calculated. The simulation was carried out for $\sigma = 1.0/\text{year}$; 128 time steps were used for the discretization of the rate process and 100 000 configurations were used to calculate the price of the bond. The error in the bond prices is negligible. The result [97] is shown in Figure 6.5, displaying a small but significant departure from affine behaviour.

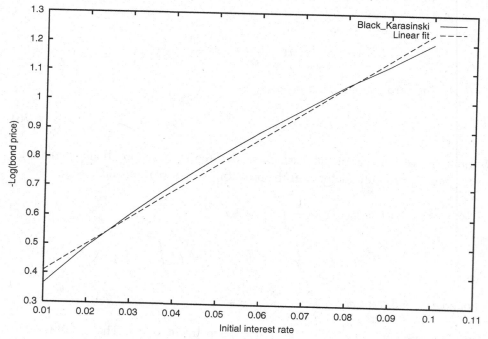

Figure 6.5 The logarithm of the bond price plotted against the initial interest rate for the Black–Karasinski model with $\sigma = 1.0/\text{year}$.

6.11 Appendix: Black–Karasinski spot rate Hamiltonian

The primary focus of this Appendix is pedagogical. The Fokker–Planck Hamiltonians have been derived using the properties of white noise in Appendix 6.8, and of which the Black–Karasinski Spot Rate Hamiltonian is a special case. The Lagrangians were then deduced from the Hamiltonians, similar to other derivations of the Lagrangian for other systems.

The question naturally arises that if the Lagrangian is known, how would one derive its Hamiltonian; the purpose of this Appendix is to carry out this derivation using quantum mechanical techniques. The reason for choosing the Black–Karasinski Hamiltonian for this exercise is because it has some similarities with the much more complex derivation, in Chapter 10, of the forward interest rates Hamiltonian from its Lagrangian. This appendix is a preparation for handling the more complex case.

From the general result derived in Eq. (6.12), the Black–Karasinski **backward** action for spot rate $r(\tau) = r_0 e^{\phi(\tau)}$, for remaining time $\tau = T - t$ and $\tau_0 = T - t_0$, is given by

$$S_B^{bk} = -\frac{1}{2} \int_0^{\tau_0} d\tau \frac{1}{\sigma^2(\tau)} \left(r_0 \frac{d\phi}{d\tau} + \alpha(r, \tau) \right)^2 \qquad (6.79)$$

Suppose for greater generality that the volatility depends on the spot rate, namely, that $\sigma = \sigma_0 e^{\nu\phi}$.

The path integral for the Black–Karasinski model is given by the following generalization of Eq. (6.78)

$$Z = \int D\phi e^{-\nu\phi} e^{S_B^{bk}} \qquad (6.80)$$

$$\int D\phi e^{-\nu\phi} \equiv \prod_{\tau=0}^{\tau_0} \int_{-\infty}^{+\infty} d\phi(t) e^{-\nu\phi(t)}$$

boundary conditions $\phi(\tau_0) = \phi_0$; $\phi(\tau = 0) = \phi_T$

To obtain the Hamiltonian, recall from the discussion of the pricing kernel in Section 5.1 that the path integral is related to the Hamiltonian by Eq. (5.8), namely

$$Z = \int D\phi e^{-\nu\phi} e^{S_B^{bk}} = <\phi_0| e^{-\tau_0 H_B^{bk}} |\phi_T> \qquad (6.81)$$

One needs to extract the Hamiltonian H_B^{bk} from the left-hand side. Since the Hamiltonian propagates the system through infinitesimal (backward) time, time is discretized into a lattice of spacing ϵ, with $\tau = n\epsilon$ and $N = \tau_0/\epsilon$. The path integral reduces to a finite $(N-1)$-fold multiple integral, analogous to what was obtained in Section 5.1, and in particular in Eq. (5.3). Discretizing the time derivative by

$d\phi/d\tau \rightarrow (\phi_{n+1} - \phi_n)/\epsilon$ gives

$$< \phi_0|e^{-\epsilon N H_B^{bk}}|\phi_N > = \prod_{n=1}^{N-1} \int d\phi_n e^{-\nu\phi_n} e^{S_B^{bk}} \tag{6.82}$$

$$S_B^{bk} \rightarrow \epsilon \sum_{n=0}^{N-1} L_B^{bk}(n)$$

$$L_B^{bk}(n) = -\frac{1}{2\epsilon\sigma_n^2}\left(r_0(\phi_{n+1} - \phi_n) + \epsilon\alpha_n\right)^2$$

As in Section 5.1, the completeness equation $\int d\phi|\phi><\phi| = \mathcal{I}$ is used $N - 1$ times to write out the expression for $e^{-\epsilon N H_B^{bk}}$, and the Hamiltonian is identified as

$$< \phi_{n+1}|e^{-\epsilon H_B^{bk}}|\phi_n > = e^{-\nu\phi_n} e^{\epsilon L_B^{bk}(n)}$$

$$= e^{-\nu\phi_n} e^{-\frac{1}{2\epsilon\sigma_n^2}\left(r_0(\phi_{n+1}-\phi_n)+\epsilon\alpha_n\right)^2} \tag{6.83}$$

and one has recovered Eq. (5.5), with a normalization that depends on the random variable ϕ. But unlike Eq. (5.5) where the Hamiltonian is known and the Lagrangian is derived from it, in Eq. (6.83) one needs to derive the Hamiltonian from the known Lagrangian.

The key feature of the Lagrangian that in general allows one to derive its Hamiltonian is that the Lagrangian contains only first-order time derivatives, and hence on discretization involves only nearest neighbours in time, thus allowing it to be represented as the matrix element of $e^{-\epsilon H_B^{bk}}$, as in Eq. (6.83). Secondly, the time derivatives appears in a quadratic form; one can therefore use Gaussian integration to re-write the right-hand side of Eq. (6.83) in the following manner

$$<\phi_{n+1}|e^{-\epsilon H_B^{bk}}|\phi_n> = e^{-\nu\phi_n} \int_{-\infty}^{+\infty} \frac{dp}{2\pi} e^{-\frac{\epsilon}{2}p^2} \exp\left\{ip\left(\frac{r_0(\phi_{n+1} - \phi_n) + \epsilon\alpha_n}{\sigma_n}\right)\right\}$$

$$= \frac{\sigma_0}{r_0} \int_{-\infty}^{+\infty} \frac{dp}{2\pi} e^{-\frac{\epsilon\sigma_n^2}{2r_0^2}p^2} e^{ip\left(\phi_{n+1}+\phi_n+\epsilon\frac{\alpha_n}{r_0}\right)} \tag{6.84}$$

where the pre-factor of $e^{-\nu\phi_n}$ has been cancelled by re-scaling the integration variable p by σ_n/r_0, with $\sigma_n = \sigma_0 e^{\nu\phi_n}$.

The Hamiltonian $H_B^{bk} = H_B^{bk}(\phi, \partial/\partial\phi)$ is a differential operator acting on the dual co-ordinate ϕ_{n+1}. Recall from the discussion at the end of Section 4.3 that the reason this choice is made is because the wavefunction $|\psi >$ is taken to be an element of the state space, and the Hamiltonian acts on the dual basis state $< x|$, and yields $<x|H|\psi> = H(x, \partial/\partial x)\psi(x)$. For this reason one has the following representation

$$<\phi_{n+1}|e^{-\epsilon H_B^{bk}}|\phi_n> = e^{-\epsilon H_{BK}(\phi_{n+1}, \partial/\partial\phi_{n+1})} <\phi_{n+1}|\phi_n>$$

$$= e^{-\epsilon H_B^{bk}(\phi_{n+1}, \partial/\partial\phi_{n+1})} \int_{-\infty}^{+\infty} \frac{dp}{2\pi} e^{ip(\phi_{n+1}-\phi_n)} \quad (6.85)$$

since $<\phi_{n+1}|\phi_n> = \delta(\phi_{n+1} - \phi_n)$.[8] Hence, ignoring overall constants, and using the property of the exponential function under differentiation, one can re-write Eq. (6.84) as

$$<\phi_{n+1}|e^{-\epsilon H_B^{bk}}|\phi_n> = \exp\left\{\frac{\epsilon\sigma^2}{2r_0^2}\frac{\partial^2}{\partial\phi_{n+1}^2} + \frac{\epsilon\alpha_n}{r_0}\frac{\partial}{\partial\phi_{n+1}}\right\} \int_{-\infty}^{+\infty} \frac{dp}{2\pi} e^{ip(\phi_{n+1}-\phi_n)}$$

$$(6.86)$$

Comparing the above equation with Eq. (6.85), and writing $\phi_{n+1} \equiv \phi$, yields the Black–Karasinski Hamiltonian as

$$H_B^{bk} = -\frac{\sigma^2}{2r_0^2}\frac{\partial^2}{\partial\phi^2} - \frac{\alpha}{r_0}\frac{\partial}{\partial\phi} \quad (6.87)$$

and is equal to the expected result given in Eq. (6.70).

The Hamiltonian is fairly general since both σ and α can be functions of the random variable ϕ, and can be used, for example, to model spot interest rates with stochastic volatility.

6.12 Appendix: Quantum mechanical spot rate models

The Black–Karasinski model is a nonlinear model, and difficult to analyze analytically. One can make a modification of the Black–Karasinski model by dropping the boundary terms from the action S_{BK} given in Eq. (6.75), and obtain the quantum mechanical model given by

$$L_1 = -\frac{1}{2\sigma^2}\left[\left(\frac{d\phi}{dt}\right)^2 + a^2(\phi - \theta)^2\right] \quad (6.88)$$

$$S_1 = \int_{t_0}^{T} L_1 dt$$

The modified Black–Karasinski model is similar to the simple harmonic oscillator of quantum mechanics, and can be used to generate an approximate expansion of

[8] From Eq. (4.35), the convention for scalar product is $<p|\phi_n> = \exp(-ip\phi_n)$, and the sign of the exponential in Eq. (6.85) reflects this choice. The definition of H_{BK} requires it to act on the dual state vector $<\phi_{n+1}|$; if one chooses to write the Hamiltonian as acting on the state vector $|\phi_n>$, $H_{BK}^\dagger$ is then obtained. Since H_{BK} is not Hermitian, this would lead to an incorrect result.

the zero coupon bonds that exhibit non-affine behaviour for the zero coupon bond.

One can also make a quantum mechanical nonlinear generalization of the Vasicek model given in Eq. (6.14) by defining the Lagrangian

$$S_2 = -\frac{1}{2\sigma^2} \int_{t_0}^{T} dt \left\{ \left[\frac{dr}{dt} - a(b-r) \right]^2 + \lambda(r-b)^4 \right\}$$

and the zero coupon bond is then given by Eq. (6.77), with S_2 being the appropriate action.

Both the spot rate actions given by S_1 and S_2 yield non-affine models for the zero coupon bonds, and are possible models for the spot interest rate.

Part III

Quantum field theory of interest rates models

7

Quantum field theory of forward interest rates

The complexity of the forward interest rates, or forward rates, is far greater than that encountered in the study of stocks and their derivatives; the reason being that a stock at a given instant in time is described by only one degree of freedom that is undergoing random evolution, whereas in the case of the interest rates it is the entire yield curve that is randomly evolving and requires **infinitely many degrees of freedom** for its description. The theory of quantum fields [106] has been developed precisely to study problems involving infinitely many (independent) degrees of freedom, and so one is naturally led to its techniques in the study of the interest yield curve.

The most widely used model of the forward rates is the HJM model [43]. The fundamental limitation of the HJM model is that all the forward rates are exactly correlated, leading, for instance, to the unreasonable possibility of hedging a 30-year Treasury Bond with a six-month Treasury Bill. Models in which the forward rates have **nontrivial correlation** are more general, and it will be seen later from the empirical studies of the forward rates that such nontrivial correlations in fact exist in the financial markets.

Field theory models are able to incorporate correlation between forward rate maturities in a parsimonious manner that is well suited to analytical and computational studies as well as to empirical implementation. This is the main motivation for studying the forward interest rates from the point of view of quantum field theory.

Treating all the forward interest rates as independent random variables has been studied in [38, 61, 95]. In references [61] and [38] a correlation between forward rates with different maturities was introduced. In [95] the forward rate was modelled as a stochastic string, and a stochastic partial differential equation in infinitely many variables was obtained. A detailed discussion of the various generalizations of the HJM model, and their relation to the field theory model of the forward rates, is given in [97].

In the approach based on quantum field theory all financial instruments are formally given as a path (functional) integral and hence is complementary to the approach based on stochastic partial differential equations. The advantages of the approach based on quantum field theory are that it offers a different perspective on financial processes, offers a variety of computational algorithms, and nonlinearities in the forward interest rates as well as its stochastic volatility can be incorporated in a fairly straightforward manner.

Quantum field theory models of the forward rates are based on taking the forward rates as a strongly correlated system with independent fluctuations for all maturities [4].

The outline of this chapter is the following. Quantum mechanics and its relation to quantum field theory is briefly reviewed for readers from disciplines other than physics. The HJM model is extended to the case with **independent** fluctuations of the forward rates for each maturity; the theory is seen to consist of a two-dimensional quantum field theory. For simplicity the linear case is studied first – which is a free (Gaussian) quantum field. The field theory model has new parameters that determine how strongly it deviates from the HJM model, and hence the HJM model can be obtained by setting these parameters to zero. A number of Gaussian models for the forward rates are studied, including the 'stiff' case that provides the best fit to the market data. A Gaussian path integration is used to derive the risk-neutral martingale measure for the linear forward interest rates models.[1]

Nonlinear generalizations of the Gaussian model are discussed, and stochastic volatility that is a function of the forward rates is introduced. The theory is further generalized to the case of linear forward rates, with nonlinear stochastic volatility being an independent quantum field.

7.1 Quantum field theory

The concept of a **quantum field** is introduced, and shown to be a natural generalization of the concept of a **particle** in quantum mechanics.

Suppose one is interested in studying how an **extended object** undergoes quantum evolution. How does one describe the quantum dynamics of such an object? Consider for example a non-relativistic (one-dimensional) string, and let its displacement from equilibrium at time t and at position x in space be denoted by $\phi(t, x)$, as shown for a particular instant t_0 in Figure 7.1.

Let the initial string position at time t_1 be given by $\phi_1(x) = \phi(t_1, x)$, and the final position at time t_2 be given by $\phi_2(x) = \phi(t_2, x)$. Suppose the string has mass

---

[1] The terms linear, Gaussian and free quantum fields are used interchangeably.

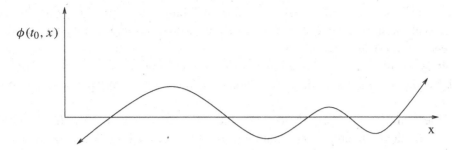

Figure 7.1 A typical string configuration

per unit length given by ρ, and string tension (energy per unit length) given by T. A general expression for the action[2] of the string is given by [106]

$$S_{\text{string}} = -\frac{1}{2} \int_{t_1}^{t_2} dt \int_{-\infty}^{+\infty} dx \rho \left(\frac{\partial \phi}{\partial t}\right)^2$$
$$-\frac{1}{2} \int_{t_1}^{t_2} dt \int_{-\infty}^{+\infty} dx \left[T \left(\frac{\partial \phi}{\partial x}\right)^2 + V(\phi) \right]$$
$$\equiv S_{\text{kinetic}} + S_{\text{potential}} \tag{7.1}$$

where $V(\phi)$ is the potential energy of the field ϕ. In analogy with quantum mechanics, **all possible string positions** are allowed to occur at each instant of the string's evolution. Hence one needs to integrate over all possible values for the string's position at each point x and for each instant t.

Let the dynamics of the field be determined by the Hamiltonian of the string given by $\hat{H}_{\text{string}}$, and which can be derived from the string action S_{string}. In analogy with Eq. (5.64) of quantum mechanics, the quantum field theory for the transition amplitude (partition function) of the string field $\phi(t, x)$ is defined by the Feynman path integral [106]

$$Z \equiv \; < \phi_2 | e^{-\tau \hat{H}_{\text{string}}} | \phi_1 > \tag{7.2}$$

$$= \prod_{t_1 < t < t_2} \prod_{-\infty < x < +\infty} \int_{-\infty}^{+\infty} d\phi(t, x) \exp(S_{\text{string}}) \tag{7.3}$$

with boundary conditions given by $\phi_1(x) = \phi(t_1, x)$ and $\phi_2(x) = \phi(t_2, x)$. The collection of infinitely many random variables $\{\phi(t, x)\}$ is called a **boson quantum field**. Unlike a classical string that has a determinate and fixed value for every x and t, the boson quantum field takes **all possible values** for each x and t.

---

[2] Physical time t is replaced by $t \to -it$, and the theory is then said to be defined in Euclidean time.

Eqs. (5.64) and (7.2) that define quantum mechanics and quantum field theory respectively look deceptively similar. At every instant t_0 there is only **one** degree of freedom $x(t_0)$ in quantum mechanics, whereas for a quantum field, there are **infinitely many** degrees of freedom, since for **each** point of space $\mathbf{x}$ that is occupied by the string, the string co-ordinate $\phi(t_0, x)$ is an independent random variable.

In the Hamiltonian formulation the state space of a single particle in quantum mechanics depends on one variable, namely is given by $|x>$, whereas for a single field ϕ, the field's state space depends on infinitely many independent variables given by the infinite tensor product $|\phi>= \otimes_{-\infty<x<+\infty} |\phi(x)>$. Hence the initial and final quantum state vectors of the (string) field in the transition amplitude in Eq. (7.2) are given by $|\phi_1>= \otimes_{-\infty<x<+\infty} |\phi(t_1, x)>$ and $|\phi_2>= \otimes_{-\infty<x<+\infty} |\phi(t_2, x)>$ respectively.[3]

One can see that quantum mechanics is a system with a finite number degrees of freedom, whereas quantum field theory is a system that has infinitely many independent degrees of freedom. This, in essence, is the difference between quantum mechanics and quantum field theory.

From a more mathematical point of view there is no measure theoretic interpretation of the expression $\prod_{t_1<t<t_2} \prod_{-\infty<x<+\infty} \int_{-\infty}^{+\infty} d\phi(t, x)$. A rigorous definition of Eq. (7.3) is to limit spacetime to a finite volume, and then discretize spacetime into a lattice so that the infinite-dimensional integration given in Eq. (7.3) is reduced to an ordinary finite-dimensional multiple integral $\prod_{m,n} \int d\phi_{m,n}$. The lattice field theory of forward rates, defined on a two-dimensional lattice, is discussed in Appendix 7.18.

A (finite) continuum limit of a nonlinear field theory defined on a finite and discrete spacetime is in general possible only if the action S defines a theory that is renormalizable [106]. Moreover, only by studying the properties of nonlinear field theories under renormalization can one decide how to construct a consistent perturbation expansion for the theory. The procedure of renormalization has as yet no mathematically rigorous definition, and, in general, the entire formalism of quantum field theory lies beyond the scope of conventional and rigorous mathematics, including stochastic calculus [65,92].

If the action S is only a quadratic function of the quantum field ϕ, the theory is said to be a free (Gaussian) quantum field, and one can take the continuum limit without having to address the problem of renormalization. The linear (Gaussian) case for the forward interest rates is analysed in some detail since the simplicity of the model allows many important features of the theory to be studied analytically.

---

[3] To rigorously define the tensor product over a continuous index x, one first needs to discretize the index $x = na$, where a is an infinitesimal, and limit the range of integer $n = 0, \pm1, \pm2, \ldots \pm N$. The tensor product is then over a finite number of state spaces, and the limit of this system is taken for $N \to \infty$ followed by the limit $a \to 0$. The discretized version of the field theory model is discussed in Appendix 7.18.

For the more complex case of nonlinear forward rates and of linear forward rates with stochastic volatility, the problem of renormalization needs to be addressed.

7.2 Forward interest rates' action

In the HJM model, as discussed in Section 6.3, the fluctuations in the forward rates at a given time t are given by 'shocks' that are delivered to the entire curve $f(t, x)$ by random variables $W_i(t)$, which do not depend on the maturity direction x. Clearly, a more general evolution of the instantaneous forward rates would be to let the whole curve evolve randomly, that is to let **all the forward rates** – that is, $f(t, x)$ for **each x and t** – fluctuate **independently**. The only constraint imposed on the random evolution of the forward rates is that, at every instant, the evolution be driven by a **risk-neutral martingale measure**, and which can be used to price financial instruments that are free from arbitrage opportunities.

At any instant t, there exist in the market forward interest rates for a duration of T_{FR} in the future. The forward rates at any instant t, namely $f(t, x)$, exists for all $t < x < t + T_{FR}$.[4] Viewed as a function of the maturity variable x, the forward rates are called the **forward rate curve**. It is important to note that the label x stands for **future time**, and not for a point of space.

Figure 7.2 shows the market data on the forward interest rates for the US\$ obtained from the Eurodollar futures for 1990–1996,[5] and will be used extensively to empirically study the various models of the forward rates.

Figure 7.2 plots the daily traded values of forward rates for only eight maturities, namely for maturities of three months, and yearly maturities from one to seven. The forward interest rates' time evolution for eight maturities has the appearance of eight points randomly evolving in time, but in a very correlated manner; for example, all the lines move up and down together and they never cross. The full forward rate curve consists of infinitely many maturities, all of which evolve randomly in a highly correlated manner.

Since at any instant t there are infinitely many forward rates, an infinite number of independent variables are required to describe its random evolution. As discussed in Section 7.1, the generic quantity describing such a system is a quantum field [106]. The forward interest rates is hence considered to be a **boson quantum field**; that is, $f(t, x)$ is taken to be an **independent** random variable for **each x and each t**. For notational simplicity both t and x are kept continuous. In Appendix 7.18 the lattice theory of the forward rates is studied when both t and x are made discrete and range over a finite set, and the continuum limit is then discussed in some detail.

---

[4] T_{FR} is greater than 30 years.
[5] See discussion in Section 8.1 on the Eurodollar market.

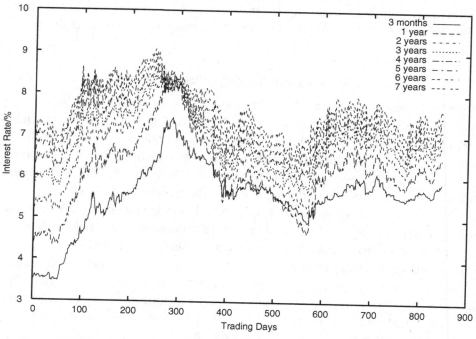

Figure 7.2 Eurodollar futures from 1990–1996, for the forward interest rates $f(t,7 \text{ years})$, $f(t, 6 \text{ years})$, ... $f(t, 1 \text{ year})$, and $f(t, 0.25 \text{ years})$.

For the sake of concreteness, consider the evolution of the forward rates starting from some initial time T_i to a future time T_f. Since all the forward rates $f(t, x)$ are always for the future, it is always true that $x > t$; hence the quantum field $f(t, x)$ is defined on a domain consisting of a parallelogram $\mathcal{P}$ that is bounded in the maturity direction by parallel lines $x = t$ and $x = T_{FR} + t$, and in the time direction by the horizontal lines $t = T_i$ and $t = T_f$ as shown in Figure 7.3. Every point inside the domain $\mathcal{P}$ represents an independent integration variable $f(t, x)$, and shows the enormous increase over the HJM random variables $W_i(t)$ given in Figure 6.1. For modelling the forward rates and Treasury Bonds, one needs to study a two-dimensional quantum field on a finite (Euclidean) domain.[6]

To define a Lagrangian $\mathcal{L}$, one needs a kinetic term, denoted by $\mathcal{L}_{\text{kinetic}}$, to describe the time evolution of the forward rates. Since it is known from the HJM model that the forward interest rates have a drift velocity $\alpha(t, x)$ and volatility $\sigma(t, x)$, these have to appear directly in the Lagrangian. The important insight of HJM [43] consists in recognizing that the combination of forward rates that occurs in finance is of the form $[\partial f(t, x)/\partial t - \alpha(t, x)]/\sigma(t, x)$; this HJM combination

---

[6] The field theory interpretation of the evolution of the forward rates, as expressed in the domain $\mathcal{P}$, is that of a (non-relativistic) quantum string moving with unit velocity in the x (maturity) direction.

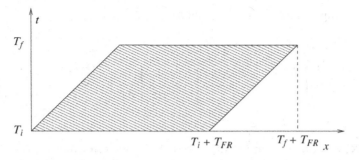

Figure 7.3 Domain $\mathcal{P}$ of the forward interest rates

of forward rates, drift velocity and volatility continues to appear throughout the treatment of the forward rates, including the linear and nonlinear cases, as well as when the volatility of the forward rates is rendered stochastic.

Another term needs to be introduced in the Lagrangian $\mathcal{L}$ for constraining the change of shape of the forward rates in the maturity direction. The analogy of this term in the case of an ordinary string is a tension term in the Lagrangian which attenuates sharp changes in the shape of the string, since the shape of the string stores potential energy. To model a similar property for the forward rates one cannot use a simple tension-like term $(\partial f / \partial x)^2$ in the Lagrangian, since, as will be shown in Section 10.7.1, this term is ruled out by the (risk-neutral) martingale condition for the forward rates.

7.3 Field theory action for linear forward rates

The existence of a martingale measure requires that the forward rates Lagrangian contain higher-order derivative terms, essentially a term of the form $(\partial^2 f / \partial x \partial t)^2$; such string systems have been studied in [83] and are said to be strings with finite **rigidity**. Rigidity yields a term in the forward rates Lagrangian, namely $\mathcal{L}_{\text{rigidity}}$, with a new parameter μ; the rigidity of the forward rates is given by $1/\mu^2$ and quantifies the strength of the fluctuations of the forward rates in the time-to-maturity direction x.

The simplest action that meets all the requirements discussed above is a **linear** (Gaussian) model for the forward rates given by

$$S[f] = \int_{T_i}^{T_f} dt \int_t^{t+T_{FR}} dx \, \mathcal{L}[f] \tag{7.4}$$

$$\equiv \int_{\mathcal{P}} \mathcal{L}[f] \tag{7.5}$$

with the Lagrangian density $\mathcal{L}[f]$ given by

$$\mathcal{L}[f] = \mathcal{L}_{\text{kinetic}}[f] + \mathcal{L}_{\text{rigidity}}[f] \tag{7.6}$$

$$= -\frac{1}{2}\left[\left\{\frac{\frac{\partial f(t,x)}{\partial t} - \alpha(t,x)}{\sigma(t,x)}\right\}^2 + \frac{1}{\mu^2}\left\{\frac{\partial}{\partial x}\left(\frac{\frac{\partial f(t,x)}{\partial t} - \alpha(t,x)}{\sigma(t,x)}\right)\right\}^2\right]$$

$$-\infty \leq f(t,x) \leq +\infty \tag{7.7}$$

The presence of the second term in the action given in Eq. (7.6) is not ruled out by the existence of a risk-neutral measure, and an empirical study [19] provides strong evidence for this term in the evolution of the forward rates.

In summary, the forward rates behave like a quantum string, with a time- and 'space-dependent' drift velocity $\alpha(t,x)$, an effective mass given by $1/\sigma^2(t,x)$, and string rigidity proportional to $1/\mu^2$. In the limit of $\mu \to 0$, it will be shown that one can recover (up to a re-scaling) the HJM model, which corresponds to an infinitely rigid string.

The drift term $\alpha(t,x)$ is completely determined by the requirement of obtaining a martingale evolution of the forward rates, with the Lagrangian having as free parameters the function $\sigma(t,x)$, and constants such as μ^2. Unlike the HJM model where a functional form is usually assumed for the volatility function $\sigma(t,x)$, in the field theory approach volatility can be kept completely arbitrary and determined from the market.

Since the field theory is defined on a finite domain $\mathcal{P}$ as shown in Figure. 7.3, to complete the definition of the model the boundary conditions need to be specified on all the four boundaries of the finite parallelogram $\mathcal{P}$.

- **Fixed (Dirichlet) initial and final conditions**
  The initial and final (Dirichlet) conditions in the time direction are given by

$$t = T_i \; ; \; T_i < x < T_i + T_{FR} \; : \; f(T_i, x) \tag{7.8}$$
$$: \text{specified initial forward rate curve}$$

$$t = T_f \; ; \; T_f < x < T_f + T_{FR} \; : \; f(T_f, x) \tag{7.9}$$
$$: \text{specified final forward rate curve}$$

- **Free (Neumann) boundary conditions**
  To specify the boundary condition in the maturity direction, one needs to impose the condition on the action given in Eq. (7.4) that it has no surface terms. The term $\int_{\mathcal{P}} \mathcal{L}_{\text{rigidity}}[f]$ in the action can be integrated by parts with respect to x, and the requirement that there are no boundary terms yields the following Neumann boundary condition

$$T_i < t < T_f, \quad \frac{\partial}{\partial x}\left(\frac{\frac{\partial f(t,x)}{\partial t} - \alpha(t,x)}{\sigma(t,x)}\right) = 0 \tag{7.10}$$

$$: x = t \text{ or } x = t + T_{FR} \tag{7.11}$$

The Neumann boundary conditions are also necessary for obtaining a Hamiltonian for the forward rates, as will be clear from the discussion in Chapter 10. Doing an integration by parts in the maturity direction using the Neumann boundary conditions yields, from Eqs. (7.5) and (7.6), the action

$$
S = -\frac{1}{2} \int_{\mathcal{P}} \left(\frac{\frac{\partial f(t,x)}{\partial t} - \alpha(t,x)}{\sigma(t,x)} \right) \left[1 - \frac{1}{\mu^2} \frac{\partial}{\partial x^2} \right] \left(\frac{\frac{\partial f(t,x)}{\partial t} - \alpha(t,x)}{\sigma(t,x)} \right) \tag{7.12}
$$

The quantum field theory of the forward rates is defined by the partition function Z, which is obtained by integrating over all configurations of $f(t,x)$, and yields the Feynman path integral

$$
Z = \int Df e^{S[f]} \tag{7.13}
$$

$$
\int Df \equiv \prod_{(t,x)\epsilon\mathcal{P}} \int_{-\infty}^{+\infty} df(t,x) \tag{7.14}
$$

$e^{S[f]}/Z$ is the probability for different field configurations to occur when the functional integral over $f(t,x)$ is performed.

The forward rates starting from some time t_0 can in principle be defined into the infinite future, that is with $T_f = \infty$. Since the forward rates $f(t,x)$ are only defined for the future, one always has $x > t$. The domain $\mathcal{P}$ of the forward rates can be extended, as shown in Figure 7.4, to a semi-infinite parallelogram that is

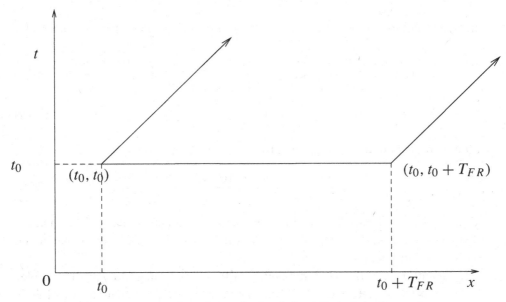

Figure 7.4 Domain of the forward rates defined for infinite future time

bounded by parallel lines $x = t$ and $x = T_{FR} + t$ in the maturity direction, and by the line $t = t_0$ in the time direction.

7.4 Forward interest rates' velocity quantum field $A(t, x)$

The action given in Eq. (7.4) is suitable for studying the formal properties of the forward rates, and is the basis, in Chapter 10, for defining the Hamiltonian of the forward interest rates. For studying the Feynman path integral, which is the focus of this chapter, it is simpler for computational purposes to change variables from the quantum field $f(t, x)$ to another quantum field $A(t, x)$.

Let $A(t, x)$ be a two-dimensional quantum field; the HJM change of variables expresses $A(t, x)$ in terms of the forward rates $f(t, x)$ as follows, namely

$$\frac{\partial f}{\partial t}(t, x) = \alpha(t, x) + \sigma(t, x)A(t, x) \tag{7.15}$$

$$f(t, x) = f(t_0, x) + \int_{t_0}^{t} dt'\alpha(t', x) + \int_{t_0}^{t} dt'\sigma(t', x)A(t', x)) \tag{7.16}$$

The quantum field $A(t, x)$ is the drift-less velocity field of the forward interest rates.

The Jacobian of the above transformation is a constant and, hence, up to a constant

$$\int Df \rightarrow \int DA \tag{7.17}$$

The action in terms of the $A(t, x)$ field, from Eqs. (7.6) and (7.15), is given by

$$S[A] = -\frac{1}{2}\int_{t_0}^{\infty} dt \int_{t}^{t+T_{FR}} dx \left\{ A^2(t, x) + \frac{1}{\mu^2}\left(\frac{\partial A(t, x)}{\partial x}\right)^2 \right\} \tag{7.18}$$

$$= \int_{\mathcal{P}} \mathcal{L}[A] \tag{7.19}$$

with Neumann boundary conditions, from Eq. (7.10), given by

$$\left.\frac{\partial A(t, x)}{\partial x}\right|_{x=t} = 0 = \left.\frac{\partial A(t, x)}{\partial x}\right|_{x=t+T_{FR}} \tag{7.20}$$

The quantum field variables at the boundary $x = t$ and $x = t + T_{FR}$, namely $A(t, t)$ and $A(t, t + T_{FR})$ take all possible values, and this results in the Neumann boundary conditions given above.

The quantum field theory is defined by a functional integral over all variables $A(t, x)$; in particular, the values of $A(t, x)$ on the boundary of $\mathcal{P}$ are unconstrained and are independent integration variables; this yields the partition function

$$Z = \int DA e^{S[A]} \tag{7.21}$$

Due to Neumann boundary conditions, repeating the derivation leading to Eq. (7.12) one has the action of velocity quantum field

$$S = -\frac{1}{2} \int_{\mathcal{P}} A(t, x) \left(1 - \frac{1}{\mu^2} \frac{\partial^2}{\partial x^2} \right) A(t, x) \tag{7.22}$$

The action $S[A]$ given in Eq. (7.22) has no derivative coupling in the time direction. The velocity quantum field $A(t, x)$ is effectively a quantum mechanical system in the maturity x direction, and, for each t, is identical to the quantum mechanical system discussed in Section 5.5.

Scaling symmetry

The change of variables from $f(t, x)$ to $A(t, x)$ has the following symmetry. Consider the transformation $\sigma(t, x) \rightarrow \zeta(t, x)\sigma(t, x)$; then a corresponding change of $A(t, x) \rightarrow \zeta(t, x)^{-1}A(t, x)$ leaves the defining Eq. (7.16) for $A(t, x)$ invariant. The change of variables from $A(t, x) \rightarrow \zeta(t, x)^{-1}A(t, x) \equiv B(t, x)$ yields an identical theory to the one written above in terms of the $A(t, x)$ field. This symmetry implies that the volatility function $\sigma(t, x)$ does not have an invariant significance, and the symmetry needs to be fixed in order to uniquely determine $\sigma(t, x)$ from market data.

7.5 Propagator for linear forward rates

Quantum fields, such as $f(t, x)$ or $A(t, x)$, themselves cannot be directly observed as they are degrees of freedom (integration variables) that are fluctuating and have no fixed value. What are observable and measurable are the **average values** of quantities that are functions of the quantum field. In particular, the measurable quantities of a quantum field are the **correlation functions** that encode the effect of the field's fluctuations at one point on the field's fluctuations at other points.

The most important correlation function for finance – the field theory analog of the variance of a single random variable – is the correlation function between the field's fluctuations at **two** different points. More precisely, the two-point correlator of the field $A(t, x)$ is given by

$$
\begin{aligned}
< (A(t, x)A(t', x')) > &= E[A(t, x)A(t', x')] \\
&= \frac{1}{Z} \int DA \; A(t, x)A(t', x')e^{S[A]} \\
&\equiv \delta(t - t')D(x, x'; t, T_{FR})
\end{aligned}
\tag{7.23}
$$

where the $\delta(t - t')$ has been factored out for future convenience. $D(x, x'; t, T_{FR})$ is called the **propagator** and is a measure of the effect that the fluctuations of the field $A(t, x)$ at point t, x has on the fluctuations of $A(t', x')$ at another point t', x'.

It is convenient to evaluate the moment generating functional for the quantum field theory, given by

$$
Z[J] = \frac{1}{Z} \int DA e^{\int_{t_0}^{\infty} dt \int_t^{t+T_{FR}} dx J(t,x)A(t,x)} e^{S[A]}
\tag{7.24}
$$

Any correlation function can be evaluated from $Z[J]$ by functional differentiation by $J(t, x)$ (discussed in A.15) and then setting $J(t, x) = 0$. In particular

$$
< A(t, x)A(t', x') > = \frac{\delta^2}{\delta J(t, x)\delta J(t', x')}Z[J]\Big|_{J=0}
\tag{7.25}
$$

The generating functional $Z[J]$ for **constant rigidity** μ has already been evaluated in Section 5.5. From Eq. (5.30)

$$
Z[J] = \exp \frac{1}{2} \int_{t_0}^{\infty} dt \int_t^{t+T_{FR}} dxdx' J(t, x)D(x, x'; t, T_{FR})J(t, x')
\tag{7.26}
$$

where the propagator $D(x, x'; t, T_{FR})$ is given in Eq. (5.38).[7] From Eq. (7.26) one sees that for the Gaussian model, similar to the normal random variable, the correlation functions between any number of fields at different points can all be expressed in terms of the propagator. This simple property of the linear (Gaussian) field theory does not extend to nonlinear theories, where it can be shown that one needs **all** the correlation functions between fields at an arbitrary number of points to fully describe the theory [106].

---

[7] The propagator $D(x, x'; t, T_{FR})$ depends **only** on the variables $x - t$ and $x' - t$ since the Lagrangian, the domain $\mathcal{P}$ and the Neumann boundary conditions are all only functions of $x - t$. This property of the propagator implies that, if the volatility function is also a function only of $x - t$, that is $\sigma(t, x) = \sigma(x - t)$, then all the properties of the future interest rates depend not on instant t but only on how far into the future one is looking at. This property is only partly realized in the financial market; see Figure 8.2.

From Eq. (5.38) the propagator is given by[8]

$$D(x, x'; t, T_{FR}) = \mu \frac{\cosh \mu \{T_{FR} - |x - x'|\} + \cosh \mu \{T_{FR} - (x + x' - 2t)\}}{2 \sinh \mu T_{FR}}$$

$$= D(x', x; t, T_{FR}) \quad : \text{ symmetric function of } x, x' \quad (7.27)$$

From Eqs. (7.22), (7.24) and (7.26), the following is a formal expression for the propagator

$$D(x, x'; t, T_{FR}; \mu) = \, < x | \frac{1}{1 - \frac{1}{\mu^2} \frac{\partial^2}{\partial x^2}} | x' > \quad : \text{ Neumann B.C.'s.} \quad (7.28)$$

The symmetry of the theory under the transformation $A(t, x) \to \zeta(t, x)^{-1} A(t, x)$ is used to fix the propagator's diagonal value to be unity, that is $D(x, x'; t, T_{FR}; \mu) \to \tilde{D}(x, x'; t, T_{FR}; \mu)$ such that $\tilde{D}(x, x; t, T_{FR}; \mu) = 1$.

For most applications $T_{FR} \to \infty$, and yields

$$D(x, x'; t) = \lim_{T_{FR} \to \infty} D(x, x'; t, T_{FR})$$

$$= \frac{\mu}{2} \left[e^{-\mu(x + x' - 2t)} + e^{-\mu|x - x'|} \right] \quad ; \quad x, x' > t \quad (7.29)$$

The first exponential in the propagator given in Eq. (7.29) is due to the boundary at $x = t$, and the second exponential is due to x taking values on an infinite interval.

The propagator has the following interpretation. If the field $A(t, x)$ has some value at point x, then the field at 'distances' $x - \mu^{-1} < x' < x + \mu^{-1}$ will tend to have the same value, whereas for other values of x' the field will have arbitrary values. Hence the fluctuations in the time-to-maturity x direction are correlated within maturity time μ^{-1}, which is the **correlation time** of the forward interest rates.

'Kink' in the propagator

An undesirable feature of the constant rigidity propagator is that – along the diagonal direction – the slope of the propagator **perpendicular** to the diagonal direction is discontinuous. The discontinuity of the slope appears as a 'kink' when the propagator is plotted against its arguments as can be seen in Figure 8.8. To analytically identify the kink, define new variables $\theta_\pm = \theta \pm \theta'$; $\theta = x - t$, $\theta' = x' - t$. The propagator is given by

$$D(\theta_+; \theta_-) = \frac{\mu}{2} \left[e^{-\mu \theta_+} + e^{-\mu|\theta_-|} \right] \quad (7.30)$$

---

[8] To make the connection with Eq. (5.38) given in Section 5.5, compare the simple harmonic oscillator action given in Eq. (5.28) with the action given in Eq. (7.22) for the velocity quantum field. This yields the following identification of the parameters: $m = 1/\mu^2$, $\omega = \mu$, $t_i = t$, $t_f = t + T_{FR}$, and $x = t$, $x' = t'$.

$\theta_- = 0$ defines the diagonal line $\theta = \theta'$ in the plot of $D(\theta, \theta')$. The slope of the propagator perpendicular to $\theta_- = 0$ (see Figure 8.10) is defined as follows

$$m = \frac{\partial D(\theta_+; \theta_-)}{\partial \theta_-}\bigg|_{\theta_-=0} \equiv \frac{\partial D(\theta_+; 0)}{\partial \theta_-} \tag{7.31}$$

Expanding the propagator about $\theta_- = 0$ yields

$$D(\theta_+; \theta_-) \simeq \frac{\mu}{2}\left[e^{-\mu\theta_+} + 1 - \mu|\theta_-| + O(\theta_-^2)\right] \tag{7.32}$$

Hence the slope m is discontinuous about $\theta_- = 0$ since[9]

$$m = -\frac{\mu^2}{2}\frac{\partial|\theta_-|}{\partial\theta_-} = \frac{\mu^2}{2}\begin{cases} -1 & \theta_- > 0 \\ +1 & \theta_- < 0 \end{cases}$$

The discontinuity in m is equal to μ^2, and can be seen in Figure 8.8.

It is shown in Chapter 8 that the field theory model with constant rigidity, while explaining some features of the market correlation of the forward rates, does not predict the correlation very well.

The basic model with constant rigidity can be generalized in various ways. The simplest generalization is to continue with the Gaussian model, but with propagators that have more structure. In Appendices 7.14 to 7.15 the following variants of the free field (Gaussian) model are studied.

1. Forward interest rates with a constraint on $A(t, t)$ arising from the special role of the spot rate $f(t, t) = r(t)$.
2. Forward interest rates with a non-constant (maturity dependent) rigidity parameter $\mu = \mu(\theta)$.
3. Forward interest rates with a fourth-order derivative in maturity time given by a 'stiffness' term $(\partial^2 A(t, x)/\partial x^2)^2$ being added to the Lagrangian with constant rigidity.
4. Forward rates with (nonlinear) psychological future time given by $z = z(\theta)$ replacing θ.

Cases 1 and 2 are related to the propagator for constant rigidity. Case 3 yields a stiff propagator that is the most important for empirical purposes, giving the best fit to market data. Case 4 introduces nonlinear psychological future time in the theory; this concept is very general, and can be applied to both linear and nonlinear theories.

All the above linear models only modify the propagator $D(x, x'; t; T_{FR})$. All the formulae that are derived for the linear models, such as the martingale measure

---

[9] From its definition

$$|\theta_-| = \begin{cases} \theta_- & \theta_- > 0 \\ -\theta_- & \theta_- < 0 \end{cases}$$

and pricing of forward rates derivatives, are valid for any Gaussian model. It will be seen in Chapter 8 that the empirical predictions of the Gaussian field theory models hinge on the propagator $D(x, x'; t, T_{FR})$, and only in the empirical studies of the forward interest rates is any reference made to the explicit form of these propagators.

7.6 Martingale condition and risk-neutral measure

The risk-neutral evolution for the forward interest rates is obtained using the martingale condition for the action $S[A]$. The general martingale condition for Treasury Bonds is given in Eq. (2.10) as

$$P(t_0, T) = E_{[t_0, t_*]}[e^{-\int_{t_0}^{t_*} r(t) dt} P(t_*, T)] \tag{7.33}$$

and has the following explicit expression in field theory

$$P(t_0, T) = \frac{1}{Z} \int Df e^{-\int_{t_0}^{t_*} r(t) dt} P(t_*, T) e^{S[f]} \tag{7.34}$$

Repeating the steps followed in the HJM model for deriving Eqs. (6.41) and (6.42) yields the field theory generalization that

$$\exp \int_{\mathcal{T}} \alpha(t, x) = \frac{1}{Z} \int DA e^{-\int_{\mathcal{T}} \sigma(t, x) A(t, x)} e^{\int_{\mathcal{P}} \mathcal{L}[A]} \tag{7.35}$$

$$= \exp \frac{1}{2} \int_{t_0}^{t_*} dt \int_{t}^{T} dx dx' \sigma(t, x) D(x, x'; t, T_{FR}) \sigma(t, x') \tag{7.36}$$

where the last equation follows from the generating functional given in Eq. (7.26).

The trapezoidal domain $\mathcal{T}$ determining the risk-neutral measure is nested inside the domain of the forward rates $\mathcal{P}$, as shown in Figure 7.5.

Dropping the time integration in Eq. (7.36) yields

$$\int_{t}^{T} dx \alpha(t, x) = \frac{1}{2} \int_{t}^{T} dx dx' \sigma(t, x) D(x, x'; t, T_{FR}) \sigma(t, x') \tag{7.37}$$

Differentiating above expression with respect to T yields the following generalization of Eq. (6.44) for the drift velocity

$$\alpha(t, x) = \sigma(t, x) \int_{t}^{x} dx' D(x, x'; t, T_{FR}) \sigma(t, x') \tag{7.38}$$

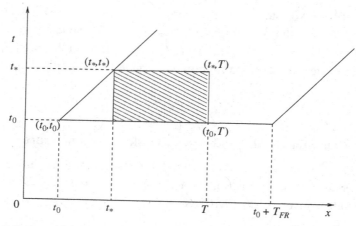

Figure 7.5 Trapezoidal domain $\mathcal{T}$ for the martingale condition contained in the domain of the forward rates

For constant volatility $\sigma(t, x) = \sigma_1$, and the exact result is[10]

$$\alpha(t, x) = \lim_{T_{FR} \to \infty} \frac{\sigma_1^2}{2} \left[1 + \frac{e^{-2\mu(T_{FR} - x + t)} - e^{-2\mu(x-t)}}{1 - e^{-2\mu T_{FR}}} \right]$$

$$= \frac{\sigma_1^2}{2} \left(1 - e^{-2\mu(x-t)} \right)$$

Eqs. (7.15) and (7.38) yield

$$f(t, x) = f(t_0, x) + \int_{t_0}^{t} dt' \sigma_i(t', x) \int_{t'}^{x} dy D(x, y; t', T_{FR}) \sigma_i(t', y)$$

$$+ \int_{t_0}^{t} dt' \sigma(t', x) A(t', x) \tag{7.40}$$

7.7 Change of numeraire

In obtaining the risk-neutral measure in Section 7.6, the martingale measure was defined by discounting the Treasury Bond by the money market account $B(t_0, t_*)$

---

[10] The drift has the limiting behaviour

$$\alpha(t, x) \simeq \begin{cases} \frac{1}{2T_{FR}} \sigma(t, x) \int_{t}^{x} dx' \sigma(t, x') & \mu \to 0 \\ \frac{1}{2} \sigma^2(t, x) & \mu \to \infty \end{cases} \tag{7.39}$$

The equation for $\alpha(t, x)$ for the case for $\mu = \infty$ is quite dissimilar from that of the HJM model given in Eq. (6.44), which, upto a rescaling by T_{FR}, is given by $\mu = 0$.

in Eq. (7.33), where

$$B(t_0, t_*) \equiv e^{\int_{t_0}^{t_*} r(t)dt} \tag{7.41}$$

$$B(t_0, t_0) = 1$$

The quantity $P(t, T)/B(t_0, t)$ is a martingale [see Eq. (A.2)], since from Eq. (7.33) the conditional expectation is

$$P(t_0, T) = E_{[t_0, t_*]}\left[e^{-\int_{t_0}^{t_*} r(t)dt} P(t_*, T) \right]$$

$$\Rightarrow \frac{P(t_0, T)}{B(t_0, t_0)} = E\left[\frac{P(t_*, T)}{B(t_0, t_*)} \bigg| \frac{P(t_0, T)}{B(t_0, t_0)} \right]$$

where $E_{[t_0, t_*]}[\ldots] \equiv E[\ldots]$.

It has been shown by Geman *et al.* [31,36,37,42] that any positive valued security can be used for discounting the Treasury Bond. In particular, one can use other Treasury Bonds with different maturities as a discounting factor instead of using the spot interest rate. The martingale condition for the changed discounting factor leads only to a change in the drift term for the action [36].

For concreteness, suppose that at time t_0 all Treasury Bonds are discounted by another Treasury Bond that matures at some fixed time t_*, that is, by $P(t_0, t_*)$. The martingale condition is now defined by

$$P(t_0, T) = P(t_0, t_*)E_*[P(t_*, T)] \tag{7.42}$$

$$\equiv P(t_0, t_*) \int DA \, P(t_*, T)e^{S_*} \tag{7.43}$$

where $E_*[\ldots]$ denotes taking the expectation value with respect to the new risk-neutral measure S_*.

After the change of numeraire $P(t, T)/P(t, t_*)$ is a martingale, since from Eq. (7.42)

$$\frac{P(t_0, T)}{P(t_0, t_*)} = E_*\left[\frac{P(t_*, T)}{P(t_*, t_*)} \bigg| \frac{P(t_0, T)}{P(t_0, t_*)} \right] \tag{7.44}$$

$$= E_*[P(t_*, T)]$$

Denote by $\alpha_*(t, x)$ the drift that corresponds to the martingale condition when discounting by the Treasury Bond $P(t_0, t_*)$. A straightforward calculation using the martingale condition determines that the drift velocity,[11] similar to Eq. (7.38),

---

[11] A Hamiltonian derivation is given in Eq. (10.71).

is given by

$$\alpha_*(t, x) = \sigma(t, x) \int_{t_*}^{x} dx' D(x, x'; t, T_{FR}) \sigma(t, x') \tag{7.45}$$

The relation of the risk-neutral probability measures e^S/Z and e^{S_*}/Z_*, obtained by discounting using the spot interest rate and a Treasury Bond respectively, can be explicitly obtained for Gaussian forward rates; a derivation is given in Appendix 7.19 using the techniques of quantum field theory. From Eq. (7.104)

$$S = S[\alpha] \quad ; \quad S_* = S[\alpha_*]$$

$$e^{S_*} = \frac{e^{-\int_{t_0}^{t_*} r(t) dt}}{P(t_0, t_*)} e^S \tag{7.46}$$

The factor relating the two actions is evaluated in the finance literature using the Radon–Nikodyn derivative [31, 36].

Eq. (7.46) is particularly useful in evaluating European options for Treasury Bonds. From Eq. (6.53), for $t_0 < t_*$ the price of a call option C is given by

$$C(t_0, t_*, T, K) = E[e^{-\int_{t_0}^{t_*} dt r(t)} (P(t_*, T) - K)_+] \tag{7.47}$$

$$= P(t_0, t_*) E_*[(P(t_*, T) - K)_+] \tag{7.48}$$

where Eq. (7.46) has been used in obtaining Eq. (7.48) above. To compute the call option using Eq. (7.48) is much simpler than doing the calculation using Eq. (7.47), since the discounting term, after a change in numeraire, is the deterministic function $P(t_0, t_*)$.[12]

7.8 Nonlinear forward interest rates

In the field theory of forward interest rates discussed so far, the forward rates were taken to be Gaussian random fields with a finite probability of being negative. As long as interest rates are well above zero, the negative valued fluctuations of the forward rates are negligible. However, if the forward rates are near zero, as has been the case for Japanese Yen since the 1990s and for the US$ since the early 2000s, it becomes important that the forward rates be a strictly positive-valued quantum field, that is $f(t, x) > 0$ for all t, x. The forward rates can then be modeled as exponential fields, and hence essentially nonlinear. Having $f(t, x) > 0$ is a major advantage of any model, since in the financial markets forward rates are always positive.

Market data discussed in Section 8.5.1 show that the volatility of volatility is small but significant, and consequently considering volatility as a stochastic

---

[12] The bond call option is computed in Sections 9.2 and 10.13 using discounting by a Treasury Bond.

quantity needs to be addressed. Volatility by its very definition is the standard deviation of the forward rates and hence is always strictly positive. The modelling of the fluctuations of volatility of the forward rates is closely tied to the positivity of the forward rates. Once the forward rates are considered to be positive-valued quantum fields, the simplest model for stochastic volatility is to consider it to be some function of the forward rates – which is consistent only if the forward rates themselves are strictly positive. Volatility can be also be generalized to a stochastic quantity by considering it to be an independent quantum field.

In the remainder of this chapter the field theory of the forward rates is extended, following the treatment given in [3], to a nonlinear theory of positive-valued forward rates, and to a theory with linear forward rates with stochastic volatility.

The following cases are analyzed

1. Nonlinear forward rates with deterministic volatility.
2. Nonlinear forward rates, with stochastic volatility a function of the forward rates.
3. Linear forward rates with an independent nonlinear quantum field for stochastic volatility.

For brevity of discussion cases 1 and 2 will sometimes be combined. The most general case of nonlinear forward rates coupled to a nonlinear stochastic volatility quantum field is not discussed.

The Lagrangians for the nonlinear forward rates discussed in this chapter are **incomplete** because the drift term $\alpha(t, x)$ is left undetermined. The reason being that a Gaussian integration was performed in obtaining the drift term given in Eq. (7.36) from its definition in Eq. (7.35); this can no longer be carried out for nonlinear forward rates. To fix the drift term it will be necessary to have a Hamiltonian formulation of the theory, which is discussed in Chapter 10.

7.9 Lagrangian for nonlinear forward rates

The forward rates are strictly positive, that is $f(t, x) > 0$. Hence, in contrast to Eq. (7.7), the forward rates are modelled as an exponential quantum field

$$f(t, x) = f_0 e^{\phi(t,x)} > 0 \; ; \quad -\infty \le \phi(t, x) \le +\infty$$

How should the Lagrangian given in Eq. (7.6) be generalized to case where the forward rates are always positive? For starters consider the volatility to be deterministic and given by some function $\sigma_0(t, x)$ that is determined from the financial markets.

The Lagrangian given in Eq. (7.6) is interpreted to be an approximate one that is valid only if all the forward rates are close to some fixed value f_0. Note

$$\frac{\partial f(t,x)}{\partial t} = f_0 e^{\phi(t,x)} \frac{\partial \phi(t,x)}{\partial t}$$

$$\simeq f_0 \frac{\partial \phi(t,x)}{\partial t} + O(\phi^2)$$

Hence the following mapping is made[13]

$$\frac{\partial f(t,x)}{\partial t} \rightarrow f_0 \frac{\partial \phi(t,x)}{\partial t}$$

Eq. (7.6) then generalizes to

$$\mathcal{L}[\phi] = \mathcal{L}_{\text{kinetic}}[\phi] + \mathcal{L}_{\text{rigidity}}[\phi] \tag{7.49}$$

$$= -\frac{1}{2}\left[\left\{ \frac{f_0 \frac{\partial \phi(t,x)}{\partial t} - \alpha(t,x)}{\sigma_0(t,x)} \right\}^2 + \frac{1}{\mu^2}\left\{ \frac{\partial}{\partial x}\left(\frac{f_0 \frac{\partial \phi(t,x)}{\partial t} - \alpha(t,x)}{\sigma_0(t,x)} \right) \right\}^2 \right]$$

The partition function of the theory is given by the Feynman path integral

$$Z = \int D\phi \, e^{S[\phi]} \tag{7.50}$$

$$\int D\phi \equiv \prod_{(t,x)\in\mathcal{P}} \int_{-\infty}^{+\infty} d\phi(t,x) \tag{7.51}$$

with Dirichlet condition for initial and final time as in Eq. (7.8), and Neumann boundary condition, similar to Eq. (7.10), given by

$$T_i < t < T_f, \quad \frac{\partial}{\partial x}\left(\frac{f_0 \frac{\partial \phi(t,x)}{\partial t} - \alpha(t,x)}{\sigma_0(t,x)} \right) = 0 \tag{7.52}$$

$$: x = t \text{ or } x = t + T_{FR}$$

The Lagrangian for the nonlinear forward rates given in Eq. (7.49) is the generalization of the Black–Karasinski spot interest rate Lagrangian given in Eq. (6.75). The Lagrangian is nonlinear due to the behaviour of the drift term $\alpha(t,x)$, as given in Eq. (10.66). Nonlinearities are also contained in the financial instruments that depend on the forward interest rates.

---

[13] This is a standard mapping in the study of Lie group valued quantum fields, with its kinetic terms being determined in a similar manner [106].

7.9.1 Fermion path integral

The case of nonlinear forward rates with deterministic volatility is analyzed to illustrate the new features of the nonlinear case. The Neumann boundary conditions given in Eq. (7.52) necessitates a change of variables from $\phi(t, x)$ to $A(t, x)$, with the velocity field satisfying the Neumann boundary conditions. This change of variables gives rise to a nontrivial Jacobian. Define the change of variables by

$$\frac{f_0 \frac{\partial \phi(t,x)}{\partial t} - \alpha(t, x)}{\sigma_0(t, x)} = A(t, x) \qquad (7.53)$$

From the analysis for the martingale condition in Chapter 10, for the special case of $\sigma_0(t, x) = \sigma_0 = $ constant, one obtains from Eq. (10.65)

$$\alpha(t, x) = -\frac{\sigma_0^2}{2f_0} D(x, x; t) + \sigma_0^2 \int_t^x dx' D(x, x'; t) e^{\phi(t,x')}$$

On solving Eq. (7.53), one obtains the following

$$\phi(t, x) = \psi[t, x; A]$$

Hence, the differentials are given by

$$d\phi(t, x) = \int dt' dx' \mathcal{J}(t, x; t', x') dA(t', x')$$

$$\text{where } \mathcal{J}(t, x; t', x'; A) = \frac{\delta \psi[t, x; A]}{\delta A(t', x')}$$

$$\Rightarrow D\phi = J[A] \, DA \quad \text{where } J = \det \left| \mathcal{J}(t, x; t', x') \right|$$

In field theory, a determinant is represented by a path integral over **fermion quantum fields** $c^\dagger(t, x), c(t', x')$ (integration variables) [106] such that

$$J[A] = \int Dc^\dagger Dc e^{S_F[c^\dagger, c, A]}$$

$$S_F = \int_{\mathcal{P}} dt dx dt' dx' c^\dagger(t, x) \mathcal{J}(t, x; t', x'; A) c(t', x') \qquad (7.54)$$

Fermions are anti-commuting integration variables that are well-defined mathematical objects, and path integrals for fermion quantum fields are discussed in [106]. Quantum fields such as the forward interest rates $f(t, x), \phi(t, x)$ and so on that take real or complex values are called **boson quantum fields** to distinguish them from fermionic fields.

The nonlinear forward rates has a partition function given by

$$Z = \int D\phi e^{S[\phi]}$$

$$= \int DAJ[A]e^{S[A]} = \int DADc^\dagger Dc \; e^{S[A]+S_F[c^\dagger,c,A]} \tag{7.55}$$

$$\text{with} \int DADc^\dagger Dc \equiv \prod_{(t,x)\in\mathcal{P}} \int_{-\infty}^{+\infty} dA(t,x) \int dc^\dagger(t,x)dc(t,x)$$

The action $S[A]$ is quadratic in the field variable $A(t,x)$, and hence contains no nonlinearities. The sum total effect of the nonlinearity of the forward interest rates is seen to be contained in the fermion path integral that generates nonlinear terms for the $A(t,x)$ field.

Nonlinear quantum field theories are notoriously difficult to analyze due to the divergences that are inherent in such theories; only those field theories that are renormalizable can be given a consistent and meaningful interpretation [106]. The renormalization of nonlinear forward interest rates has yet to be addressed and solved.

7.10 Stochastic volatility: function of the forward rates

For forward interest rates that are positive valued, volatility can be modelled as a stochastic quantity by considering it to be a function of the forward rates [1, 2, 103]. The standard models using this approach consider that volatility is given by[14]

$$\sigma(t,x,f(t,x)) = \sigma_0(t,x)e^{\nu\phi(t,x)} \tag{7.56}$$

$$\sigma_0(t,x) \quad : \text{deterministic function}$$

In the limit of zero rigidity $\mu \to 0$, the following variants of the HJM models are special cases of Eq. (7.56), and have been discussed from an empirical point of view in [2].

1. Ho and Lee (1986) Model : $\sigma(t,x,f(t,x)) = \sigma_0$: deterministic
2. CIR (1985) : $\sigma(t,x,f(t,x)) = \sigma_0 f^{\frac{1}{2}}(t,x)$
3. Courtadon (1982) : $\sigma(t,x,f(t,x)) = \sigma_0 f(t,x)$
4. Vasicek (1977) : $\sigma(t,x,f(t,x)) = \sigma_0 \exp(-\lambda(x-t))$: deterministic
5. Linear proportional HJM (1992) : $\sigma(t,x,f(t,x)) = [\sigma_0 + \sigma_1(x-t)]f(t,x)$

---

[14] Using no arbitrage arguments, it can be shown that $\nu \geq -1/2$ [2,97].

The following is a straightforward generalization of the nonlinear Lagrangian given in Eq. (7.49)

$$\mathcal{L}[\phi] = \mathcal{L}_{\text{kinetic}}[\phi] + \mathcal{L}_{\text{rigidity}}[\phi] \tag{7.57}$$

$$= -\frac{1}{2}\left[\left\{\frac{f_0\frac{\partial\phi(t,x)}{\partial t} - \alpha(t, x)}{\sigma_0(t, x)e^{\nu\phi(t,x)}}\right\}^2 + \frac{1}{\mu^2}\left\{\frac{\partial}{\partial x}\left(\frac{f_0\frac{\partial\phi(t,x)}{\partial t} - \alpha(t, x)}{\sigma_0(t, x)e^{\nu\phi(t,x)}}\right)\right\}^2\right]$$

with Dirichlet condition for initial and final time as in Eq. (7.8), and Neumann boundary condition, similar to Eq. (7.10), given by

$$T_i < t < T_f, \quad \frac{\partial}{\partial x}\left(\frac{f_0\frac{\partial\phi(t,x)}{\partial t} - \alpha(t, x)}{\sigma_0(t, x)e^{\nu\phi(t,x)}}\right) = 0 \tag{7.58}$$

$$: x = t \text{ or } x = t + T_{FR}$$

The quantum field theory is defined by

$$Z = \int D\phi\, e^{-\nu\phi} e^{S[\phi]}$$

$$\int D\phi\, e^{-\nu\phi} = \prod_{(t,x)\in\mathcal{P}} \int_{-\infty}^{+\infty} d\phi(t, x) e^{-\nu\phi(t,x)} \tag{7.59}$$

The parameter ν is a measure of the extent to which volatility is stochastic, and has a natural $\nu \to 0$ limit to the case of the nonlinear theory of forward rates with deterministic volatility $\sigma_0(t, x)$.

The parameter ν can in principle be determined from data. In the fortunate circumstance that ν is a small parameter, one can do a perturbative expansion of the nonlinear theory about the Gaussian theory, a procedure that is well studied and understood in field theory, and use perturbation theory to self-consistently estimate ν from market data.

7.11 Stochastic volatility: an independent quantum field

Consider the case where volatility $\sigma(t, x)$ is stochastic and modelled as an **independent** quantum field.[15] Since one can only measure the behaviour of the forward rates, all effects of stochastic volatility are manifested only through the observed properties of the forward rates.

---

[15] The HJM model [43] has been developed by [1, 103] (and references cited therein) to account for stochastic volatility. Amin and Ng [2] studied the market data of Eurodollar options to obtain the implied forward rates volatility; they concluded that many features of the market, and in particular of the (stochastic) volatility of the forward rates curve, could not be fully explained in the HJM framework.

For simplicity, consider the forward interest rates as a linear quantum field, and from Eq. (7.7)

$$f(t, x): \quad -\infty \le f(t, x) \le +\infty$$

Since the volatility function $\sigma(t, x) > 0$ is always positive, introduce another **boson quantum field** $h(t, x)$ such that

$$\sigma(t, x) = \sigma_0 e^{h(t,x)}, \quad -\infty \le h(t, x) \le +\infty$$

The system now consists of **two interacting quantum fields**, namely $f(t, x)$ and $h(t, x)$. The interacting system's Lagrangian should have the following features.

- A parameter ξ that quantifies the extent to which the field $h(t, x)$ is stochastic. A limit of $\xi \to 0$ would, in effect, 'freeze' all the fluctuations of the field $h(t, x)$, and reduce it to a deterministic function.
- A parameter κ to control the fluctuations of $h(t, x)$ in the maturity direction similar to the parameter μ that controls the fluctuations of the forward rates $f(t, x)$ in the maturity direction x.
- A parameter ρ with $-1 \le \rho \le +1$ that quantifies the correlation of the forward rates' quantum field $f(t, x)$ with the volatility quantum field $h(t, x)$.
- A drift term for volatility, namely $\beta(t, x)$, which is analogous to the drift term $\alpha(t, x)$ for the forward rates.

The Lagrangian for the interacting system is not unique; there is a wide variety of choices that one can make to fulfil all the conditions given above. A possible Lagrangian for the interacting system, written by analogy with the Lagrangian for the case of stochastic volatility for a single security [5], is given by

$$\mathcal{L} = -\frac{1}{2(1-\rho^2)} \left(\frac{\frac{\partial f}{\partial t} - \alpha}{\sigma} - \rho \frac{\frac{\partial h}{\partial t} - \beta}{\xi} \right)^2 - \frac{1}{2} \left(\frac{\frac{\partial h}{\partial t} - \beta}{\xi} \right)^2$$

$$- \frac{1}{2\mu^2} \left(\frac{\partial}{\partial x} \left(\frac{\frac{\partial f}{\partial t} - \alpha}{\sigma} \right) \right)^2 - \frac{1}{2\kappa^2} \left(\frac{\partial}{\partial x} \left(\frac{\frac{\partial h}{\partial t} - \beta}{\xi} \right) \right)^2 \tag{7.60}$$

with action

$$S[f, h] = \int_{\mathcal{P}} \mathcal{L}$$

One needs to specify the boundary conditions for the interacting system. The initial and final conditions for the forward rates $f(t, x)$ given in Eq. (7.8) continue to hold for the interacting case. The volatility quantum field's boundary conditions are the following.

Fixed (Dirichlet) initial and final conditions

The initial value is specified from data, that is

$$T_i < x < T_i + T_{FR}, \quad \sigma(T_i, x)$$

: specified initial volatility curve

$$T_f < x < T_f + T_{FR}, \quad \sigma(T_f, x)$$

: specified final volatility curve

The boundary condition in the x-direction for the forward rates $f(t, x)$ – as given in Eq. (7.10) – continues to hold for the interacting case, and for the volatility field is similarly given by the Neumann boundary condition.

Free (Neumann) boundary conditions

$T_i < t < T_f$:

$$\frac{\partial}{\partial x}\left(\frac{\partial h(t, x)}{\partial t} - \beta(t, x)\right)\Big|_{x=t} = 0 = \frac{\partial}{\partial x}\left(\frac{\partial h(t, x)}{\partial t} - \beta(t, x)\right)\Big|_{x=t+T_{FR}} \quad (7.61)$$

On quantizing the volatility field $\sigma(t, x)$ the boundary condition for the forward rate $f(t, x)$ given in Eq. (7.10) is rather unusual. The martingale measure yields that the drift velocity α is a (quadratic) functional of the volatility field $\sigma(t, x)$; hence the boundary condition Eq. (7.10) for the case of stochastic volatility is a form of **interaction** between the $f(t, x)$ and $\sigma(t, x)$ fields.

The requirement that the system have a well-defined Hamiltonian dictates that path-integration measure for the quantum field $h(t, x)$ needs to be defined as follows

$$\int Df Dh e^{-h} = \prod_{(t,x)\in\mathcal{P}} \int_{-\infty}^{+\infty} df(t, x)dh(t, x)e^{-h(t,x)} \quad (7.62)$$

The partition function for the forward interest rates with stochastic volatility is defined by the Feynman path integral as

$$Z = \int Df Dh e^{-h} e^{S[f,h]} \quad (7.63)$$

Similar to nonlinear forward interest rates, the problem of renormalization has to be solved for the field theory of forward rates with stochastic volatility to be mathematically consistent [106].

The (observed) market value of a financial instrument, say $\mathcal{O}[f, h]$, is expressed as the **average value** of the instrument – denoted by $< \mathcal{O}[f, h] >$ – taken over all possible values of the quantum fields $f(t, x)$ and $h(t, x)$, with the probability density given by the (appropriately normalized) exponential of the action. In symbols

$$< \mathcal{O}[f, h] > = \frac{1}{Z} \int Df Dh e^{-h} \, \mathcal{O}[f, h] e^{S[f,h]} \qquad (7.64)$$

Stochastic volatility is reduced to a deterministic function in the following limit: ξ, ρ and $\kappa \to 0$. The $1/\xi^2$ kinetic term of the $h(t, x)$ field in the action given in Eq. (7.61) has the limit (up to irrelevant constants)

$$\lim_{\xi \to 0} \prod_{t, x \in \mathcal{P}} \exp\left\{ -\frac{1}{2} \int_{\mathcal{P}} \left(\frac{\frac{\partial h}{\partial t} - \beta}{\xi} \right)^2 \right\} \to \prod_{t, x \in \mathcal{P}} \delta\left(\frac{\partial h}{\partial t} - \beta \right)$$

which implies that

$$< \sigma(t, x) > = \sigma_0 < e^{h(t,x)} >$$

$$\Rightarrow \sigma_0 \exp\left\{ \int_{t_0}^{t} dt' \beta(t', x) \right\} + O(\xi, \kappa, \rho)$$

7.12 Summary

The quantum field theory of the forward interest rates is based on each forward interest rate being an independent degree of freedom; in particular, the correlation of fluctuations in the field theory of the forward rates typically have a finite range, whereas in the HJM model **all** the fluctuations are exactly correlated. The finite correlations in the time-to-maturity direction that exist for the forward rates can be efficiently captured using a variety of field theory models.

Gaussian models of the forward rates have the important property of being analytically tractable. Five different variations of the Gaussian model were discussed, showing the flexibility and versatility of the Gaussian models.

The field theory model was further developed to account for nonlinear forward rates. The nonlinearities give rise, for the first time in financial modelling, to a fermion path integral that results from a nonlinear change of variables. Stochastic volatility of the forward rates was firstly considered as a nonlinear function of the forward rates, and was then introduced as an independent quantum field. In both cases, the path integral has natural extensions that account for nonlinear forward rates, and for their stochastic volatility.

For the case of deterministic volatility with linear forward rates, it was shown in Section 7.4 that, in effect, the two-dimensional quantum field theory reduced to a one-dimensional quantum mechanical problem due to the specific nature of the Lagrangian. However, the quantum field theory of nonlinear forward rates was seen to be irreducibly two dimensional.

Although the formalism of stochastic calculus can be extended to the case of free Gaussian quantum fields, it cannot be extended to account for nonlinear quantum field theories. The concept of renormalization, that is essential for understanding nonlinear field theories, has no counterpart in stochastic calculus.

The methodology of quantum field theory provides a computationally tractable and independent perspective for studying and understanding the (nonlinear) random processes that drive the forward interest rates.

7.13 Appendix: HJM limit of the field theory

The action $S[A]$ given in (7.18) allows **all** the degrees of freedom of the field $A(t, x)$ to fluctuate independently and can be thought of as a 'string' with string rigidity equal to $1/\mu^2$; in this language the forward rate curve in the HJM model is a string with **infinite rigidity**.

From Eq. (7.29) the constant rigidity propagator has the following limits

$$D(x, x'; t) \rightarrow \begin{cases} \mu + O(\mu^2) & \mu \rightarrow 0 \\ \frac{1}{2}\mu e^{-\mu|x-x'|} \rightarrow \delta(x - x') & \mu \rightarrow \infty \end{cases} \tag{7.65}$$

As expected, in the limit of $\mu \rightarrow 0$ all the fluctuations in the x direction are 'frozen' in that they are exactly correlated; in other words the values of $A(t, x)$ for different maturities are all the same, and this is the limit that reproduces the HJM model. The 'freezing' of all quantum fluctuations can be directly seen from the action given in Eq. (7.18); when $\mu = 0$ any configuration of $A(t, x)$ with variations in the x direction gives an infinite negative contribution to the action, and hence is eliminated from the path integral; the only random configurations that survive are the ones for which all the variables $A(t, x)$ have the same value in the x direction, and yield the HJM model.[16]

For Eq. (7.65) the limits of $T_{FR} \rightarrow \infty$ and $\mu \rightarrow 0$ cannot be interchanged since the convergence is not uniform. When $\mu \rightarrow 0$, the simplest procedure to obtain the HJM model is to set the field theory propagator to unity. In other words

$$\lim_{\text{QFT} \rightarrow \text{HJM}} D(x, x'; t, T_{FR}) \rightarrow 1$$

Define

$$j(t) = \frac{1}{\sqrt{T_{FR}}} \int_t^{t+T_{FR}} dx J(t, x) \tag{7.66}$$

---

[16] The other limit of $\mu \rightarrow \infty$ corresponds to all the fluctuations of the $A(t, x)$ being completely de-correlated.

From Eqs. (7.26) and (7.66) the HJM limit of the propagator yields

$$\lim_{\mu \to 0} Z[j] = \exp \frac{1}{2} \int_{t_0}^{\infty} dt \, j^2(t)$$

which is the result obtained earlier, in Eq. (6.35), for the HJM model.

The K-factor HJM model is

$$\frac{\partial f(t, x)}{\partial t} = \alpha(t, x) + \sum_{i=1}^{K} \sigma_i(t, x) W_i(t)$$

with white noise given by $< W_i(t) W_j(t') > = \delta_{i-j} \delta(t - t')$. The forward rates correlator is hence given by

$$< \left[\frac{\partial f(t, x)}{\partial t} - \alpha(t, x) \right] \left[\frac{\partial f(t', x')}{\partial t} - \alpha(t', x') \right] > = \delta(t - t') H(t, x, x')$$

where

$$H(t, x, x') = \sum_{i=1}^{K} \sigma_i(t, x) \sigma_i(t, x')$$

The Lagrangian describing the random process of a K-factor HJM model is hence given by

$$\mathcal{L}[A] = -\frac{1}{2} A(t, x) H^{-1}(t, x, x') A(t, x')$$

where $H^{-1}(t, x, x')$ denotes the inverse of the function $H(t, x, x')$. For the K-factor HJM model, the expression $H^{-1}(t, x, x')$ is singular, and one has to introduce a vanishingly small rigidity parameter to obtain a well-defined field theory.

7.14 Appendix: Variants of the rigid propagator

The constant rigidity model of the forward interest rates is based on the term in the Lagrangian of the form $(\partial A(t, x)/\partial x)^2$ which constrains the fluctuations of the field $A(t, x)$ in the maturity direction. Various generalizations of this constant rigidity can be made, and two of these are considered below.

7.14.1 Constrained spot rate

All the forward rates have been treated on par in defining the constant rigidity model. However, it is known that the spot interest rate $r(t) = f(t, t)$ has a special role in the financial markets, since central banks frequently intervene to change the spot rate to suit changes in government policies. Hence, it is reasonable to assume that the spot interest rate $f(t, t)$ should be treated in a special manner.

Since what is meaningful is the rate of change of spot rate $r(t)$, it is constrained to fluctuate about some arbitrary value c. Eq. (7.38) shows that $\alpha(t, t) = 0$, and hence $(\partial f/\partial t)(t, t) = \sigma(t, t)A(t, t)$. In other words, constraining the rate of change of the spot rate is equivalent to constraining the velocity field $A(t, t)$.

The propagator for constant rigidity μ, namely $D(x, x'; t, T_{FR})$, is modified to incorporate the special behaviour of the spot rate. Introduce a new term in the action to constrain the boundary field variable $A(t, t)$, namely

$$e^S \rightarrow \frac{1}{\sqrt{2\pi a}} \exp\left\{-\frac{1}{2a}\int_t^\infty dt\left(A(t, t) - c\right)^2\right\} e^S$$

The parameter a controls the degree of randomness of the variable $A(t, t)$, and in the limit of $a \rightarrow 0$, the value $A(t, t)$ is fixed at c. The constrained propagator $D_C(x, x'; t, T_{FR})$ does not depend on the parameter c as it only changes the value of $\alpha(0)$. For calculational purposes it is easiest to assume that $c = 0$.

The constraint can be implemented by modifying the action to

$$e^S \rightarrow \int_{-\infty}^\infty d\xi e^S e^{i\xi A(t,t)} e^{-a^2\xi^2/2} \qquad (7.67)$$

Using Gaussian integrations, one can derive that the constrained propagator is given by

$$D_C(x, x'; t, T_{FR}) = D(x, x'; t, T_{FR}) - \frac{D(t, x; t, T_{FR})D(t, x'; t, T_{FR})}{D(t, t; t, T_{FR}) + a} \qquad (7.68)$$

The constrained propagator is only a function of $\theta = x - t$ since the constraint is imposed at $\theta = 0$. In the limit of $a \rightarrow +\infty$, as expected $D_C \rightarrow D(x, x'; t, T_{FR})$.

7.14.2 Non-constant rigidity

Another modification of the correlation structure is to make μ a function of the maturity time θ [9, 97]. The dependence of μ on maturity has a direct physical meaning; if one imagines that the forward rates' rigidity increases with increasing maturity, this in turn implies that μ decreases as a function of θ. The analytically tractable function $\mu = \mu_0/(1 + k\theta)$ is chosen for rigidity as it declines to zero as θ becomes large, and contains constant rigidity μ_0 as a limiting case. The action is given by

$$S_M = -\frac{1}{2}\int_{t_0}^{t_1} dt \int_0^{T_{FR}} d\theta \left(A^2 + \left(\frac{1 + k\theta}{\mu_0}\frac{\partial A}{\partial \theta}\right)^2\right) \qquad (7.69)$$

This is a quadratic action, and is simplified by performing integration by parts and setting the boundary term to zero, since Neumann boundary conditions are assumed. The propagator for this action is found to be [9, 97]

$$
D_M(\theta, \theta'; T_{FR}) = \frac{\mu_0^2 \alpha}{2\lambda\alpha(\alpha + 1/2)(1 - (1 + kT_{FR})^{-2\alpha})}
$$
$$
\times \left(\frac{\alpha + 1/2}{\alpha - 1/2}(1 + kT_{FR})^{-2\alpha}(1 + k\theta)^{\alpha - 1/2} + (1 + \lambda\theta)^{-\alpha - 1/2} \right)
$$
$$
\times \left(\frac{\alpha + 1/2}{\alpha - 1/2}(1 + kT_{FR})^{-2\alpha}(1 + k\theta')^{\alpha - 1/2} + (1 + k\theta')^{-\alpha - 1/2} \right) \quad (7.70)
$$

where $\alpha = \sqrt{\frac{1}{4} + \frac{\mu_0^2}{4k^2}}$. The bound on the θ variable is explicitly kept at T_{FR}. To compare this model with the HJM model the limit of $\mu_0 \to 0$ has to be taken before the limit of $T_{FR} \to \infty$, and the limits cannot be interchanged since the convergence of the limits is not uniform.

The following are two of the expected limits

$$
D_M(\theta, \theta'; T_{FR}) = \begin{cases} D(\theta, \theta'; T_{FR}; \mu_0) & k \to 0 \\ \frac{1}{T_{FR}} & \mu_0 \to 0 \end{cases} \quad (7.71a)
$$

If the limit $T_{FR} \to \infty$ is taken first, the propagator becomes, for $\theta > \theta'$

$$
D_{M1}(\theta, \theta') = \frac{\mu_0^2(\alpha - 1/2)}{2k\alpha(\alpha + 1/2)}(1 + k\theta)^{-\alpha - 1/2}
$$
$$
\times \left(\frac{\alpha + 1/2}{\alpha - 1/2}(1 + k\theta')^{\alpha - 1/2} + (1 + k\theta')^{-\alpha - 1/2} \right) \quad (7.72)
$$

which exhibits a θ dependence in the non-HJM limit of $\mu_0 \to 0$. Hence, $D_{M1}(\theta, \theta')$ cannot be made equivalent to HJM model since the limit of $T_{FR} \to \infty$ has been taken before the limit of $\mu_0 \to 0$ in $D_M(\theta, \theta'; T_{FR})$.

7.15 Appendix: Stiff propagator

It was shown in the last paragraph of Section 7.5 that the constant rigidity propagator shows a 'kink' around its diagonal co-ordinates.[17] All the variants of the propagator based on a rigidity term, including the propagators discussed in Appendices 7.14.1 and 7.14.2 continue to show the same 'kink'.

However, as can be seen from Figure 8.4, the surface of the empirical propagator is extremely smooth and showing no such kinks. The shape of the empirical

---

[17] Recall a 'kink' in the θ, θ' plot of the propagator $D(\theta, \theta')$ is a discontinuity, along the diagonal, in the derivative of the propagator along the direction orthogonal to the diagonal axis.

propagator is taken to be an indication in [6] that the evolution of the forward rates curve needs to be attenuated by a higher power of the derivative of the forward rates in the maturity direction than the second-order derivative provided by the constant rigidity model. The higher power of the derivative **stiffens** the fluctuations of the forward rates curve, and produces the desired smooth behaviour of the empirical propagator, as is demonstrated in this Appendix.

A **stiffness** term of the form $(\partial^2 A(t, x)/\partial x^2)^2$ is added to the Lagrangian[18] that strongly constrains any fluctuations of the forward rates in the maturity direction. Hence, extending Eq. (7.18), the following is an action, called the stiff action and Lagrangian, with both the rigidity and stiffness terms ($\mu, \lambda > 0$)

$$
\begin{aligned}
S_Q[A] = -\frac{1}{2} \int_{t_0}^{\infty} dt \int_{t}^{t+T_{FR}} dx \left\{ A^2(t, x) + \frac{1}{\mu^2} \left(\frac{\partial A(t, x)}{\partial x} \right)^2 \right. \\
\left. + \frac{1}{\lambda^4} \left(\frac{\partial^2 A(t, x)}{\partial x^2} \right)^2 \right\} \\
= \int_{\mathcal{P}} \mathcal{L}_Q[A]
\end{aligned}
\tag{7.73}
$$

with $A(t, x)$ satisfying Neumann boundary conditions as in Eq. (7.20), and which yields, on integrating by parts – similar to Eq. (7.22) – the following

$$
S_Q = -\frac{1}{2} \int_{\mathcal{P}} A(t, x) \left(1 - \frac{1}{\mu^2} \frac{\partial^2}{\partial x^2} + \frac{1}{\lambda^4} \frac{\partial^4}{\partial x^4} \right) A(t, x)
\tag{7.74}
$$

Stiffness $\lambda (\neq \infty)$ introduces a **quartic derivative** term in the action that attenuates, even more strongly than the μ rigidity term, all the high-frequency fluctuations of the forward rates in the maturity direction. The propagator is now given by[19]

$$
G(x, x'; t) = \lambda^4 < x | \frac{1}{\lambda^4 + (\lambda^2/\mu)^2 p^2 + p^4} | x' > \quad : \text{Neumann B.C.'s.}
\tag{7.75}
$$

$$
\text{where} \quad p^2 \equiv -\frac{\partial^2}{\partial x^2}
$$

Note that

$$
\lambda^4 + (\lambda^2/\mu)^2 p^2 + p^4 = (p^2 + \alpha_+)(p^2 + \alpha_-)
$$

$$
\text{with} \quad \alpha_{\pm} = \frac{\lambda^4}{2\mu^2} \left[1 \pm \sqrt{1 - 4(\frac{\mu}{\lambda})^4} \right]
$$

---

[18] Given the need to have a positive term in the Lagrangian, a third-order derivative is ruled out.
[19] Henceforth $T_{FR} \to \infty$.

Hence

$$\frac{1}{\lambda^4 + (\lambda^2/\mu)^2 p^2 + p^4} = \left(\frac{1}{\alpha_+ - \alpha_-}\right)\left[\frac{1}{p^2 + \alpha_-} - \frac{1}{p^2 + \alpha_+}\right] \quad (7.76)$$

Recall from Eq. (7.28) that

$$< x|\frac{1}{p^2 + \alpha_\pm}|x' > = \frac{1}{\alpha_\pm}D(x, x'; t; \sqrt{\alpha_\pm})$$

Define new variables

$$\theta_\pm = \theta \pm \theta'$$
$$\theta = x - t \; ; \; \theta' = x' - t \quad (7.77)$$

Hence, from Eqs. (7.75), (7.76) and (7.77)

$$G(\theta_+; \theta_-) = \left(\frac{\lambda^4}{\alpha_+ - \alpha_-}\right)\left[\frac{1}{\alpha_-}D(\theta_+; \theta_-; \sqrt{\alpha_-}) - \frac{1}{\alpha_+}D(\theta_+; \theta_-; \sqrt{\alpha_+})\right]$$
$$(7.78)$$

and, from Eq. (7.29)

$$D(\theta_+; \theta_-; \sqrt{\alpha_\pm}) = \frac{\sqrt{\alpha_\pm}}{2}\left[e^{-\sqrt{\alpha_\pm}\theta_+} + e^{-\sqrt{\alpha_\pm}|\theta_-|}\right] \quad (7.79)$$

$\lambda \rightarrow \infty$ gives the limiting behaviour of $\alpha_+ \simeq \lambda^4/\mu^2$ and $\alpha_- \simeq \mu^2$. Hence the propagator has the following limit

$$\lim_{\lambda \to \infty} G(\theta_+; \theta_-; \mu, \lambda) \rightarrow D(\theta_+; \theta_-; \mu)$$

and, as expected, reduces to the case of constant rigidity.

The solution for $\alpha_\pm$ yields three distinct cases, namely, when $\alpha_\pm$ is real, complex or degenerate. Note for all three cases $\lambda \neq \infty$, and giving rise to a qualitatively different propagator than the variants based on the rigidity term.

Case I: $\mu < \sqrt{2}\lambda$; $\alpha_\pm$ real

Choose the following parametrization

$$\alpha_\pm = \lambda^2 e^{\pm b}$$
$$e^{\pm b} = \frac{\lambda^2}{2\mu^2}\left[1 \pm \sqrt{1 - 4\left(\frac{\mu}{\lambda}\right)^4}\right] \; ; \; b \geq 0$$

In this parametrization, from Eqs. (7.78) and (7.79)

$$G_b(\theta_+; \theta_-) = \frac{\lambda}{2\sinh(2b)} \left[e^{-\lambda\theta_+\cosh(b)\}} \sinh\{b + \lambda\theta_+ \sinh(b)\} \right.$$
$$\left. + e^{-\lambda|\theta_-|\cosh(b)} \sinh\{b + \lambda|\theta_-|\sinh(b)\} \right] \quad (7.80)$$

The limit of $\mu \ll \lambda$ corresponds to $e^b \simeq \lambda^2/\mu^2$.

Case II: $\lambda < \mu/\sqrt{2}$; $\alpha_\pm$ complex

Use the parametrization

$$\alpha_\pm = \lambda^2 e^{\pm i\phi}$$

$$e^{\pm i\phi} = \frac{\lambda^2}{2\mu^2} \left[1 \pm i \sqrt{4\left(\frac{\mu}{\lambda}\right)^4 - 1} \right]$$

Hence

$$G_\phi(\theta_+; \theta_-) = \frac{\lambda}{2\sin(2\phi)} \left[e^{-\lambda\theta_+\cos(\phi)} \sin\{\phi + \lambda\theta_+ \sin(\phi)\} \right.$$
$$\left. + e^{-\lambda|\theta_-|\cos(\phi)} \sin\{\phi + \lambda|\theta_-|\sin(\phi)\} \right] \quad (7.81)$$

where $\cos(\phi) > 0$.

Case III: $\alpha_\pm = \lambda^2$: degenerate

Case I and Case II are separated by the degenerate case of $\alpha_\pm = \lambda^2$ corresponding to $\mu = \sqrt{2}\lambda$, which is equivalent to $b = 0 = \phi$. The propagator is

$$G_d(\theta_+; \theta_-) = \frac{\lambda}{4} \left[e^{-\lambda\theta_+}\{1 + \lambda\theta_+\} + e^{-\lambda|\theta_-|}\{1 + \lambda|\theta_-|\} \right] \quad (7.82)$$

For $|\theta_-| \simeq 0$, due to a nontrivial cancellation of the terms linear in $|\theta_-|$, the propagator given in Eq. (7.82) has the limit $G_d(\theta_+; \theta_-) \simeq -(1/2)\lambda^2\theta_-^2 + O(\theta_-^3)+$ function of θ_+, showing that the kink that was encountered for the constant (and non-constant) rigidity propagator has disappeared. This result is now proven in general.

Absence of 'kink' in the stiff propagator

The derivations carried out for the kink in Section 7.5 is repeated for the propagator $G(\theta_+; \theta_-)$, with the result being valid for all three cases; the explicit expression from Case I is used for convenience. The slope orthogonal to the line $\theta_- = 0$, as shown in Figure 8.10, is defined as

$$m_Q = \frac{\partial G_b(\theta_+; \theta_-)}{\partial \theta_-}|_{\theta_-=0} \equiv \frac{\partial G_b(\theta_+; 0)}{\partial \theta_-}$$

Expanding the propagator $G(\theta_+; \theta_-)$ about $\theta_- = 0$ yields[20]

$$G_b(\theta_+; \theta_-) \simeq \frac{\lambda}{2\sinh(2b)} \left(1 - \lambda|\theta_-|\cosh(b) + \frac{1}{2}\lambda^2\theta_-^2 \cosh^2(b) + O(\theta_-^3)\right)$$

$$\times \left(\sinh(b)\{1 + \lambda|\theta_-|\cosh(b)\} + \frac{1}{2}\lambda^2\theta_-^2 \sinh^2(b) + O(\theta_-^3)\right) \qquad (7.83)$$

$$= \frac{\lambda}{2\sinh(2b)} \left[\sinh(b) - \frac{1}{2}\lambda^2\theta_-^2 \sinh(b)\left(\cosh^2(b) - \sinh(b)\right)\right] + O(\theta_-^3)$$

The nontrivial cancellation of the term linear in $|\theta_-|$ in Eq. (7.83) gives a final result that is a function of θ_-^2 and consequently has a continuous derivative in the limit of $|\theta_-| \to 0$; hence

$$m_Q = r_Q\theta_-$$
$$\to 0 \quad \text{as} \quad \theta_- \to 0$$

showing that there is no kink for the stiff propagator; in fact $m_Q = 0$ also implies that along the full length of line defined by $\theta_- = 0$ the propagator has a maxima. The curvature orthogonal to the diagonal line $\theta_- = 0$ is given by

$$r_Q = \frac{\partial^2 G_b(\theta_+; \theta_-)}{\partial\theta_-^2}\bigg|_{\theta_-=0}$$

$$= \frac{\lambda}{2\sinh(2b)} \left[-\lambda^2 \sinh(b)\{\cosh^2(b) - \sinh(b)\}\right] \qquad (7.84)$$

$r_Q < 0$ follows from the fact that $b \geq 0$, confirming that the value of the propagator along $\theta_- = 0$ is a maximum.

In the limit of $\lambda \to \infty$, since one can no longer carry out the Taylors expansion around $\theta_- = 0$, the cancellation of the term linear in $|\theta_-|$ becomes invalid, and the propagator $G(\theta_+; \theta_-)$ develops the expected kink.

7.16 Appendix: Psychological future time

The forward interest rates model can be further improved by noting that the predicted correlation structure for field theory models depends only on variable $\theta = x - t$, where t is present time and x is future time. The variable θ is a measure of how far in the future is future time x. If one is studying nature, then there is no ambiguity in what is meant by future time $-\theta$ must grow linearly for any parametrization of the future. But in finance 'future time' is determined by how investors

---

[20] Up to a function of θ_+ that does not affect the singularity structure of the propagator near $\theta_- = 0$.

and fund managers view the future. Based on their subjective and personal views about how the financial system will perform in the future, practitioners of finance form their views – in the present – as to what the future holds, such as the risks inherent in future lending and borrowing, the present price of a futures contract and so on.

It is well known that one can predict very little about the very long-term behaviour of the market; the best that one can do is have some credible models for the next year or two. Hence it is expected that in taking positions far from the immediate future, the sense of future time should grow more slowly than calendar time. For instance, the future time interval between say the next four years and 12 years is eight calendar years, but in the minds of the investors and brokers, this could appear to be equal to the interval between the present and two years in the immediate future.

Future time θ is replaced by some nonlinear function $z = z(\theta)$ [9, 97], to be determined from the market. The nonlinear maturity variable $z = z(\theta)$ measures **psychological future time** in an investor's minds that corresponds to calendar future time given by θ. The specification of psychological future time $z(\theta)$ is an **independent** ingredient of the field theory model, and needs to be specified in conjunction with the Lagrangian. Some general features of function $z(\theta)$ is that it is invertible, namely $\theta = \theta(z)$ is well defined, and that $z(0) = 0$; $z(\infty) = \infty$. The independent variables are now $t, z(\theta)$ instead of t, x. The forward rates from the market are always given for $f(t, \theta)$, and so both future calendar time θ as well as psychological future time $z(\theta)$ are necessary to connect with the market. The defining equation for psychological future time, similar to Eq. (7.15), is given by

$$\frac{\partial f}{\partial t}(t, \theta) = \alpha(t, z(\theta)) + \sigma(t, z(\theta))A(t, z(\theta)) \; ; \; \theta = x - t \qquad (7.85)$$

where $f(t, \theta)$ depends only on calendar time $\theta = x - t$. An important feature of the defining equation above is that **both** future times, namely $\theta = x - t$ and psychological time $z(\theta)$ occur in the theory.[21]

The stiff Lagrangian for psychological future time is written as

$$S_z = -\frac{1}{2} \int_{t_0}^{t_1} dt \int_{z(0)}^{z(\infty)} dz \left(A^2 + \frac{1}{\mu^2} \left(\frac{\partial A}{\partial z} \right)^2 + \frac{1}{\lambda^4} \left(\frac{\partial^2 A}{\partial z^2} \right)^2 \right) \qquad (7.86)$$

---

[21] The field theory for psychological future time can be defined entirely in terms of forward rates $\bar{f}(t, z(\theta))$. However, for imposing the martingale condition, one needs to apply the functional differential operator $\delta/\delta \bar{f}(t, z(\theta))$ on Treasury Bonds $P(t, T) = \exp(- \int_0^{T-t} d\theta f(t, \theta))$ as given in Eq. (10.60). Hence it is necessary to specify the relation between $\bar{f}(t, z(\theta))$ and $f(t, \theta)$, and in effect one would recover Eq. (7.85).

The propagator for S_z is $G(z, z'; \mu, \lambda)$ as in Eq. (7.78) and the martingale condition for psychological future time is given by Eq. (10.61) as

$$\alpha(t, z) = \sigma(t, z) \int_{z(0)}^{z} dz' G(z, z') \sigma(t, z')$$

For the case of constant rigidity, as expected, the propagator is given by $D(z, z'; \mu) = (\mu/2)(e^{-\mu|z-z'|} + e^{-\mu(z+z')})$.

Introducing psychological future time $z(\theta)$ is different from giving a maturity dependence to the rigidity $\mu = \mu(\theta)$. To see this, write the action, for rigidity function $\mu(\theta) \equiv \mu_0(dg(\theta)/d\theta)$, as

$$S_M = -\frac{1}{2} \int_{t_0}^{t_1} dt \int_0^{\infty} d\theta \left(A^2 + \frac{1}{\mu_0^2} \left(\frac{\partial A}{\partial g} \right)^2 \right)$$

With a change of variables from θ to g the action is given by

$$S_M = -\frac{1}{2} \int_{t_0}^{t_1} dt \int_{g(0)}^{g(\infty)} dg \left(\frac{d\theta}{dg} \right) \left(A^2 + \frac{1}{\mu_0^2} \left(\frac{\partial A}{\partial g} \right)^2 \right) \neq S_z$$

Hence, unlike the case of psychological future time $z(\theta)$, the Lagrangian for nonlinear variable $g(\theta)$ has an additional Jacobian factor $dg(\theta)/d\theta$.

Examples for the possible choices for psychological future time are (i) $z = \tanh \beta(\theta - \theta_0)$, (ii) $z = \theta^\nu$ and so on.

The introduction of nonlinear future time $z(\theta)$ is a new way of thinking of the interest rates models. In the framework of field theory, empirical data can be used to gain an insight into market psychology that results in the formation of $z(\theta)$, and which constitutes subjective future time for the market players.

7.17 Appendix: Generating functional for forward rates

All the financial instruments are expressed in terms of the forward rates $f(t, x)$, and not in terms of the velocity field $A(t, x)$, which has been introduced primarily as an efficient tool for computations. The generating functional of the forward rates is directly computed, and is useful for evaluating various financial instruments.

The generating functional for the forward rates is defined by

$$Z_f[J] = \frac{1}{Z} \int Df e^{\int_{t_0}^{\infty} dt \int_t^{t+T_{FR}} dx J(t,x) f(t,x)} e^{S[f]} \tag{7.87}$$

with the path integral defined over the semi-infinite domain given in Figure 7.4. Change integration variables from $f(t, x)$ to the velocity field $A(t, x)$ in Eq. (7.87) for evaluating $Z_f[J]$. Recall from Eq. (7.16) that

$$
f(t, x) = f(t_0, x) + \int_{t_0}^{t} dt' \alpha(t', x) + \int_{t_0'}^{t} dt' \sigma(t', x) A(t', x)
$$

The equation above, after some simplifications, gives the following

$$
\int_{t_0}^{\infty} dt \int_{t}^{t+T_{FR}} dx J(t, x) f(t, x) = F_0
$$

$$
+ \int_{t_0}^{\infty} dt \int_{t}^{\infty} d\tau \int_{\tau}^{\tau+T_{FR}} dx J(\tau, x) \sigma(t, x) A(t, x)
$$

$$
\text{with } F_0 = \int_{t_0}^{\infty} dt \int_{t}^{t+T_{FR}} dx J(t, x) f(t_0, x)
$$

$$
+ \int_{t_0}^{\infty} dt \int_{t}^{t+T_{FR}} dx J(t, x) \int_{t_0}^{t} dt' \alpha(t', x)
$$

Since $\int Df \to \int DA$, from the generating functional $Z[A]$ given in Eq. (7.26) and the equation above

$$
Z_f[J] = e^{F_0} \frac{1}{Z} \int DA e^{\int_{t_0}^{\infty} dt \int_{t}^{\infty} d\tau \int_{\tau}^{\tau+T_{FR}} dx J(\tau,x)\sigma(t,x)A(t,x)} e^{S[A]} = e^{F_0} e^{W}
$$

$$
\Rightarrow W = \frac{1}{2} \int_{t_0}^{\infty} dt \left[\int_{t}^{\infty} d\tau \int_{\tau}^{\tau+T_{FR}} dx J(\tau, x) \sigma(t, x) \right]
$$

$$
\times \left[\int_{t}^{\infty} d\tau' \int_{\tau'}^{\tau'+T_{FR}} dx' J(\tau', x') \sigma(t, x') \right] D(x, x'; t) \qquad (7.88)
$$

Eq. (7.88) is useful for directly evaluating the forward interest rates' various Libor options and other derivatives.

7.18 Appendix: Lattice field theory of forward rates

A rigorous treatment of the quantum field theory of the forward interest rates that is defined over a continuous domain $\mathcal{P}$ is presented. The main idea is to truncate the full functional integral, that requires an independent integration for every point of the continuous domain $\mathcal{P}$, into a finite-dimensional multiple integral. The continuous domain $\mathcal{P}$ is replaced by a finite set of lattice points constituting a discrete domain $\hat{\mathcal{P}}$ that is obtained by discretizing the continuous variables t, x into a finite lattice. One can then rigorously discuss the continuous field theory as the limit of the lattice theory.

Discretize the domain $\mathcal{P}$ into a lattice of discrete points. Let $(t, x) \to (m\epsilon, na)$, where ϵ is an infinitesimal time step and a is an infinitesimal in the future maturity time direction. The semi-infinite domain $\mathcal{P}$ given in Figure 7.3 is truncated into a finite discretize domain $\hat{\mathcal{P}}$, with an upper limit in the time direction given by $M\epsilon$. Let $N = T_{FR}/a$ and $m_0 = t_0/\epsilon$ as shown in Figure 10.1.

The discrete and finite domain $\hat{\mathcal{P}}$ is bounded in the time direction by $m = m_0$ and $m = M$, and in the maturity direction by $na = m\epsilon$ and $na = m\epsilon + Na$. The integers take values in the discrete domain $\hat{\mathcal{P}}$, and are given by

$$\hat{\mathcal{P}} = \{m = m_0, m_0 + 1, \ldots, M - 1; na = m\epsilon, m\epsilon + a, \ldots, m\epsilon + Na\}$$

The forward rates and velocity field yield on discretization

$$f(t, x) \rightarrow f(m\epsilon, na) \equiv f_{mn}$$
$$A(t, x) \rightarrow A(m\epsilon, na) \equiv A_{mn}$$

and similarly for α and σ. The functional integral yields

$$\int DA \rightarrow \prod_{m=m_0}^{M-1} \prod_{n=m}^{N+m} \int_{-\infty}^{+\infty} dA_{mn} \qquad (7.89)$$

Boundary conditions on A_{mn} variables

The A_{mn} variables are always chosen, both for the linear and nonlinear forward rates, such that they satisfy the Neumann boundary conditions in the maturity direction; this in turn implies, as discussed in Eqs. (5.26) and (5.27), that the A_{mn} variables at the maturity boundaries are integration variables. In the time direction, the initial condition on the forward rates, namely $f(t_0, x)$ is specified, and hence does not constrain the A_{mn} variables; the future values of the forward rates, as in the case of pricing the value of an option, are integrated over, hence again do not constrain any of the A_{mn} variables. In summary, **all** the lattice variables A_{mn}, for every point on the lattice, are independent integration variables. The velocity quantum field $A(t, x)$ is the natural set of variables for any numerical study of the forward rates' path integral.

The functional integral over the field $A(t, x)$ has been reduced to a **finite-dimensional** multiple integral over the A_{mn} variables, which in the case above consists of $N(M - m_0)$-independent variables; hence all the techniques useful for evaluating finite-dimensional integrals can be used for performing the integration over A_{mn}, both for the linear and nonlinear cases.

To achieve the correct normalization, one in fact need not keep track of the constants that correctly normalize $\int DA$ in (7.89). Instead, one simply redefines the action by

$$e^{S[A]} \rightarrow e^{S[A]}/Z$$

$$Z \equiv \int DAe^{S[A]} \tag{7.90}$$

All the constants in $\int DA$ cancel out; and, more importantly, the expression e^S/Z is correctly normalized to be interpreted as a probability distribution, and hence can be used for Monte Carlo studies of this theory. The lattice action is the starting point for any of the simulations that are required of the model, including the pricing of path-dependent derivatives; there are well-known numerical algorithms developed in physics for numerically studying quantum fields [44].

Linear forward rates

Discretizing Eq. (7.15) yields the lattice change of variables from f_{mn} to A_{mn}

$$\frac{\partial f}{\partial t}(t, x) = \alpha(t, x) + \sigma(t, x)A(t, x)$$

$$\Rightarrow f_{m+1n} = f_{mn} + \epsilon\alpha_{mn} + \epsilon\sigma_{mn}A_{mn}$$

$$f_{m_0,n} : n = m_0, m_0 + 1, \ldots, N + m_0 : \text{ initial forward rate curve}$$

Consider the action for the case of constant rigidity given in Eq. (7.18)

$$S[A] = -\frac{1}{2}\int_{t_0}^{\infty} dt \int_{t}^{t+T_{FR}} dx \left\{ A^2(t, x) + \frac{1}{\mu^2}\left(\frac{\partial A(t, x)}{\partial x}\right)^2 \right\} \tag{7.91}$$

Using finite differences to discretize derivatives in the action yields

$$S_{\text{lattice}}[A] = -\frac{\epsilon^2}{2}\sum_{m=m_0}^{M-1}\left\{ \sum_{n=m}^{N+m} A_{mn}^2 + \frac{1}{(\epsilon\mu)^2}\sum_{n=m}^{N+m-1}(A_{mn+1} - A_{mn})^2 \right\} \tag{7.92}$$

Discretizing the path-integration measure yields

$$\int DA = \prod_{(t,x)\in\mathcal{P}}\int dA \equiv \lim_{\epsilon\to0, a\to0, M\to\infty}\prod_{mn}\int dA_{mn}$$

and the lattice partition function is given by

$$Z_{\text{lattice}} = \int DA_{mn}e^{S_{\text{lattice}}[A]}$$

All the variations of the Gaussian models that are discussed in Appendices 7.14–7.16 can be similarly analyzed.

HJM limit

To get some idea of the lattice theory, the case of $\mu \to 0$ is studied to see how the HJM model emerges. For $\mu \to 0$, the second term in the lattice action in Eq. (7.92) gives a product of δ functions, and hence

$$\lim_{\mu \to 0} e^{S[A]} = e^{S_0} \prod_{m=m_0}^{M-1} \prod_{n=m}^{N+m-1} \delta(A_{mn+1} - A_{mn})$$

$$S_0 = -\frac{\epsilon}{2} \sum_{m=m_0}^{M-1} \sum_{n=m}^{N+m} A_{mn}^2 \tag{7.93}$$

Consider evaluating a typical expression like Z in (7.26). For each m, there are N integration variables A_{mn}; from Eq. (7.93) it can be seen that there are only $(N-1)$ δ functions, leaving, for every m, only one variable, say A_{mm} unrestricted. For simplicity, take $\epsilon = a$; performing the A_{mn} integrations yields

$$Z = \prod_{m=m_0}^{M-1} \int dA_{mm} e^{S_0}$$

$$S_0 = -\frac{\epsilon}{2}(N+1) \sum_{m=m_0}^{M-1} A_{mm}^2$$

Defining $W(m) = \sqrt{N+1} A_{mm}$, one can see from Eq. (6.32) that the HJM model is recovered. For the HJM limit take the continuum limit by letting $\mu \to 0$, $M \to 0$ limit, and hence

$$S_0 \to -\frac{1}{2} \int_{t_0}^{\infty} dt \, W^2(t)$$

$$W(t) = \frac{1}{\sqrt{T_{FR}}} \int_t^{t+T_{FR}} dx \, A(t, x) \tag{7.94}$$

Nonlinear field theories: Fermion path integral

The change of variables from the nonlinear variable $\phi(t, x)$ to $A(t, x)$, as in Section 7.9.1, is made for the purpose of satisfying the Neumann boundary conditions; in fact the fundamental utility of the velocity field $A(t, x)$, and both in its continumm and lattice versions, results from the Neumann boundary conditions.

The nonlinear path integral is given from Eq. (7.55) as

$$Z = \int DA Dc^\dagger Dc\, e^{S[A] + S_F[c^\dagger, c, A]}$$

$$\int DA Dc^\dagger Dc \equiv \prod_{(t,x)\in\mathcal{P}} \int_{-\infty}^{+\infty} dA(t, x) \int dc^\dagger(t, x) dc(t, x)$$

where $c^\dagger(t, x), c(t, x)$ are fermionic quantum fields.

The nonlinear lattice path integral is given on the two-dimensional lattice as

$$Z = \prod_{(mn)\in\tilde{\mathcal{P}}} \int dA_{mn} dc^\dagger_{mn} dc_{mn} e^{S[A] + S_F[c^\dagger, c, A]} \tag{7.95}$$

with the action $S[A]$ given by Eq. (7.92); $S_F[c^\dagger, c, A]$ is obtained by discretizing the action given in Eq. (7.54). One can generalize $S[A]$ to the case of the stiff action as well as to the nonlinear psychological time variable $z(\theta)$.

For simulating nonlinear forward rates given in Eq. (7.95), since the action $S[A]$ involves only derivatives it can be treated like other similar local quantum fields. The fermion path integral is nonlocal and nonlinear in the field variable $A(t, x)$, and needs special algorithms similar to the ones used for simulation of fermions in lattice gauge theory [106].

Path-dependent options

The payoff function of an Asian option at time t_0 on a zero coupon bond $P(t, T)$ with exercise time t_* is given by

$$g[P(*, T)] = \left[\frac{1}{t_* - t_0} \int_{t_0}^{t_*} dt\, P(t, T) - K \right]_+$$

Another example is the price of a European call option on a coupon bond $\mathcal{B}(t, T)$ given in (2.7); the payoff function is given by

$$g[\mathcal{B}] = (\mathcal{B}(t_*, T) - K)_+$$

The payoff function $g[A]$ in both the cases above is path dependent. Expressing all the zero coupon bonds in terms of the quantum field $A(t, x)$, the prices of such path-dependent options at time t_0 are given by

$$C(t_0, t_*, T, K) = \frac{1}{Z} \int DA\, e^{-\int_{t_0}^{t_*} dt\, r(t)} g[A] e^{S[A]}$$

The computation above can be performed numerically [44], and requires the lattice formulation of functional integral over $A(t, x)$.

7.19 Appendix: Action S_* for change of numeraire

A field theory derivation is given of the result quoted in Eq. (7.46), namely that

$$e^{S_*} = \frac{e^{-\int_{t_0}^{t_*} r(t)dt}}{P(t_0, t_*)} e^S \tag{7.96}$$

Recall from Eq. (7.12) that the action is given by[22]

$$S \equiv S[\alpha] = -\frac{1}{2} \int_{\mathcal{P}} \left(\frac{\frac{\partial f(t,x)}{\partial t} - \alpha(t, x)}{\sigma(t, x)} \right) D^{-1}(x, x'; t) \left(\frac{\frac{\partial f(t,x')}{\partial t} - \alpha(t, x')}{\sigma(t, x')} \right) \tag{7.97}$$

and

$$S_* \equiv S[\alpha_*] = -\frac{1}{2} \int_{\mathcal{P}} \left(\frac{\frac{\partial f(t,x)}{\partial t} - \alpha_*(t, x)}{\sigma(t, x)} \right) D^{-1}(x, x'; t) \left(\frac{\frac{\partial f(t,x')}{\partial t} - \alpha_*(t, x')}{\sigma(t, x')} \right) \tag{7.98}$$

where

$$\left(1 - \frac{1}{\mu^2} \frac{\partial}{\partial x^2} \right) D(x, x'; t) = \delta(x - x') \; : \; \text{Neumann B.C.'s.}$$

$$\Rightarrow D^{-1}(x, x'; t) = \left(1 - \frac{1}{\mu^2} \frac{\partial}{\partial x^2} \right) \delta(x - x') \tag{7.99}$$

The drift velocities, from Eqs. (7.38) and (7.45), are

$$\alpha(t, x) = \sigma(t, x) \int_t^x dx' D(x, x'; t) \sigma(t, x')$$

$$\alpha_*(t, x) = \sigma(t, x) \int_{t_*}^x dx' D(x, x'; t) \sigma(t, x') \tag{7.100}$$

Define

$$\delta\alpha \equiv \alpha - \alpha_* = \sigma(t, x) \int_t^{t_*} dx' D(x, x'; t) \sigma(t, x')$$

$$\Rightarrow \left(\frac{\delta\alpha}{\sigma} \right)(t, x) = \int_t^{t_*} dx' D(x, x'; t) \sigma(t, x') \tag{7.101}$$

---

[22] Without loss of generality take $T_{FR} \to \infty$, and hence

$$\int_{\mathcal{P}} = \int_{t_0}^{t_*} dt \int_t^\infty dx$$

Hence

$$S_* = -\frac{1}{2}\int_{\mathcal{P}}\left(\frac{\frac{\partial f(t,x)}{\partial t} - \alpha(t,x) + \delta\alpha}{\sigma(t,x)}\right) D^{-1}(x,x';t)\left(\frac{\frac{\partial f(t,x')}{\partial t} - \alpha(t,x') + \delta\alpha}{\sigma(t,x')}\right)$$

$$= S - \int_{\mathcal{P}}\left(\frac{\frac{\partial f(t,x)}{\partial t} - \alpha(t,x)}{\sigma(t,x)}\right) D^{-1}(x,x';t)\left(\frac{\delta\alpha}{\sigma}\right)(t,x')$$

$$-\frac{1}{2}\int_{\mathcal{P}}\left(\frac{\delta\alpha}{\sigma}\right)(t,x)D^{-1}(x,x';t)\left(\frac{\delta\alpha}{\sigma}\right)(t,x') \tag{7.102}$$

Eqs. (7.101) and (7.99) yield the following simplification

$$\left(1 - \frac{1}{\mu^2}\frac{\partial}{\partial x^2}\right)\left(\frac{\delta\alpha}{\sigma}\right)(t,x) = \left(1 - \frac{1}{\mu^2}\frac{\partial}{\partial x^2}\right)\int_t^{t_*} dx' D(x,x';t)\sigma(t,x')$$

$$= \int_t^{t_*} dx'\delta(x - x')\sigma(t,x')$$

$$= \begin{cases} \sigma(t,x); & x \in [t,t_*] \\ 0; & x \notin [t,t_*] \end{cases}$$

$$= \sigma(t,x)\Theta(x - t_*)$$

Hence, from above and Eq. (7.102)

$$S_* = S - \int_{t_0}^{t_*} dt \int_t^\infty dx\left\{\frac{\partial f(t,x)}{\partial t} - \alpha(t,x)\right\}\Theta(x - t_*)$$

$$-\frac{1}{2}\int_{t_0}^{t_*} dt \int_t^\infty dx\left[\int_t^{t_*} dx' D(x,x';t)\sigma(t,x')\right]\sigma(t,x)\Theta(x - t_*)$$

$$\tag{7.103}$$

$$= S - \int_{t_0}^{t_*} dt \int_t^{t_*} dx\frac{\partial f(t,x)}{\partial t}$$

since the last two terms in Eq. (7.103) cancel out exactly.
From the identity

$$\int_{t_0}^{t_*} dt \int_t^T dx\frac{\partial f(t,x)}{\partial t} = \int_{t_0}^{t_*} dt f(t,t) + \int_{t_*}^T dx f(t,x) - \int_{t_0}^T dx f(t,x)$$

and using $f(t,t) = r(t)$ gives the expected result

$$S_* = S - \int_{t_0}^{t_*} dt r(t) + \int_{t_0}^{t_*} dx f(t,x)$$

$$\Rightarrow e^{S_*} = \frac{e^{-\int_{t_0}^{t_*} r(t)dt}}{P(t_0,t_*)}e^S \tag{7.104}$$

The derivation given above goes through unchanged for a general quadratic (Gaussian) forward rates' action, given by generalizing Eq. (7.97) to

$$S = S[\alpha] \hspace{6cm} (7.105)$$

$$= -\frac{1}{2} \int_{\mathcal{P}} \left(\frac{\frac{\partial f(t,x)}{\partial t} - \alpha(t,x)}{\sigma(t,x)} \right) G^{-1}(x,x';t) \left(\frac{\frac{\partial f(t,x')}{\partial t} - \alpha(t,x')}{\sigma(t,x')} \right)$$

where $G^{-1}(x,x';t)$ is any arbitrary function of the future maturities with

$$\int_t^\infty dx'' G^{-1}(x,x'';t) G(x'',x';t) = \delta(x-x')$$

8

Empirical forward interest rates and field theory models

The validity of any model of the forward rates is ultimately an empirical question [26, 34], and whether the model is useful depends on it being computationally tractable. The computational and empirical aspects are closely related, since only computationally tractable models can produce testable results.

In this chapter, an empirical study of the forward interest rates is undertaken, and due to its computational simplicity only the Gaussian field theory models that have been discussed in Chapter 7 will be analyzed. Nonlinear theories of the forward rates will not be discussed and are left for future studies.

The empirical study of the field theory model of the forward rates is based on data from Eurodollar futures. A brief review of the empirical properties of the forward rates based on the findings in [19, 73, 74] will be presented. The rest of this chapter is focussed on the correlation structure of the instantaneous changes in forward interest rates as a function of maturity [6, 9, 11].

The theoretical framework for the empirical study is the formulation of the forward rates as a two-dimensional Gaussian quantum field theory. Recall the key feature of the field theory models is that the forward rates $f(t, x)$ are strongly correlated in the maturity direction $x > t$. To fully calibrate the field theory models, all parameters in the Lagrangians need to be fixed, which consists of numerical quantities like the parameters μ and λ that control the fluctuations in the maturity direction, and the (complete) volatility function $\sigma(t, x)$ that is a measure of the degree to which the forward rates are stochastic.

In the field theory approach, $\sigma(t, x)$ is fixed from the market.[1] In general, to fix an arbitrary function from data is a difficult exercise. It will turn out that due to a special property of Gaussian field theories, $\sigma(t, x)$ can be

---

[1] In the HJM model, the volatility function $\sigma(t, x)$ is assumed to have a specific mathematical form that is taken to reflect the expected behaviour of the forward rates.

determined independently of all the Gaussian models, and hence will give a model-independent result.

The main conclusion of the study of the Gaussian models is that only the stiff propagator, together with an appropriate psychological future time variable, fits all aspects of the market behaviour of the forward rates. The reason for studying the other Gaussian models is primarily pedagogical.

8.1 Eurodollar market

All the empirical analysis is exclusively based on data from the Eurodollar futures market, and hence the main features of this market are briefly discussed [63, 100].

Eurodollars refer to US$ bank deposits in commercial banks outside the US. These commercial banks are either non-US banks or US banks outside the US. The deposits are made for a fixed time, the most common being 90- or 180-day time deposits, and are exempt from certain US government regulations that apply to time deposits inside the US.

The Eurodollar deposit market constitutes one of the largest financial markets. The Eurodollar market is dominated by London, and the interest rates offered for these US$ time deposits are often based on Libor, the **London Interbank Offer Rate**. The Libor is a simple interest rate derived from a Eurodollar time deposit of 90 days. The minimum deposit for Libor is a par value of $1 000 000. Libor are interest rates for which commercial banks are willing to lend funds in the interbank market. The Libor spot market is active in maturities ranging from a few days to ten years, with the greatest depth in the three- and six-months time deposits. Libor contracts are commercial, not sovereign, and as such better reflect the term structure available for corporations. The Libor is the benchmark for many derivatives markets in interest rates, as well as in the hedging of Treasury Bonds.

In the Eurodollar time deposit market, cash deposits are traded between commercial banks, with varying maturities, principal and interest rates. On the contract's initiation date the principal is sent by lender to borrower, and on the maturity date the principal plus interest is sent by the borrower to lender.

Eurodollar futures

Eurodollar futures contracts are amongst the most important instrument for short-term contracts and have come to dominate this market. The Eurodollar futures contract, like other futures contracts, is an undertaking by participating parties to loan or borrow a fixed amount of principal at an interest rate fixed by Libor and executed at a specified future date. Eurodollar deposits are non-transferable and preclude any delivery. Hence the Eurodollar futures is settled entirely by cash payments with no delivery of an actual good. Once the final settlement price has been determined, traders with open positions settle with cash through the standard marking-to-market procedure, and the contract expires.

Since 1981, the Eurodollar futures cash settlement price has been determined in the following manner.

- Contracts are settled by cash, with no transfer of time deposits.
- There is no flexibility in delivery or timing for either investor who is long or short.
- The expiration date is the second London business day before the third Wednesday of the maturity month.
- On expiration, the clearing house determines the Libor for a 90-day deposit at two times – termination of trading and at a randomly chosen time 90 minutes before closing. The clearing house randomly selects 12 reference banks from a list of 20 participating banks. Each bank provides a quote to the clearing house for the 90-day Eurodollar deposit. The clearing house automatically eliminates the two highest and the two lowest bids, and takes the arithmetic mean of the remaining eight quotes to be the Libor at that time.
- The average of the computation performed at the two times is taken to be the three-month Libor, and is finally determined by rounding off to the nearest basis point (100 basis points equal an annual interest of 1%).
- At maturity, the Eurodollars futures converges to 100(1-Libor).

Eurodollar futures contracts extends from a few days up to ten years in the future. The contract is extensively used for hedging, interest rate swaps, Forward Rate Agreements (FRAs) and other interest rate derivatives. The open positions on Eurodollar futures in 1999 was nearly 750 000, representing a par value of $750 billion. The liquidity in the Eurodollar market has grown since then.

The Eurodollar futures trade on the International Monetary Market (IMM) of the Chicago Mercantile Exchange (CME), and on SIMEX (Singapore International Monetary Exchange), LIFFE (London International Financial Futures Exchange) and on TIFFE (Tokyo International Financial Futures Exchange). Contracts entered in one exchange can be offset by those in other exchanges.

Libor resolution

The Eurodollar futures prices are in percentages of $1 million face value on a 90-day time deposit. The market has a resolution of 1 basis point, and this is worth (1 year = 360 days)

$$1\,000\,000 \times \frac{1}{100} \times \frac{1}{100} \times \frac{90}{360} = \$25$$

8.1.1 Libor forward rates curve

Eurodollar futures as expressed by Libor extends to up to ten years into the future, and hence there are underlying forward interest rates driving all Libors with different maturities. The interest earned on a Eurodollar time deposit is a simple interest based on Libor. For a futures contract entered into at time t for a 90-day deposit of \$1 million from future time T to $T + \tau$ ($\tau = 90/360$ year), the principal plus simple interest that will accrue – on the maturity – to an investor long on the contract is given by

$$P + I = 1 + \tau L(t, T) \tag{8.1}$$

where $L(t, T)$ is the (annualized) three-month (90-day) Libor. Let the forward interest rates for the three-month Libor be denoted by $f(t, x)$. One can express the principal plus interest based on the compounded forward interest rates and obtain

$$P + I = e^{\int_T^{T+\tau} dx f(t,x)} \tag{8.2}$$

Hence the relationship between Libor and its forward rates is given by

$$1 + \tau L(t, T) = e^{\int_T^{T+\tau} dx f(t,x)}$$
$$\Rightarrow L(t, T) = \frac{e^{\int_T^{T+\tau} dx f(t,x)} - 1}{\tau} \tag{8.3}$$

Given the data on Libor obtained from the Eurodollar futures contracts, the Libor forward rates can be extracted. Since Libor is determined on a daily basis, the data for the forward rates are discrete in time and are given daily. Future time is also discrete with the Libor given at 90-day intervals.

Libor forward interest rates contain the risk inherent in commercial lending, and have a spread above the interest rates that are determined by risk-free Treasury Bonds; in the finance literature this is known as the TED (Treasury Eurodollar) spread [63, 100].

8.2 Market data and assumptions used for the study

A key feature in the empirical testing of the forward rates' field theory models is to obtain evidence for the nontrivial correlation of the forward rates. The market price of the Eurodollars futures provides data that is ideally suited for this purpose.

The empirical analysis in this and later chapters uses the daily data for the Libor forward rates from 1990 to 1996 that are plotted in Figure 7.2. Significant historical data for contracts on deposits up to seven years into the future were used in the empirical analysis. This is the same data set used in [19] and [9, 11].

It is assumed that the Eurodollar futures Libor prices are equal to the forward rates. More precisely, from Eq. (8.3)

$$L(t, T) = \frac{e^{\int_T^{T+\tau} dx f(t,x)} - 1}{\tau} \tag{8.4}$$

$$\simeq f(t, T) + O(\tau) \tag{8.5}$$

The errors in setting Libor equal to the forward interest rates are negligible, given the other errors that arise in the empirical study; the justification for this assumption is discussed in [19]. In sum, the Libor is identified with the instantaneous forward interest rates.

Market data on the Libor are given for daily time t in the form of $L(t, T_i - t)$, and have fixed dates of maturity T_i (March, June, September and December). The shortest maturity time is $\theta_{min} = 3$ months, and the spot rate is taken to be $r(t) = f(t, \theta_{min})$. Consider a future time falling within these fixed maturity times, say $\theta = x - t$, with $T_i - t \le \theta \le T_{i+1} - t$; to obtain $f(t, \theta)$ with fixed θ, a simple linear interpolation[2] between the values of the Libor [9, 11, 73, 74] at the two neighbouring maturities is carried out, resulting in the values of $f(t, \theta)$ for fixed θ as shown in Figure 8.1. The linear interpolation is necessary, since the forward rate data are provided for maturity times $T_i - t$, whereas for the empirical study the data for the forward rates are required for constant θ.

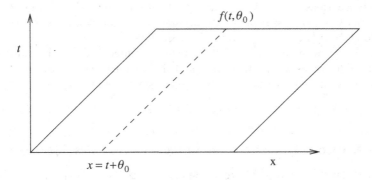

Figure 8.1 Forward rates on lines of constant θ obtained by linear interpolation from Libor forward rates specified at constant maturity time.

It is assumed in the empirical study that the volatility function, as well as the correlation functions of the forward rates, depend only on the variable θ and do

---

[2] It is assumed that the Libor rates are smooth; the assumption is reasonable as one would intuitively expect that the forward rate, say three years into the future would not be too different from that of three years and one month into the future. The loss in accuracy due to linear interpolation is unimportant if the time interval of t between specifications of the forward rates is small, since the random changes that are being studied have much larger errors than those introduced by the linear interpolation.

not have any explicit dependence on the time parameter t. In other words, it is assumed that $\sigma(t, \theta) = \sigma(\theta)$ with no explicit dependence on t, and similarly for the correlation functions. This assumption is made only for simplicity, since carrying out an analysis when these quantities are subject to changes in time is more difficult and does not add much to the understanding of the models. Furthermore, the limit of $T_{FR} \to \infty$ will always be assumed.

The empirical analysis consists of taking averages over the time series data for the various combinations of the forward rates. One fundamental assumption in the empirical analysis is to treat these time averages as being equivalent to averaging over the stochastic fluctuations of the forward rates. This assumption is called the ergodic hypothesis in statistical physics. In particular, all the correlation functions of the forward rates will be calculated by computing the corresponding time averages.

It should be noted that in full generality one expects time-dependent behaviour in the market for $\sigma(t, \theta)$, as can be seen from Figure 8.2. There are periods during which the assumption of dependence on only θ is valid, and the market can then change its behaviour and enter another regime where all the quantities are changed to new values and so on; this is further discussed in Section 8.4.

Treasury Bond data

One can also use Treasury Bond tick data from the GovPx database to study the forward rates. The problem in doing so is the lack of accuracy of the forward rates obtained from the Treasury Bond data. The main reason for this is that, while one can obtain reasonably accurate yields for a few maturities, the differentiation required to get the forward rates from the yields introduces too many inaccuracies. This is somewhat unfortunate since Treasury Bonds represent risk-free instruments while a small credit risk exists for Eurodollar deposits.

8.3 Correlation functions of the forward rates' models

Since all data on the forward interest rates are available only at discrete moments the parameter of time t needs to be discretized for empirically studying the field theory models.

Recall, from Eq. (7.15), the fundamental relation

$$\frac{\partial f}{\partial t}(t, x) = \alpha(t, x) + \sigma(t, x)A(t, x) \tag{8.6}$$

where, from Eq. (7.38), the martingale condition yields for the drift

$$\alpha(t, x) = \sigma(t, x)\int_t^x dx' D(x, x'; t)\sigma(t, x') \tag{8.7}$$

Discretize time into a lattice of points with spacing ϵ, with $t = n\epsilon$. Hence, from Eq. (8.6)

$$\delta f(t, x) \equiv f(t + \epsilon, x) - f(t, x)$$
$$= \epsilon \alpha(t, x) + \epsilon \sigma(t, x) A(t, x) \tag{8.8}$$

Recall from Eq. (7.23) that

$$E(A(t, x)) = 0 \tag{8.9}$$
$$E(A(t, x) A(t', x')) = \frac{1}{Z} \int DA e^{S[A]} A(t, x) A(t', x') \tag{8.10}$$
$$= \delta(t - t') D(x, x'; t) \tag{8.11}$$

Henceforth, all correlators will be expressed in co-ordinates $\theta = x - t$, $\theta' = x - t$ as these are the requisite ones for the empirical study. On discretizing time, the equal time expectation value of the fields at two maturities is, for $\delta(0) = 1/\epsilon$, given by

$$E(A(t, \theta) A(t, \theta')) \equiv < A(t, \theta) A(t, \theta') > = \frac{1}{\epsilon} D(\theta, \theta') \tag{8.12}$$

Hence, from Eqs. (8.8) and (8.12)

$$< \delta f(t, \theta) > = \epsilon \alpha(\theta)$$
$$< \delta f(t, \theta) \delta f(t, \theta') >_c \equiv < \delta f(t, \theta) \delta f(t, \theta') > - < \delta f(t, \theta) > < \delta f(t, \theta') >$$
$$= \epsilon^2 \sigma(\theta) \sigma(\theta') < A(t, \theta) A(t, \theta') >$$
$$= \epsilon \sigma(\theta) \sigma(\theta') D(\theta, \theta') \tag{8.13}$$
$$\Rightarrow < \delta f^2(t, \theta) >_c = \epsilon \sigma^2(\theta) D(\theta, \theta) \tag{8.14}$$

using the notation $< \delta f^2(t, \theta) >_c \equiv < [\delta f(t, \theta)]^2 >_c$

The drift velocity in the market is not the one given by the martingale measure (see Section 8.5 for a discussion on this point); hence $\alpha(t, x)$ cannot be determined from the empirical evolution of the forward rates, but instead can only be determined from the prices of interest rate derivatives. The connected correlation function $< \delta f(t, \theta) \delta f(t, \theta') >_c$, on the other hand, is completely independent of drift velocity $\alpha(\theta)$, and consequently directly models the actual market evolution of the forward rates; for this reason the connected correlator can be used to test the field theory model's validity by comparing it with the observed behaviour of the forward rates.

8.4 Empirical correlation structure of the forward rates

The correlator given in Eq. (8.14), namely $< \delta f^2(t, \theta) >_c = \epsilon \sigma^2(\theta) D(\theta, \theta)$, is the defining equation for the volatility, except for the factor of $\epsilon D(\theta, \theta)$ that depends on the model for the propagator. To be able to compare the volatilities of different Gaussian models, its determination has to be made model independent. The scaling symmetry discussed at the end of Section 7.4 allows the field $A(t, \theta)$ to be re-scaled so that $D(\theta, \theta') \to \tilde{D}(\theta, \theta')$ such that $\tilde{D}(\theta, \theta) = 1/\epsilon$. The re-scaled frame yields the usual definition of volatility of the forward rates, given as follows

$$< \delta f^2(t, \theta) >_c = \sigma^2(t, \theta) \tag{8.15}$$

The kurtosis for a single maturity time is defined by

$$\kappa(t, \theta) = \frac{< \left[\delta f(t, \theta) \right]^4 >}{\sigma^4(t, \theta)} - 3 \tag{8.16}$$

The normalization of kurtosis has been chosen so that it is zero if the forward rates are completely Gaussian.

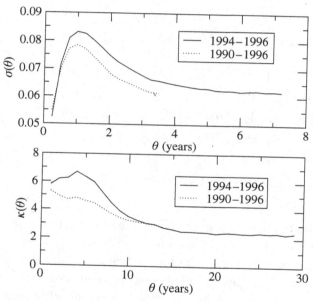

Figure 8.2 Empirically determined volatility function $\sigma(\theta) = \sqrt{< \delta f^2(t, \theta) >_c}$ and kurtosis $\kappa(t, \theta) = < [\delta f(t, \theta)]^4 > /\sigma^4(t, \theta) - 3$ of the forward rates. The functions are given for two distinct time periods showing a gradual change in their values. (*Source*: J.-P. Bouchaud and M. Potters, *Theory of Financial Risks*.)

The volatility and kurtosis of forward rates observed in the Eurodollar market for the period of 1990–1996 are shown Figure 8.2. Data show that kurtosis is far from zero, providing evidence that the empirical forward rates are non-Gaussian. The data in the figure have been broken up into two periods, namely 1990–1996 and 1994–1996, and show that both the volatility and kurtosis functions gradually change from period to period. Within a given period, one can assume that σ depends only on future time, that is $\sigma(t, \theta) = \sigma(\theta)$, but this assumption cannot be indefinitely extended.

The value of σ for the period 1990–1996 is used in all the computations discussed in this chapter.

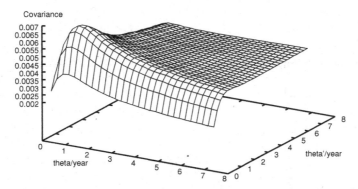

Figure 8.3 Empirical covariance $< \delta f(t, \theta)\delta f(t, \theta') >_c$

The empirical covariance of the forward rates is given by $<\delta f(t, \theta)\delta f(t, \theta')>_c$, and is plotted in Figure 8.3. An important quantity in the analysis of forward rates $f(t, \theta)$ is the normalized correlation (or scaled covariance) between the instantaneous changes in the forward rates for different maturities θ, θ', and is defined by

$$\mathcal{C}(\theta, \theta') = \frac{\langle \delta f(t, \theta)\delta f(t, \theta')\rangle_c}{\sqrt{\langle \delta f^2(t, \theta)\rangle_c}\sqrt{\langle \delta f^2(t, \theta')\rangle_c}} \qquad (8.17)$$

The normalized empirical correlation function, also called the **empirical propagator**, is estimated from the market using the Eurodollar futures data, and is shown in Figure 8.4, with another perspective given in Figure 8.5.

The figures clearly show that the normalized correlator has an extremely smooth surface with no discontinuities or 'kinks'. Furthermore the normalized correlation function everywhere has a value greater than about 0.65, showing that all the forward rates are **highly correlated**, with no two forward rates being de-correlated – no matter how large is their separation in maturity time. It is to explain this smooth and highly correlated behaviour of the forward rates that the stiff Gaussian interest rate model was introduced in [6].

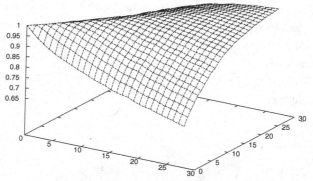

Figure 8.4 Empirical correlation $\mathcal{C}(\theta, \theta') = \dfrac{\langle \delta f(t,\theta) \delta f(t,\theta') \rangle_c}{\sqrt{\langle \delta f^2(t,\theta) \rangle_c} \sqrt{\langle \delta f^2(t,\theta') \rangle_c}}$.

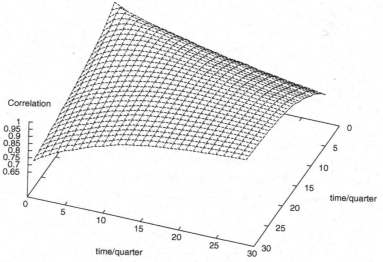

Figure 8.5 Empirical correlation $\mathcal{C}(\theta, \theta') = \dfrac{\langle \delta f(t,\theta) \delta f(t,\theta') \rangle_c}{\sqrt{\langle \delta f^2(t,\theta) \rangle_c} \sqrt{\langle \delta f^2(t,\theta') \rangle_c}}$: a different view.

8.4.1 Gaussian correlation functions

Due to the manner in which volatility appears in the Gaussian field theory models, as expressed in Eqs. (8.13) and (8.14), the entire volatility function can be completely factorized out of the normalized correlation functions. Hence, the Gaussian field theory models yield for the correlator

$$\mathcal{C}_{QFT}(\theta, \theta') = \frac{D(\theta, \theta')}{\sqrt{D(\theta, \theta) D(\theta', \theta')}} \tag{8.18}$$

Since for free (Gaussian) quantum fields the normalized correlation is **independent** of $\sigma(\theta)$, no assumption of the form of volatility needs to be made. This is the reason for using the scaled covariance rather than the covariance itself to perform the empirical study. Parameters such as μ, λ and so on, that need calibration in the Gaussian models, can be fitted from data, independent of the value of $\sigma(\theta)$.

HJM correlation functions

The covariance for the two factor HJM model is given by

$$\delta f(t, \theta) = \epsilon \alpha(\theta) + \sum_{i=1}^{2} \sigma_i(\theta) W_i(t)$$

$$\Rightarrow C(\theta, \theta') = <\delta f(t, \theta) \delta f(t, \theta') >_c$$

$$= \sigma_1(\theta) \sigma_1(\theta') + \sigma_2(\theta) \sigma_2(\theta') \tag{8.19}$$

The functional form for σ_1 and σ_2 needs to be obtained from the data, which is impractical as it is not possible to numerically estimate an entire function. The usual specification of $\sigma_1(\theta) = \sigma_0$ and $\sigma_2(\theta) = \sigma_1 e^{-\lambda \theta}$ is seen to be unable to explain many features of the covariance in Figure 8.3 such as the peak at one year or the sharp reduction in the covariance as the maturity goes to zero.

For the one factor HJM model, the normalized correlation structure is constant as all changes in the forward rates are perfectly correlated. In other words, $D(\theta, \theta') = 1$. For the two-factor HJM model, the predicted correlation structure, from Eq. (8.19) is given by

$$\mathcal{C}_{HJM}(\theta, \theta') = \frac{\sigma_1(\theta) \sigma_1(\theta') + \sigma_2(\theta) \sigma_2(\theta')}{\sqrt{\sigma_1^2(\theta) + \sigma_2^2(\theta)} \sqrt{\sigma_1^2(\theta') + \sigma_2^2(\theta')}} = \frac{1 + g(\theta) g(\theta')}{\sqrt{1 + g^2(\theta)} \sqrt{1 + g^2(\theta')}} \tag{8.20}$$

The normalized correlation structure requires the empirical determination of the function $g(\theta) = \sigma_1(\theta)/\sigma_2(\theta)$, something which is quite impractical.

In conclusion, the one-factor HJM model is insufficient to characterize the data, while the two-factor HJM model has too much freedom.

8.5 Empirical properties of the forward rates

In a series of publications [18, 19, 73, 74] Bouchaud and collaborators have uncovered a number of regularities in the behaviour of the forward interest rates. They have analyzed the forward rates for the US$ [19, 73] and other currencies [74], and the discussion in this section is largely drawn from these references.

The two key properties of the forward interest rates that need to be explained are (i) its average value $< f(t, x) >$ and (ii) its volatility $\sigma^2(t, x) \equiv< \delta f^2(t, x) >_c$.

Average value of the forward interest rates

Extensive analysis of the data shows that [19, 73, 74][3]

$$< f(t, x) >= r(t) + \text{constant } \sqrt{\theta} \; ; \; \theta = x - t \tag{8.21}$$

It is shown in [19, 73, 74] that all models of the forward rates based on a martingale measure are unable to explain the square root behaviour of the forward rates' average. Recall from Eq. (7.40) that for Gaussian models the martingale measure yields, using $< A(t, x) >= 0$, the following

$$< f(t, x) >= f(t_0, x) + \int_{t_0}^{t} dt' \sigma_i(t', x) \int_{t'}^{x} dy D(x, y; t', T_{FR}) \sigma_i(t', y)$$

The expectation value of $f(t, x)$ for a martingale evolution is equal to a quadratic function of its volatility. It is shown in [19, 74] that a quadratic term such as σ^2 is too small to account for the observed behaviour of $< f(t, x) >$; moreover, it is not possible to produce the square root law from the expression above.[4] This result is not unexpected, since the martingale measure is an evolution that is risk free, and this is not what transpires in the market. The forward rates encode all forms of risks that are inherent in the borrowing and lending of money, and hence the drift velocity of the forward rates in the financial market is not expected to be given by the no arbitrage requirement.[5]

It is shown in [19, 73, 74] that the concept of **value at risk** offers a possible explanation of the square root behaviour of the forward rates. Money lenders dominate short-term lending and borrowing in the money market. Money lenders agree, at time t, to loan money at the forward rate $f(t, \theta)$ for a loan that will run from time $t + \theta$ to $t + \theta + d\theta$. These money lenders themselves will borrow money from banks at the prevailing short-term rate at time $t + \theta$, namely at $r(t + \theta)$. Money lenders who loan out money at rate $f(t, \theta)$ bet that that at time $t + \theta$ the spot rate $r(t + \theta)$ is going to be **below** $f(t, \theta)$, since they lose money if $r(t + \theta) > f(t, \theta)$.

The money lender is willing to be on the losing side of the bet with likelihood of p; one expects that $p \simeq 0.1$; in other words the money lender is willing to lose only about 10% of the time. With this risk analysis, the money lender will fix the forward rate $f(t, \theta)$ to have a value such that the likelihood of the spot rate $r(t + \theta)$ having a value greater than $< f(t, \theta) >$ is only p. This procedure

---

[3] One may think that having a prediction on how the forward rates will evolve allows for the possibility of arbitrage. However the square root law only holds for the average of the forward rates and hence profit is not certain [19].

[4] One can include a term linear in the volatility σ to represent the market premium for risk, but this does not change the conclusions [73].

[5] The rationale for evolving the forward rates with a risk-free martingale measure is that it can be used for pricing derivative instruments of forward rates that are free from arbitrage possibilities. The martingale measure is not supposed to predict the actual evolution of the forward rates that takes place in the debt markets.

for valuing the forward interest rates reflects the value at risk for the money lender.

Let the conditional probability that the spot interest rate r at time t has a value r' at t' be given by $P(r', t'; r, t)$. The value at risk is then expressed by [19, 73, 74]

$$\int_{<f(t,\theta)>}^{\infty} dr' P(r', t + \theta; r, t) = p \tag{8.22}$$

For small time periods non-Gaussian effects are important in the evolution of the spot interest rate. However, for durations on the scale of months, one has [18]

$$P(r', t'; r, t) = \frac{1}{\sqrt{2\pi \sigma_r^2 (t' - t)}} e^{-\frac{1}{2\sigma_r^2(t'-t)}(r'-r)^2}$$

Hence from Eq. (8.22) and above, defining $\xi = (r' - r)/\sigma_r^2 \sqrt{\theta}$, yields[6]

$$\frac{1}{\sqrt{2\pi}} \int_{(<f(t,\theta)>-r)/\sigma_r \sqrt{\theta}}^{\infty} d\xi e^{-\frac{1}{2}\xi^2} = p = \frac{1}{\sqrt{2\pi}} \int_A^{\infty} d\xi e^{-\frac{1}{2}\xi^2}$$

$$\Rightarrow < f(t, x) > = r(t) + A\sigma_r \sqrt{\theta} \; ; \; A = \sqrt{2}\text{erfc}^{-1}(2p) \tag{8.23}$$

The formula above is plotted against data in Figure 8.6, and is seen to match the market data quite well for $p \simeq 0.16$, and with σ_r being the empirical volatility of the spot rate [18, 19]; Figure 8.6 also shows that the HJM model gives an incorrect result. The result derived above is valid for many of the major currencies such as the British, Japanese and Australian, with the appropriate spot rate volatility σ_r. Further refinements of the square root behaviour of the forward rates are considered in [74].

Forward rates' volatility 'hump' and anticipated trends

Recall the volatility of the forward rates is defined by $\sigma^2(t, x) \equiv < \delta f^2(t, x) >_c$. The volatility for all periods have a characteristic hump (a gradual maxima) at about one year in the future as can be seen from Figure 8.2. A similar volatility hump also occurs in the implied volatility for caps and floors on interest rates [51].

Intuitively, the volatility hump can be seen to arise from the anticipated evolution of the spot rate $r(t)$. Money lenders estimate the future evolution of the spot interest rate based on averaging over its past behaviour. Any small change in the value of the spot rate triggers a change in the market's expectation of the future evolution of the spot interest rate. The future volatility of the spot rate gets multiplied by the maturity time, enhancing its anticipated value [73, 74]. The money

---

[6] The erfc function is the related to function defined in Eq. (3.24).

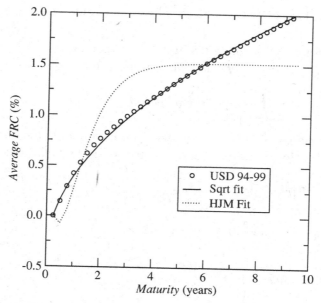

Figure 8.6 Empirically determined average value of the forward rates curve (FRC) $<f(t, \theta) - f(t, \theta_{min})>$ as a function of future time $\theta = x - t$. (*Source*: J.-P. Bouchaud and M. Potters, *Theory of Financial Risks*.)

lenders do not project their expectations for more than a year as this is beyond the horizon of their lending and borrowing – spot rates far into the future being fixed by policy makers and macroeconomic considerations. Hence the volatility peaks at about a year in the future. For future time, of greater than about a year and a half, volatility slowly decreases to a constant value, reflecting the anticipated risks in long-term borrowing and lending. This qualitative discussion has been made more rigorous in [19, 73, 74].

A good fit to the empirical value of the volatility for US$ forward rates for the period of 1994–1999 is as follows [73]

$$\sigma(\theta) = 0.061 - 0.014e^{-1.55(\theta - \theta_{min})} + 0.074(\theta - \theta_{min})e^{-1.55(\theta - \theta_{min})} \quad (8.24)$$

where the linearly increasing term results from the projection of the anticipated trends of the spot rate.

8.5.1 The Volatility of volatility of the forward rates

The volatility of the forward rates is a central measure of the degree to which the forward rates fluctuate. The question naturally arises as to whether the volatility

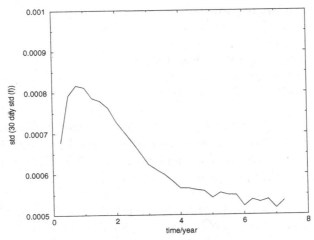

Figure 8.7 The volatility of volatility of the forward rates, denoted by std in the diagram, is based on a 30-day moving average.

itself should considered to be a randomly fluctuating quantity. The volatility of volatility is an accurate measure of the degree to which volatility is random. Market data for the Eurodollar futures provide an estimate of the forward rates for the US dollar, and also yield the volatility of volatility of the forward rates.

Figure 8.7 plots the 30-day moving average of the volatility of volatility for the forward rates, and shows that it contributes about 0.0006–0.0007 per year to the forward rates, as well as the characteristic peak at one and a half years that occurs for volatility itself. The volatility of the forward rates is approximately 0.01 per year; hence the fluctuations in the volatility of volatility are about 6–7% of the volatility. Although this number may look insignificant, one needs to remember that the effective volatility is given by the square of volatility being multiplied by the square root of the duration of a contract, and hence the effective volatility of volatility for contracts with a long maturity period could become significant.

It can be concluded from data that the volatility of the forward rates should be treated as a fluctuating quantum (stochastic) field.

8.6 Constant rigidity field theory model and its variants

This section summarizes the results of an empirical study of the models that are derived from the underlying constant rigidity model, based on the $(\partial A(t, x)/\partial x)^2$

term in the Lagrangian. In particular, these models are

1. Constant rigidity
2. Psychological future time for constant rigidity
3. Constrained constant rigidity
4. Non-constant rigidity

Although all these models have similar shortcomings, studying them gives insight into the flexibility of Gaussian field theories, and prepares the ground for studying the more complex stiff Lagrangian model. The common features of all the rigidity-based models are the following [11].

- The parameters of all the models are fitted to the market using the Levenberg–Marquardt method [71, 84], and the best fit is obtained by minimizing the square of the error.
- The error of the fit for these models range from about 4.2% to 2.4%. The lower figures for the error are achieved at the cost of more parameters and more complicated propagators.
- To estimate the error bounds, the data were split into 346 data sets of 500 contiguous days of data and the estimation done for each of the sets.
- For the constrained propagator, the Levenberg–Marquardt method shows that the fitted values of μ and a are very small, of the order of 10^{-7}/year for μ and 10^{-13}/year^2 for a; the fit of both the parameters is very unstable but the ratio a/μ^2 is stable.
- For the non-constant (maturity-dependent) rigidity case it is found that μ_0 is very unstable but always very small (less than 10^{-2}/year), while the 90% confidence interval for k is (0.099, 0.149)

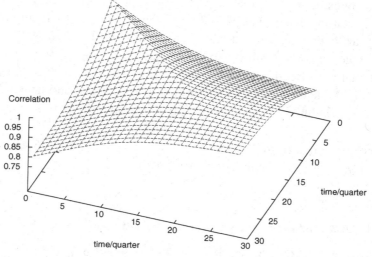

Figure 8.8 Fitted correlation $C_R(\theta, \theta') = \sqrt{\dfrac{e^{-\mu\theta}\cosh\mu\theta'}{e^{-\mu\theta'}\cosh\mu\theta}}$ for constant rigidity model

Table 8.1 *Summary of constant and non-constant rigidity models of the forward rates (μ, $\sqrt{a}$, λ, β are all measured in units of (year)$^{-1}$)*

Model	μ	a/μ^2	k	β	$\sqrt{\text{Error}}$
Constant rigidity μ	0.06	–	–	–	4.23%
Constrained spot rate	–	9.4	–	–	3.54%
$\mu(\theta) = \mu/(1 + k\theta)$	0.011	–	0.1	–	3.35%
μ ; $z = \tanh(\beta\theta)$	0.48	–	–	0.31	2.46%

The summary of the best fit for these models is given in Table 8.1.

The major shortcomings of all the rigidity-based models are the following.

1. The main problem can be seen from a comparison between the best fit for the constant rigidity μ given in Figure 8.8 with the market's correlation structure in Figure 8.4. The constant rigidity model has an artificial 'kink' along the diagonal values of the normalized correlation function that is absent in the market; this problem persists for all the cases considered, and is somewhat alleviated when μ is combined with nonlinear psychological time. The kink is due to the term $(\partial A(t, x)/\partial x)^2$ in the Lagrangian.
2. Another shortcoming is that the correlator is largely independent of the value of θ, being dominated by $|\theta - \theta'|$; this is inconsistent with the empirical correlator given in Figure 8.5 that increases as θ increases. The nonlinear variable partly corrects this problem, but not fully.
3. Due to the kink along the diagonal, a major feature of the correlation function, namely the curvature of the correlation function that determines how rapidly the correlation falls off as one moves orthogonal to the diagonal, cannot be studied. It will be seen in the next section that curvature plays a crucial role in the behaviour of the correlation function, providing a fairly precise measure of psychological future time.

Constant rigidity

From Eq. (7.29)

$$D(\theta, \theta') = \frac{\mu}{2}\left(e^{-\mu(\theta-\theta')} + e^{-\mu(\theta+\theta')}\right) = \mu e^{-\mu\theta}\cosh\mu\theta' \ ; \ \theta > \theta' \quad (8.25)$$

The predicted correlation structure for this model follows from (8.25) and is given by

$$C_R(\theta, \theta') = \sqrt{\frac{e^{-\mu\theta}\cosh\mu\theta'}{e^{-\mu\theta'}\cosh\mu\theta}} \ ; \ \theta > \theta' \quad (8.26)$$

The best fit for constant rigidity is shown in Figure 8.8.

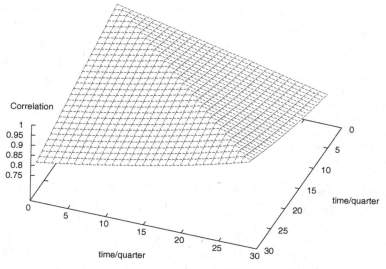

Figure 8.9 Fitted correlation $\mathcal{C}_{RZ}(\theta, \theta') = \sqrt{\dfrac{e^{-\mu z(\theta)}\cosh \mu z(\theta')}{e^{-\mu z(\theta')}\cosh \mu z(\theta)}}$ with $z(\theta) = \tanh \beta\theta$

Psychological future time

A psychological future (maturity) time variable $z(\theta)$ that is close to θ for θ small, and converges to 1 for large θ, is given by $z(\theta) = \tanh \beta\theta$ [11]. This choice is a bit extreme since for $\theta >> \beta^{-1}$ the nonlinear maturity time variable becomes a constant ($z \simeq 1$) and hence psychological time ceases to flow. The variable $z(\theta)$ nevertheless shows the dramatic effect that a nonlinear maturity variable has on the correlation function. The correlation function is given, from Eq. (8.27), by

$$\mathcal{C}_{RZ}(\theta, \theta') = \sqrt{\frac{e^{-\mu z(\theta)}\cosh \mu z(\theta')}{e^{-\mu z(\theta')}\cosh \mu z(\theta)}} \; ; \quad \theta > \theta' \qquad (8.27)$$

The shape of the fitted function is shown in Figure 8.9, and is the best fit for the models that are variations of the constant rigidity Lagrangian.

Constrained spot rate

The propagator $D(\theta, \theta')$ for this model is given by

$$D(\theta, \theta') = \mu e^{-\mu\theta}\left(\cosh \mu\theta' - \frac{\mu e^{-\mu\theta'}}{\mu + a}\right); \quad \theta > \theta' \qquad (8.28)$$

The normalized correlation function is given by

$$C_C(\theta, \theta') = \sqrt{\frac{e^{-\mu\theta}\left(\cosh \mu\theta' - \frac{\mu e^{-\mu\theta'}}{\mu+a}\right)}{e^{-\mu\theta'}\left(\cosh \mu\theta - \frac{\mu e^{-\mu\theta}}{\mu+a}\right)}}; \quad \theta > \theta' \tag{8.29}$$

The graph of the best-fit propagator is almost identical to the constant rigidity case shown in Figure 8.8 [11].

Non-constant rigidity

The normalized correlation function for a rigidity that depends on the maturity variable θ, from Eq. (7.72), is given by

$$C_M(\theta, \theta') = \left(\frac{(\alpha + 1/2)(1 + k\theta')^{2\alpha} + \alpha - 1/2}{(\alpha + 1/2)(1 + k\theta)^{2\alpha} + \alpha - 1/2}\right)^{1/2}; \quad \theta > \theta' \tag{8.30}$$

The correlation structure has a limit of $\mu_0 \to 0$ given by the following

$$\lim_{\mu_0 \to 0} C_M(\theta, \theta') = \sqrt{\frac{1 + k\theta'}{1 + k\theta}}; \quad \theta > \theta'$$

and due to the very small value of μ_0 for the fitted function, this is a very good approximation for the fit.[7] The obtained fit for the correlation function [11] look almost identical to the constant rigidity case given in Figure 8.8.

8.7 Stiff field theory model

The stiff propagator based on a quartic derivative term in the forward rates' Lagrangian yields a very smooth propagator, and, as discussed in Appendix 7.15, is free from the 'kink' along the diagonal that has plagued the rigid Gaussian models. It is consequently to be expected that this model should give the best empirical results [6]. The normalized correlation function is given by

$$C_Q(\theta, \theta') = \frac{G(\theta, \theta')}{\sqrt{G(\theta, \theta)G(\theta', \theta')}} \tag{8.31}$$

The propagator has three branches, as shown in Appendix 7.15, and the real branch is given, from Eq. (7.80), as follows

$$G_b(\theta_+; \theta_-) \equiv \frac{\lambda}{2 \sinh(2b)}[g_+(\theta_+) + g_-(\theta_-)] \tag{8.32}$$

---

[7] The normalized correlator given by $C_M(\theta, \theta')$ does not belong to the HJM class of models, since in the limit of $\mu_0 \to 0$ it does not go to a constant.

where

$$g_+(\theta_+) = e^{-\lambda \theta_+ \cosh(b)} \sinh\{b + \lambda \theta_+ \sinh(b)\} \tag{8.33}$$

$$g_-(\theta_-) = e^{-\lambda |\theta_-| \cosh(b)} \sinh\{b + \lambda |\theta_-| \sinh(b)\} \tag{8.34}$$

$$\theta_\pm = \theta \pm \theta' \tag{8.35}$$

In this representation

$$C_Q(\theta_+; \theta_-) = \frac{g_+(\theta_+) + g_-(\theta_-)}{\sqrt{[g_+(\theta_+ + \theta_-) + g_-(0)][g_+(\theta_+ - \theta_-) + g_-(0)]}} \tag{8.36}$$

The diagonal axis is a line of maxima for the correlator since

$$\left.\frac{\partial C_Q(\theta_+; \theta_-)}{\partial \theta_-}\right|_{\theta_-=0} \equiv \frac{\partial C_Q(\theta_+; 0)}{\partial \theta_-} = 0 \tag{8.37}$$

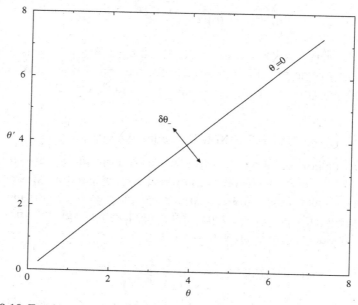

Figure 8.10 For $\theta_- = \theta - \theta'$, the figure shows that the diagonal axis is given by $\theta_- = 0$. The direction of change in θ_- for constant θ_+, namely $\delta\theta_-$, is orthogonal to the diagonal, as shown in the figure.

The propagator $G_b(\theta_+; \theta_-)$ has a finite curvature perpendicular to the diagonal, as shown in Eq. (7.84), and hence one can compute the curvature of $C_Q(\theta, \theta')$ and compare it with the data. The curvature orthogonal to the diagonal axis is defined as follows

$$R_Q(\theta_+) = -\frac{\partial^2 C_Q(\theta_+; \theta_-)}{\partial \theta_-^2}\bigg|_{(\theta_-=0)} \equiv -\frac{\partial^2 C_Q(\theta_+; 0)}{\partial \theta_-^2} \tag{8.38}$$

The calculation is performed in Appendix 8.9, and Eq. (8.47) shows that

$$R_Q(\theta_+) = \frac{|g_-''(0)|}{g_+(\theta_+) + g_-(0)} - \frac{|g_+''(\theta_+)|[g_+(\theta_+) + g_-(0)] + [g_+'(\theta_+)]^2}{[g_+(\theta_+) + g_-(0)]^2} \tag{8.39}$$

$|g_-''(0)|$ is, up to an irrelevant scale, the curvature of the stiff propagator that was shown in Eq. (7.84) to be a finite constant. For all propagators with a kink, $|g_-''(0)|$ is infinite along the diagonal, and hence would invalidate the curvature calculation.

The expression for $R_Q(\theta_+)$ shows that as θ_+ increases, which in effect means that one is moving on the diagonal axis away from the origin, the curvature (slowly) **increases** because the denominator of the first term decreases, while at the same time the second (negative) term becomes smaller; this behaviour of the curvature is plotted as the dashed line in the inset of Figure 8.11 and is seen to slope upwards.

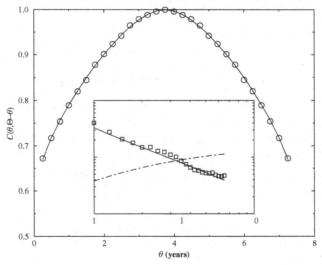

Figure 8.11 Fitted stiff correlation

$$C_Q(\theta+; \theta_-) = -\frac{g_+(z_+) + g_-(z_-)}{\sqrt{[g_+(z_+ + z_-) + g_-(0)][g_+(z_+ - z_-) + g_-(0)]}}$$

with nonlinear psychological time variable $z_\pm(\theta_+; \theta_-) \equiv z(\theta) \pm z(\theta')$. The inset shows a plot of $\log(\theta_+)$ versus (a) dashed line showing curvature $\log(R_Q(\theta_+))$ with an (incorrect) upwards slope and (b) the curvature with the nonlinear future (maturity) time $\log([z'(\theta_+)]^2 R_Q(2z(\theta_+/2)))$ correctly sloping downwards.

The curvature calculation predicts that the model's normalized correlator should fall off more rapidly as one moves on the diagonal away from the origin. If one looks carefully at Figures 8.4 and 8.5, one can see that the empirical correlator shows the **opposite** behaviour. As one moves away from the origin on the diagonal axis the curve flattens out, showing that the curvature is **decreasing** as θ_+ increases. Hence, as it stands, the stiff correlator cannot explain the empirical behaviour of the forward rates.

8.7.1 Psychological future time

On empirically studying the curvature, one finds a power law fall-off for the curvature given by $C_Q(\theta_+) \simeq 1/\theta_+^{1.3}$. Since the curvature for the stiff propagator increases very slowly, one could try and rectify the problem by multiplying the propagator with a pre-factor that cancels the gradual rise in curvature and instead makes it fall off with a power law. Psychological future time variable $z(\theta)$ plays precisely this role.

The defining equation for psychological maturity time $z(\theta)$ is given by Eq. (7.85) as

$$\frac{\partial f}{\partial t}(t, \theta) = \alpha(t, z(\theta)) + \sigma(t, z(\theta))A(t, z(\theta)) \; ; \; \theta = x - t$$

that yields for the normalized correlator

$$C_{Qz}(\theta, \theta') = \frac{G(z(\theta), z(\theta'))}{\sqrt{G(z(\theta), z(\theta))G(z(\theta'), z(\theta'))}} \tag{8.40}$$

The co-ordinates of the correlator of the forward rates $C_{Qz}(\theta, \theta')$ do not change when going to psychological future time. Instead, only the description of this correlator by the field theory model changes, and, consequently, the left-hand side of above equation depends only on the calendar time variables θ, θ', whereas the right-hand side depends only on the psychological time variables $z(\theta), z(\theta')$. Writing the correlator more explicitly, similar to Eq. (8.41), gives

$$C_{Qz}(\theta+; \theta_-) = \frac{g_+(z_+) + g_-(z_-)}{\sqrt{[g_+(z_+ + z_-) + g_-(0)][g_+(z_+ - z_-) + g_-(0)]}} \tag{8.41}$$

$$z_\pm(\theta+; \theta_-) \equiv z(\theta) \pm z(\theta') \tag{8.42}$$

It is shown in Eq. (8.49) that the curvature in the nonlinear variable is

$$-\frac{\partial^2 C_{Qz}(\theta_+; 0)}{\partial \theta_-^2} = [z'(\theta_+)]^2 R_Q(2z(\theta_+/2)) \tag{8.43}$$

The curvature with psychological future time is exactly factored into the result of the original model multiplied by a factor of $[z'(\theta_+)]^2$; hence any model with a

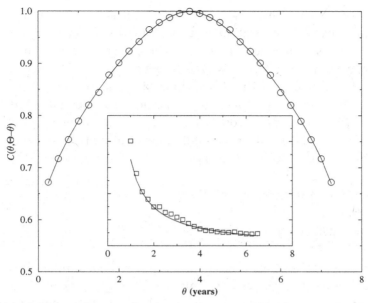

Figure 8.12 Figure showing the fitted propagator with the inset showing the empirical values and theoretical curve for the curvature $[z'(\theta_+)]^2 R_Q(2z(\theta_+/2))$.

kink along the diagonal, such as all variants based on rigidity, will always give a divergent result for curvature that cannot be rendered finite by any non-divergent choice of psychological future time.

Since one expects a power law to fall off for the curvature, the ansatz $z(\theta) = \theta^\eta$ is used for fitting the data. Using the fact that $R_Q(2z(\theta_+/2))$ varies very slowly as a function of θ_+, one can make the following approximation

$$[z'(\theta_+)]^2 \propto \frac{1}{\theta_+^{1.3}}$$

$$\Rightarrow \theta_+^{2\eta-2} \propto \frac{1}{\theta_+^{1.3}} \Rightarrow \eta \simeq 0.35 \qquad (8.44)$$

The best fit yields $\eta = 0.34$ showing that the psychological time variable almost completely dominates the curvature of the correlator.[8] What is striking is that the psychological future time variable yields a curvature for the stiff Gaussian model that matches the curvature of the empirical correlator over the **entire** range of the data, as shown in Figure 8.12 [6].

---

[8] One can verify that if one uses nonlinear maturity time $z(\theta) = \theta^{0.34}$ in the constant rigidity model, the curvature orthogonal to the diagonal still remains infinite, and, consequently, leads to no significant improvement.

Psychological future time $z(\theta) = \theta^{0.34}$ is significantly different (slower) than calendar time, and influences all financial instruments. Psychological future time may have regimes similar to volatility in that η changes over a long period of calendar time, since it is affected by market sentiment.

η is a scaling exponent and is always dimensionless. The units for λ and μ are fixed in Eq. (7.86) so that λz and μz are dimensionless; since $z = \theta^{\eta}$, define $\lambda z = [\tilde{\lambda}\theta]^{\eta}$, $\mu z = [\tilde{\mu}\theta]^{\eta}$ so that new constant $\tilde{\lambda}$, $\tilde{\mu}$ always have dimensions of (time)$^{-1}$. θ is measured in years, and the result of the empirical study is summarized below. $[\tilde{\mu} = \tilde{\lambda}/(2\cosh(b))^{1/2\eta} = 0.403/\text{year}]$

Best Fit of the Stiff Propagator's Parameters
$\tilde{\lambda} = 1.790/\text{year};$ $\tilde{\mu} = 0.403/\text{year};$ $b = 0.845;$ $\eta = 0.34$
Root mean square error for the entire fit: 0.40%

The stiff propagator, together with nonlinear psychological future time, matches data with a root mean square error of only 0.40%, and graphically looks identical to the empirical correlator given in Figure 8.4. Figure 8.11 shows a plot of the model's correlator on the diagonal line that is orthogonal to the $\theta_- = 0$ diagonal – this is the longest stretch for comparing the model's correlator with its empirical value; the agreement with the data is almost exact. What is noteworthy is that, even though the nonlinear maturity variable $z(\theta)$ was introduced to address the behaviour of the correlator in the neighbourhood of the diagonal axis, it continues to give the correct behaviour for the correlator even far from the diagonal region.

The existence of the boundary at $x = t$, or $\theta = 0$, is reflected in the θ_+ term in the propagator; if one removes this term, and in effect assumes that the forward rates exist for all $-\infty \leq x \leq +\infty$, then the fit deteriorates with the root mean square error increasing from 0.40% to 0.53%. The existence of the boundary at $x = t$ can hence be seen to have an important effect on the correlation of the forward rates.

8.8 Summary

A detailed analysis of data on the forward interest rates shows that the quantum field theory model of correlated forward rates agrees quite well with the market. All the variants of the constant rigidity model are seen to be inadequate for explaining the behaviour of the forward rates. The main result of the empirical study is that the forward rates evolve with a rigidity and a stiffness term in the Lagrangian, and together with psychological future time, provide an excellent fit to the observed correlation functions of the forward interest rates.

The empirical study focussed on only the two point correlation function, namely the normalized covariance of the forward rates. For Gaussian (linear) field theories, all the higher correlation functions involving changes in forward rates with three or more different maturities can be expressed in terms of the propagator. The empirical result given for the forward rates kurtosis in Figure 8.2, and preliminary calculations [97] show that the forward rates are in fact nonlinear. The empirical methods employed in this chapter need to be extended to nonlinear theories of forward interest rates that are discussed in Chapter 7.

8.9 Appendix: Curvature for stiff correlator

From Eq. (8.41) the normalized correlator is

$$C_\varrho(\theta_+; \theta_-) = \frac{g_+(\theta_+) + g_-(\theta_-)}{\sqrt{[g_+(\theta_+ + \theta_-) + g_-(0)][g_+(\theta_+ - \theta_-) + g_-(0)]}}$$

From Eq. (8.38) the curvature orthogonal to the diagonal axis is defined by

$$R_\varrho(\theta_+) = -\frac{\partial^2 C_\varrho(\theta_+; 0)}{\partial \theta_-^2}$$

A straightforward but tedious calculation shows that

$$R_\varrho(\theta_+) = -\frac{g_-''(0)}{g_+(\theta_+) + g_-(0)} + \frac{g_+''(\theta_+)[g_+(\theta_+) + g_-(0)] - \left(g_+'(\theta)\right)^2}{[g_+(\theta_+) + g_-(0)]^2}$$

$$(8.45)$$

where

$$g_+'(\theta) = -\lambda e^{-\lambda\theta_+ \cosh(b)} \sinh[\lambda\theta_+ \sinh(b)] < 0$$
$$g_+''(\theta_+) = -\lambda^2 e^{-\lambda\theta_+ \cosh(b)} \sinh[\lambda\theta_+ \sinh(b) - b] < 0$$

Eq. (7.84) gives

$$g_-''(0) = -\lambda^2 \sinh(b)\left(\cosh^2(b) - \sinh(b)\right) < 0 \qquad (8.46)$$

Hence

$$R_\varrho(\theta_+) = \frac{|g_-''(0)|}{g_+(\theta_+) + g_-(0)} - \frac{|g_+''(\theta_+)|[g_+(\theta_+) + g_-(0)] + \left(g'(\theta)\right)^2}{[g_+(\theta_+) + g_-(0)]^2}$$

$$(8.47)$$

Psychological future time

Note that

$$\frac{\partial C_{Qz}(\theta_+; \theta_-)}{\partial \theta_-} = \frac{\partial z_-}{\partial \theta_-} \frac{\partial C_Q(\theta_+; \theta_-)}{\partial z_-} + \frac{\partial z_+}{\partial \theta_-} \frac{\partial C_Q(\theta_+; \theta_-)}{\partial z_-+}$$

$$= \frac{\partial z_-}{\partial \theta_-} \frac{\partial C_Q(\theta_+; \theta_-)}{\partial z_-} \quad \text{since} \quad \frac{\partial z_+}{\partial \theta_-} = 0 \qquad (8.48)$$

Furthermore

$$\frac{\partial^2 C_{Qz}(\theta_+; \theta_-)}{\partial \theta_-^2}\bigg|_{(\theta_-=0)} = \left[\frac{\partial z_-}{\partial \theta_-}\right]^2\bigg|_{(\theta_-=0)} \frac{\partial^2 C_{Qz}(\theta_+; \theta_-)}{\partial z_-^2}\bigg|_{(z_-=0)}$$

since, from Eq. (8.37) $\partial C_{Qz}(\theta_+; 0)/\partial z_- = 0$. Using that $[\partial z_-/\partial \theta_-]|_{(\theta_-=0)} = z'(\theta_+)$, the curvature in the nonlinear co-ordinate from Eq. (8.38) is given by

$$\frac{\partial^2 C_{Qz}(\theta_+; 0)}{\partial \theta_-^2} = [z'(\theta_+)]^2 \frac{\partial^2 C_{Qz}(\theta_+; \theta_-)}{\partial z_-^2}\bigg|_{(z_-=0)} = [z'(\theta_+)]^2 \frac{\partial^2 C_{Qz}(z_+; 0)}{\partial z_-^2}$$

Since the correlator $C_{Qz}(z_+; z_-)$ is the same function of z_+, z_- as the correlator $C_Q(\theta_+; \theta_-)$ is of θ_+, θ_-, the result is given by

$$-\frac{\partial^2 C_{Qz}(\theta_+; 0)}{\partial \theta_-^2} = [z'(\theta_+)]^2 R_Q(z_+)\bigg|_{(z_-=0)} = [z'(\theta_+)]^2 R_Q(2z(\theta_+/2)) \quad (8.49)$$

since $z_+|_{z_-=0} = 2z(\theta_+/2)$, and $R_Q(z)$ is given by Eq. (8.39).

9

Field theory of Treasury Bonds' derivatives
and hedging

The quantum field theory of forward interest rates developed in Chapter 7 is applied to the pricing of futures and options on Treasury Bonds. The 'Greeks' for the European option on Treasury Bonds are derived to highlight the new features that emerge from the field theory of forward rates.

For Gaussian field theory models of the forward rates all the formulae derived for futures and options are similar to the HJM model; the main difference lies in the volatility of the derivatives, which now contains new couplings due to the nontrivial correlation of the forward rates.

The concept of hedging of Treasury Bonds is generalized to the case of field theory; a hedged port folio is defined to be one for which the variance of its final random value is a minimum; a more precise definition will be given in Section 9.7. The special case of the HJM-model is shown to emerge in the limit of zero rigidity. Due to its computational tractability, all derivations are carried out only for the Gaussian models.

9.1 Futures for Treasury Bonds

Let $\mathcal{F}(t_0, t_*, T_i)$ denote the price at time t_0 of a futures contract that matures at time t_* on a zero coupon bond maturing at time T, with $t_0 < t_* < T$. From Eqs. (6.46), (6.48) and (6.49)

$$
\begin{aligned}
\mathcal{F}(t_0, t_*, T) &\equiv E_{[t_0, t_*]}[P(t_*, T)] \\
&= F(t_0, t_*, T)e^{\Omega_\mathcal{F}} ; \quad e^{\Omega_\mathcal{F}} = e^{\Omega}e^{-\int_\mathcal{R} \alpha(t, x)} \qquad (9.1)
\end{aligned}
$$

where the domain $\mathcal{R}$ over which the drift velocity is integrated is given in Figure 9.1. The drift velocity for field theory has been computed in Eq. (7.38), and using Eq. (7.26) to compute Ω yields

$$e^{\Omega} = \frac{1}{Z} \int DA e^{-\int_{\mathcal{R}} dx \sigma(t,x) A(t,x)} e^{\int_{\mathcal{P}} \mathcal{L}[A]} \tag{9.2}$$

$$= \exp\left\{\frac{1}{2}\int_{t_0}^{t_*} dt \int_{t_*}^{T} dx dx' \sigma(t,x) D(x,x'; t, T_{FR}) \sigma(t,x')\right\} \tag{9.3}$$

The drift velocity in Eq. (7.38) yields the generalization of (6.52)

$$\Omega_{\mathcal{F}}(t_0, t_*, T) = \Omega - \int_{\mathcal{R}} \alpha(t,x)$$

$$= -\int_{t_0}^{t_*} dt \int_t^{t_*} dx \sigma(t,x) \int_{t_*}^{T} dx' D(x,x'; t, T_{FR}) \sigma(t,x') \tag{9.4}$$

where the integration for evaluating $\Omega_{\mathcal{F}}(t_0, t_*, T)$ is the trapezoidal domain $\mathcal{T}$ given in Figure 7.5.

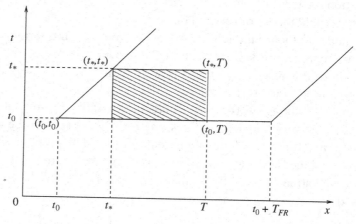

Figure 9.1 Domain $\mathcal{R}$ for the futures and option prices

For exponential volatility function $\sigma_2 e^{-\lambda(x-t)}$ the expression for $\Omega_{\mathcal{F}}(t_0, t_*, T)$ can be obtained exactly. For constant volatility $\sigma_1 \neq 0$ ($T_{FR} \to \infty$)

$$\Omega_{\mathcal{F}}(t_0, t_*, T) = -\frac{\sigma_1^2}{4\mu^2}\left(1 - e^{-\mu(T-t_*)}\right)\left\{e^{-2\mu(t_*-t_0)} - 1 + 2\mu(t_* - t_0)\right\} \tag{9.5}$$

9.2 Option pricing for Treasury Bonds

The European bond option price, from Eq. (6.53), is given by

$$C(t_0, t_*, T, K) = E_{[t_0, t_*]}\left[e^{-\int_{t_0}^{t_*} dt r(t)}(P(t_*, T) - K)_+\right] \tag{9.6}$$

Since the European bond option is one of the most important derivatives of the Treasury Bond, a field theory derivation of its price is given below.

Field theory derivation of European bond option price

Recall from Eq. (6.55) that the payoff function can be written as

$$(P(t_*, T) - K)_+ = \frac{1}{2\pi} \int_{-\infty}^{+\infty} dG dp e^{ip(G + \int_{t_*}^T dxf(t_*, x))} (e^G - K)_+$$

$$= \int_{-\infty}^{+\infty} dG \Psi(G, t_*, T)(e^G - K)_+$$

Following the discussion in Section 7.7, the price of the option is computed with the action S_* obtained by discounting the future value of the payoff function with the bond $P(t_0, t_*)$. From Eq. (7.48), the current price of the option is given by

$$C(t_0, t_*, T, K) = P(t_0, t_*) E_*[(P(t_*, T) - K)_+] \tag{9.7}$$

For domain $\mathcal{R}$ defined in Figure 9.1

$$\int_{t_*}^T dxf(t_*, x) = \int_{\mathcal{R}} \alpha_*(t, x) + \int_{t_*}^T dxf(t_0, x) + \int_{\mathcal{R}} \sigma(t, x) A(t, x)$$

From the expression above and the payoff function given in Eq. (9.7) one needs to compute

$$E_*[e^{ip \int_{\mathcal{R}} \sigma(t,x) A(t,x)}] = \frac{1}{Z} \int DA e^{ip \int_{\mathcal{R}} \sigma(t,x) A(t,x)} e^{S_*} = e^{-\frac{q^2}{2} p^2}$$

$$\Rightarrow q^2 = \int_{t_0}^{t_*} dt \int_{t_*}^T dx dx' \sigma(t, x) D(x, x'; t) \sigma(t, x') \tag{9.8}$$

Using the expression given in Eq. (7.45) for $\alpha_*(t, x)$, one obtains

$$\int_{\mathcal{R}} \alpha_*(t, x) = \int_{t_*}^T dx \int_{t_0}^{t_*} dt \alpha_*(t, x)$$

$$= \int_{t_*}^T dx \int_{t_0}^{t_*} dt \sigma(t, x) \int_{t_*}^x dx' D(x, x'; t) \sigma(t, x') = \frac{q^2}{2} \tag{9.9}$$

Collecting the results yields

$$\Psi(G, t_*, T) = \frac{1}{2\pi} \int_{-\infty}^{+\infty} dp e^{ip(G + \frac{q^2}{2} + \int_{t_*}^T dxf(t_0, x)) - \frac{q^2}{2} p^2}$$

Performing the Gaussian integration over p yields the desired result.

The computation above gives the result that

$$\Psi(G, t_*, T) = \frac{1}{\sqrt{2\pi q^2}} e^{-\frac{1}{2q^2}\left(G + \int_{t_*}^T dxf(t_0, x) + \frac{q^2}{2}\right)^2} \tag{9.10}$$

with volatility q^2 given in Eq. (9.8).[1] For field theory model with constant volatility

$$q^2 = \frac{\sigma_1^2}{4\mu^2}\left[4\mu(t_* - t_0)\{e^{-\mu(T-t_*)} - 1 + \mu(T - t_*)\}\right.$$
$$\left. + (1 - e^{-2\mu(t_*-t_0)})(1 - e^{-\mu(T-t_*)})^2\right]$$

For the European bond option given by

$$C(t_0, t_*, T, K) = \int_{-\infty}^{+\infty} dG \Psi(G, t_*, T)(e^G - K)_+ \qquad (9.11)$$

the final answer can be read off by a direct comparison of Eq. (9.10) with the case of the option price for a single equity given in Eq. (3.23). The bond option is given by

$$C(t_0, t_*, T, K) = P(t_0, t_*)[F(t_0, t_*, T)N(d_+) - KN(d_-)] \qquad (9.12)$$
$$F \equiv F(t_0, t_*, T) = \exp\left\{-\int_{t_*}^{T} dx f(t_0, x)\right\}; \quad d_\pm = \frac{1}{q}\left[\ln\frac{F}{K} \pm \frac{q^2}{2}\right]$$

The expression for the price of the European option for a zero coupon bond is very similar to the one for equity derived by Black and Scholes.

9.3 'Greeks' for the European bond option

The hedging parameters for bond options obtained from the field theory model, the so-called 'Greeks', are studied in some detail. The major difference in the price of the option arises in the expression for the volatility q. Hence all the 'Greeks' are similar to options on equities, except for the theta and vega of the option price, since these measure the change in the option price when the volatility changes.

For evaluating the hedging parameters the following identities are useful (the prime stands for differentiation)

$$N'(d_\pm) = \frac{1}{\sqrt{2\pi}}\exp\left\{-\frac{1}{2}d_\pm^2\right\}; \quad FN'(d_+) = KN'(d_-) \qquad (9.13)$$

Delta

The delta of the option is given by the variation of its value with a change in the underlying security. In the case of a bond, the underlying security is the the forward price of the bond, namely F and hence

$$\Delta \equiv \frac{\partial C}{\partial F} = P(t_0, t_*)N(d_+)$$

---

[1] One recovers the result given for the HJM model given in Eq. (6.61) by setting $D(x, x'; t) \to 1$.

Gamma

The second variation of the option price with respect to the underlying security yields gamma, which is given by

$$\Gamma \equiv \frac{\partial^2 C}{\partial F^2} = \frac{1}{qF} N'(d_+)$$

Theta

The sensitivity of the option with respect to the changes in present time t_0 is given by

$$\Theta \equiv \frac{\partial C}{\partial t_0} = f(t_0, t_0)C + P(t_0, t_*)FN'(d_+)\frac{\partial q}{\partial t_0}$$

Eq. (9.8) yields

$$\frac{\partial q}{\partial t_0} = -\frac{1}{2q} \int_{t_*}^T dx dx' \sigma_i(t_0, x) D(x, x'; t_0) \sigma_i(t_0, x')$$

and since $r(t_0) = f(t_0, t_0)$

$$\Theta = r(t_0)C - \frac{1}{2q}FN'(d_+) \int_{t_*}^T dx dx' \sigma_i(t_0, x) D(x, x'; t_0) \sigma_i(t_0, x')$$

Vega

Vega is the variation of the option price with respect to changes in the forward rates' volatilities $\sigma(t_0, x)$. There are infinitely many volatilities at time t_0, one for each x, namely $\sigma(t_0, x)$, and each volatility can be varied independently. Consequently vega, for each x, is given by

$$\mathcal{V}(t_0, x) \equiv \frac{\delta C}{\delta \sigma(t_0, x)} = \frac{\delta q}{\delta \sigma(t_0, x)} \frac{\partial C}{\partial q} = P(t_0, t_*)\frac{\delta q}{\delta \sigma(t_0, x)} FN'(d_+) \quad (9.14)$$

The concept of functional differentiation is discussed in Eq. (A.15), and yields

$$\frac{\delta \sigma(t', x')}{\delta \sigma(t, x)} = \delta(t - t')\delta(x - x')$$

Hence, from Eq. (9.8)

$$\frac{\delta q}{\delta \sigma(t_0, x_0)} = \frac{2}{q} \int_{t_0}^{t_*} dt \int_{t_*}^T dx dx' D(x, x'; t)\sigma(t, x')\delta(t - t_0)\delta(x - x_0)$$

$$= \frac{1}{q} \int_{t_*}^T dx D(x, x_0; t_0)\sigma(t_0, x)$$

where $\int_{t_0}^{t_*} dt f(t)\delta(t - t_0) = \frac{1}{2}f(t_0)$.

In practice one hedges against only a finite number of volatilities; for this purpose, discretize future time x into a finite number of points $x = n\epsilon$ spaced by a time interval ϵ equal to one month or three months. One then has a finite collection of volatilities σ_n for which one computes the vegas.

The option price also varies with changes in the parameters of the Lagrangian for the forward rates, namely the rigidity parameter μ, and the other parameters such as a, β, λ introduced for the modified propagators. Generically denote all the parameters by μ_i where i is an integer that ranges over some finite interval. Note all these parameters enter into the pricing of the option only through the volatility parameter q, and hence, similar to Eq. (9.14), there are a new set of vegas given by

$$\mathcal{V}_i(t_0) \equiv \frac{\partial C}{\partial \mu_i} = \frac{\partial q}{\partial \mu_i}\frac{\partial C}{\partial q} = P(t_0, t_*)\frac{\partial q}{\partial \mu_i}FN'(d_+) \qquad (9.15)$$

where

$$\frac{\partial q}{\partial \mu_i} = \frac{1}{q}\int_{t_0}^{t_*} dt \int_{t_*}^{T} dxdx'\sigma(t, x)\frac{\partial D(x, x'; t; \mu_1, \mu_2, \ldots)}{\partial \mu_i}\sigma(t, x')$$

Constant volatility field theory model

Consider the simplest case of constant volatility, where $\sigma(x, t_0) = \sigma_0 = \text{constant}$. For the case of the stiff propagator, vega depends on three parameters, namely σ_0, the rigidity parameter μ and the stiffness parameter λ. The vega computed for the change in the option price under the variations of σ_0, μ, λ is

$$\mathcal{V}_\sigma(t_0) = \frac{\partial C}{\partial \sigma_0}; \quad \mathcal{V}_\mu(t_0) = \frac{\partial C}{\partial \mu}; \quad \mathcal{V}_\lambda(t_0) = \frac{\partial C}{\partial \lambda}$$

9.4 Pricing an interest rate cap

A **cap** is an interest rate option for reducing a borrowers exposure to interest rate fluctuations, and guarantees a maximum interest rate for borrowing over a fixed time. Interest rate caps are fixed using Libor, discussed in Section 8.1. Recall, from Eq. (8.3), that the relation between the forward interest rates and Libor is given by (recall $\tau = 90$ days)

$$L(t, t_*) = \frac{e^{\int_{t_*}^{t_*+\tau} dxf(t,x)} - 1}{\tau} \qquad (9.16)$$

A **cap** gives its holder the option to fix the maximum (simple) interest equal to K for a fixed period from t_* to $t_* + \tau$ if the Libor $L_* \equiv L(t_*, t_*)$ is greater than the cap rate K. The interest rate cap is similar to a put option on a Treasury Bond (as will be shown later), and an interest rate **floor** is similar to call option on a Treasury Bond; the price of a floor can be obtained from the expression for the price of a cap using put–call parity arguments [51]. A cap over a duration longer than τ is made from the sum of caps $\text{Cap}(t_0, t_n, K_n)$ over a series of times t_n to $t_n + \tau$ with cap rate K_n, where $t_n = t_0 + n\tau$ with $n = 1, 2, \ldots, N$.

Let the principal amount be V. The cap is exercised at time t_*, and the payments are made, in arrears, at time $t_* + \tau$; hence the payoff function at time $t_* + \tau$ is given by

$$g(t_* + \tau) = \tau V (L_* - K)_+ \tag{9.17}$$

Recall from Section 8.1.1 that Libor is a simple interest rate, and L_* is fixed at time t_* for a duration of τ. Hence, by discounting the payoff with L_* for an interval τ, the payoff function at time t_* is given by [51,52], from Eq. (9.16)

$$g(t_*) = \frac{\tau V}{1 + \tau L_*}(L_* - K)_+ = V \left(1 - \frac{1 + \tau K}{1 + \tau L_*} \right)_+ \tag{9.18}$$

$$= \tilde{V} \left(X - e^{-\int_{t_*}^{t_* + \tau} dx f(t_*, x)} \right)_+$$

where $\quad \tilde{V} = V(1 + \tau K) \quad$ and $\quad X \equiv \dfrac{1}{1 + \tau K}$

The result above states that an interest rate cap is equivalent to a European put option on a Treasury Bond $P(t_*, t_* + \tau) = \exp\{-\int_{t_*}^{t_* + \tau} dx f(t_*, x)\}$ [51].

Similar to the case of the European call option given in Eq. (9.6), the price of the cap at time $t_0 < t_*$ is given by the average value of the discounted payoff function

$$\text{Cap}(t_0, t_*, X) = \tilde{V} E_{[t_0, t_*]} \left[e^{-\int_{t_0}^{t_*} dt r(t)} \left(X - e^{-\int_{t_*}^{t_* + \tau} dx f(t, x)} \right)_+ \right]$$

$$= \tilde{V} P(t_0, t_*) \int_{-\infty}^{+\infty} dG \, \Psi_{\text{cap}}(G, t_*)(X - e^G)_+ \tag{9.19}$$

where, similar to Eq. (9.10)

$$\Psi_{\text{cap}}(G, t_*) = \frac{1}{\sqrt{2\pi q_{cap}^2 (t_* - t_0)}} e^{-\frac{1}{2q_{cap}^2 (t_* - t_0)}(G + \int_{t_*}^{t_* + \tau} dx f(t_0, x) + \frac{q_{cap}^2 (t_* - t_0)}{2})^2}$$

and q_{cap}^2 is given, similar to Eq. (9.8), by[2]

$$q_{cap}^2 = \frac{1}{t_* - t_0} \int_{t_0}^{t_*} dt \int_{t_*}^{t_* + \tau} dx dx' \sigma(t, x) D(x, x'; t) \sigma(t, x') \tag{9.20}$$

The domain of integration for q_{cap}^2, with $T = t_* + \tau$, is $\mathcal{R}$ as in Figure 9.1.

---

[2] q_{cap} has been defined to match the conventional definition by scaling out remaining time $t_* - t_0$ [51].

Similar to Eq. (9.12) for the call option, the **price of a cap** is given by

$$\text{Cap}(t_0, t_*, X) = \tilde{V} P(t_0, t_*)[X N(-d_-) - F_{\text{cap}} N(-d_+)] \qquad (9.21)$$

with

$$F_{\text{cap}} = \exp\left\{-\int_{t_*}^{t_*+\tau} dx f(t_0, x)\right\}; \quad d_{\pm} = \frac{1}{q_{\text{cap}}\sqrt{t_* - t_0}}\left[\ln\frac{F_{\text{cap}}}{X} \pm \frac{q_{\text{cap}}^2(t_* - t_0)}{2}\right]$$

The price of the cap **at money** is given by $X = F_{\text{cap}}$, and which yields $d_{\pm} = \pm q_{\text{cap}}\sqrt{t_* - t_0}/2$. Hence, at the money

$$
\begin{aligned}
\text{Cap}(t_0, t_*)\Big|_{\text{At the money}} &= \tilde{V} P(t_0, t_*)[N(-d_-) - N(-d_+)] \\
&= \tilde{V} P(t_0, t_*)[N(d_+) - N(d_-)] \\
&= \tilde{V} P(t_0, t_*)\left[N\left(\frac{1}{2}q_{\text{cap}}\sqrt{t_* - t_0}\right)\right. \\
&\quad \left. - N\left(-\frac{1}{2}q_{\text{cap}}\sqrt{t_* - t_0}\right)\right]
\end{aligned}
\qquad (9.22)
$$

9.4.1 Black's formula for interest rate caps

Black's model for interest caps is briefly reviewed in order to compare it with the field theory model. Black's formula is based on the assumption that the spot interest rate is a log normal random variable [51, 52]. The payoff function for Black's formula at time t_*, similar to Eq. (9.18), is

$$g_B(t_*) = \frac{\tau V}{1 + \tau f_*}(R(t_*) - R_X)_+; \quad f_* = \int_{t_*}^{t_*+\tau} dx f(t_*, x) \qquad (9.23)$$

$R(t_*)$ is the spot interest rate at time t_*, and R_X is the cap strike price on the interest rate.

Black's formula for the value of the cap at time $t_0 < t_*$ is [52]

$$\text{Cap}_B(t_0, t_*, R_X) = \frac{V\tau}{1 + \tau f_0} P(t_0, t_*)[f_0 N(d_+^B) - R_X N(d_-^B)] \qquad (9.24)$$

$$f_0 = \int_{t_0}^{t_0+\tau} dx f(t_0, x); \quad d_{\pm}^B = \frac{1}{\sigma_B\sqrt{t_* - t_0}}\left[\ln\frac{f_0}{R_X} \pm \frac{\sigma_B^2(t_* - t_0)}{2}\right]$$

Clearly the price of a cap derived from an arbitrage free model, as in Eq. (9.22), and Black's formula do not agree in general. However, **at the money**, Black's formula is

$$\text{Cap}_B(t_0, t_*)\Big|_{\text{ATM}} = \frac{V\tau f_0}{1 + \tau f_0} P(t_0, t_*)\left[N(\frac{1}{2}\sigma_B\sqrt{t_* - t_0}) - N\left(-\frac{1}{2}\sigma_B\sqrt{t_* - t_0}\right)\right]$$

Comparing the equation above with Eq. (9.22) yields the result that the price of the option, at the money, for both Black's and the field theory model, shows that the volatility of Black's model is precisely the same as the one computed from the field theory model. In other words

$$q_{\text{cap}} = \sigma_B \qquad (9.25)$$

The result given in Eq. (9.25) is central to any analysis of data on caps and floors. The price of a cap is quoted in the market by specifying its effective volatility based on Black's formula, and, hence, due to Eq. (9.25), one can directly equate the volatility obtained from the field theory model with the volatility of Black's model.

9.5 Field theory hedging of Treasury Bonds

The pricing of derivatives is only meaningful if a strategy for hedging these instruments is also provided. All forms of financial instruments are subject to risks due to the unpredictable behaviour of the financial markets. There are many ways of defining risk [18]. Recall from the discussion in Section 2.6 that hedging is a general term for the procedure of **reducing**, and if possible completely eliminating, the **risk** to the future value of a financial instrument – due to its random fluctuations – by including the instrument being hedged in a portfolio together with other related instruments.

For bonds, the main risks are changes in interest rates and the risk of default. In this chapter, all bonds are taken to be default free so that the only source of risk is taken to be the (random) changes in the interest rates.

The **risk** of an instrument, evolving over some time interval $[t_0, t_*]$ is defined to be the **variance of its final value**, at time t_*. This definition of risk is valid for both finite and instantaneous hedging. Hence, when a certain instrument is hedged, one is trying to create a portfolio of the hedged and hedging instruments that minimizes the final variance of the portfolio. A perfectly hedged portfolio in this formulation is one with zero final variance.

In the HJM model the forward rates are perfectly correlated; one can therefore hedge a 30-year Treasury Bond with a six month Treasury Bill – clearly something that the market does not support. In the case of a K-factor HJM model, perfect hedging (i.e., a zero variance portfolio) is achievable once any K-independent hedging instruments are used. However, the difficulties introduced by an infinite number of factors in the HJM models has resulted in very little literature on this important subject [14].

The primary focus of the next two sections is on hedging (the fluctuations of) zero coupon Treasury Bonds, and a hedged portfolio will be formed that will include either other bonds with different maturities or futures contracts on bonds.

Field theory models address the theoretical limitations of finite factor term structure models by allowing nontrivial correlations between forward rates of every maturity. In principle, since every forward rate fluctuates independently, one can hedge against the fluctuations of any number of forward rates.

The field theory model offers computationally expedient hedge parameters for fixed-income derivatives and provides a methodology for uniquely fixing the number and maturity of bonds to be included in a hedged portfolio.

In the One Factor HJM model – at some instant t – all the forward rates are driven by a single stochastic variable, and hence, as discussed in [52], a Treasury Bond can be exactly hedged. In contrast, in the field theory of the forward rates, since at each instant t there are infinitely many degrees of freedom (random variables) driving the yield curve, exact hedging is not possible; the best that one can do is to hedge against the fluctuations of a finite number of forward rates, say $f(t, x_1), f(t, x_2), \ldots, f(t, x_N)$.

9.6 Stochastic delta hedging of Treasury Bonds

Consider the hedging of a zero coupon Treasury Bond $P(t, T)$ against fluctuations in the spot rate $r(t)$. A portfolio $\Pi(t)$ that is composed of $P(t, T)$ and other instruments needs to be formed so that the fluctuations in the value of the portfolio can ideally be made exactly zero, or at least made substantially less than the fluctuations in the instrument that one is hedging, namely $P(t, T)$.

The simplest case is if $P(t, T)$ is hedged with another bond $P(t, T_1)$ with maturity $T_1 \neq T$ [52]. Form the portfolio

$$\Pi(t) = P(t, T) + n_1(t) P(t, T_1) \tag{9.26}$$

where n_1 is the hedging parameter. Note the portfolio needs to be re-balanced every instant since the values of the bonds are changing with time.

For a change in time t by a finite amount δt, during which the spot rate changes by an amount δr, the change in the value of the portfolio is given by

$$\delta \Pi(t) = \frac{\partial \Pi(t)}{\partial t} \delta t + \frac{\partial \Pi(t)}{\partial r} \delta r + \frac{1}{2} \frac{\partial^2 \Pi(t)}{\partial r^2} (\delta r)^2 + \cdots$$

For the portfolio to be free from the risk of losing value, due to a change in the spot interest rate, the portfolio is required to be independent of small changes in the spot rate r. **Delta hedging** of this portfolio requires setting the coefficient of the δr term above to zero, namely

$$\frac{\partial}{\partial r} \Pi(t) = 0 \tag{9.27}$$

which yields for the hedge parameter, from Eq. (9.26)

$$n_1 = -\frac{\partial P(t, T)}{\partial r} \bigg/ \frac{\partial P(t, T_1)}{\partial r} \tag{9.28}$$

In the HJM model [52] the Treasury Bond can be expressed directly as a function of the spot rate $r(t)$; this is possible because there is only one random variable for each time t that is driving the forward rates. [These and related questions are addressed in Appendices 9.13 and 9.14.] Jarrow and Turnbull [52] further assume an exponential form for the volatility, namely $\sigma(x, t) = \sigma_2 \exp -\beta(x - t)$, and solve Eq. (9.28) to obtain the hedging parameter n_1.

In field theory, for each time t, there are infinitely many degrees of freedom driving the forward rates, and hence one can never delta hedge by satisfying Eq. (9.27). The best that can be done is to satisfy delta hedge only on average, and this scheme is called **stochastic delta hedging**. To implement stochastic delta hedging one needs to form the conditional expectation value of the portfolio $\Pi(t)$, keeping the spot rate $r(t)$ fixed – namely $E\big[\Pi(t)|r(t)\big]$. Define the conditional probability of a Treasury Bond by

$$B(r; t; T) = E\big[P(t, T)|r\big] \tag{9.29}$$

with the mathematical expression being given in Eq. (9.48). Eq. (9.26) gives

$$E\big[\Pi(t)|r(t)\big] = B(r; t, T) + n_1(t)B(r; t, T_1) \tag{9.30}$$

Stochastic delta hedging is defined by generalizing Eq. (9.27) to

$$\frac{\partial}{\partial r} E\big[\Pi(t)|r(t)\big] = 0 \tag{9.31}$$

Hence, from Eqs. (9.31) and (9.30) stochastic delta hedging yields

$$n_1 = -\frac{\partial B(r; t, T)}{\partial r} \bigg/ \frac{\partial B(r; t, T_1)}{\partial r} \tag{9.32}$$

As can be seen from above, changes in the hedged portfolio $\Pi(t)$, for delta hedging in field theory, are only **on the average** insensitive to the fluctuations in the spot rate r.

The hedging weight n_1 is evaluated explicitly for the field theory of forward rates in Appendix 9.13, and the result, from Eq. (9.57), is given by

$$n_1 = -\left[\frac{\int_t^T dx D(t, x; t, T_{FR})\sigma(x, t)}{\int_t^{T_1} dx D(t, x; t, T_{FR})\sigma(x, t)}\right] \frac{B(r; t, T)}{B(r; t, T_1)} \tag{9.33}$$

For $T_1 = T$ the hedging coefficient is given by $n_1 = -1$, as indeed one expects, since the best way to hedge a bond is to short it, leading to zero fluctuations.

However, since this is a trivial solution to hedging, one always assumes that $T_1 \neq T$.

To hedge against the $\Gamma = \partial^2 \Pi(t)/\partial r^2$ fluctuations, one needs to form a portfolio with two hedging bonds

$$E\big[\Pi(t)|r(t)\big] = B(r; t, T) + n_1(t)B(r; t, T_1) + n_2(t)B(r; t, T_2)$$

and to minimize the change in the value of $E\big[\Pi(t)|r(t)\big]$ with respect to both delta and gamma fluctuations.

Finite time delta hedging can be defined by hedging against the fluctuations of the spot rate $r(t_*)$ at some finite future time $t_* > t$. One would then need to compute $B(r; t_*, T)$ as has been done in Appendix 9.13.

Suppose a Treasury Bond needs to be hedged against the fluctuations of N forward rates, namely $f(t_*, x_i); i = 1, 2, \ldots, N$. The conditional probability for $P(t_*, T)$ given in Eq. (9.49), with the N forward rates fixed at $f(t_*, x_i) = f_i; i = 1, 2, \ldots, N$ is

$$B(f_1, f_2, \ldots, f_N; t_*, T) = E\big[P(t_*, T)|f_1, f_2, \ldots, f_N\big]$$

A hedged portfolio with bonds of varying maturities $T_i \neq T$ is

$$\Pi(t) = P(t, T) + \sum_{i=1}^{N} n_i P(t, T_i)$$

and the stochastic delta hedging conditions are given by

$$\frac{\partial}{\partial f_j} E\big[\Pi(t_*)|f_1, \ldots, f_N\big] = 0; \quad j = 1, 2, \ldots, N$$

$$\Rightarrow \frac{\partial B(f_1, f_2, \ldots, f_N; t_*, T)}{\partial f_j} + \sum_{i=1}^{N} n_i \frac{\partial B(f_1, f_2, \ldots, f_N; t_*, T_i)}{\partial f_j} = 0$$

One can solve the above system of N simultaneous equations to determine the N hegding parameters n_i. The volatility of the hedged portfolio can be reduced by increasing N.

Stochastic delta hedging against N forward rates for large N tends to be complicated, and a closed-form solution is difficult to obtain.

9.7 Stochastic hedging of Treasury Bonds: minimizing variance

The main shortcoming of stochastic delta hedging is that one does not have any control on the effectiveness of the hedging procedure, which is determined by the variance of the instantaneous change in the value of the hedged portfolio.

The value of the portfolio itself is deterministic at time t, and a measure of of the effectiveness of hegding is given by the **volatility** of the **instantaneous change in the portfolio**, namely by $\text{Var}[d\Pi(t)/dt]$.

Recall that in the Black–Scholes analysis for the price of an option on a security, one forms a hedged portfolio $\Pi_{BS} = C - (\partial C/\partial S)S$ such that $d\Pi_{BS}/dt$ is **deterministic**, and which implies that there are no fluctuations in its value. In other words, for the Black–Scholes portfolio

$$\text{Var}\left[\frac{d\Pi_{BS}}{dt}\right] = 0 \tag{9.34}$$

Since for the field theory case one cannot set the variance of $d\Pi(t)/dt$ to be exactly zero due to the infinity of degrees of freedom (random variables at each instant), the next best thing is to **minimize the variance**. Stochastic hedging is hence defined by demanding that the variance of the instantaneous change in the value of the hedged portfolio be a minimum – to be made more precise later – and this will yield a procedure for uniquely fixing all the hedging parameters. The discussion on stochastic hedging of Treasury Bonds is based on the results of [12].

The hedged portfolio $\Pi(t)$ at time t is given by

$$\Pi(t) = P(t, T) + \sum_{i=1}^{N} \Delta_i P(t, T_i) \tag{9.35}$$

where Δ_i are the hedging weights and denote the amount of the ith bond $P(t, T_i)$ that is included in the hedged portfolio. The value of bonds $P(t, T)$ and $P(t, T_i)$ are determined by observing their market values at time t, and hence are not stochastic. As mentioned earlier, it is the **instantaneous change** in the portfolio value that is stochastic.

Often it is more convenient to hedge the zero coupon bond using futures contracts as these are more liquid. Let $\mathcal{F}_i$ denote the price, at time t, of a futures contract $\mathcal{F}(t, t_F, T_i)$, expiring at time t_F, on a zero coupon bond maturing at time T_i and given in Eq. (9.1). The hedged portfolio in terms of the futures contract is given by

$$\Pi(t) = P + \sum_{i=1}^{N} \Delta_i \mathcal{F}_i \tag{9.36}$$

$$\mathcal{F}_i \equiv \mathcal{F}(t, t_F, T_i) = e^{-\int_{t_F}^{T_i} dx f(t,x)} e^{\Omega_{\mathcal{F}}}$$

where the $\mathcal{F}_i$s are the observed market prices.

Stochastic hedging of a zero coupon bond $P(t, T)$ is accomplished by minimizing the variance $d\Pi(t)/dt$ of the hedged portfolio, which in turn uniquely fixes the hedging weights Δ_i.[3] The hedging strategy is hence given by

$$\text{Var}\left[\frac{d\Pi(t)}{dt}\right] : \text{ minimize}$$

$$\frac{\partial}{\partial \Delta_i}\text{Var}\left[\frac{d\Pi(t)}{dt}\right] = 0 \Rightarrow \text{ fixes } \Delta_i\text{'s}$$

For the instantaneous case, the computation will always minimize the value of $\text{Var}[d\Pi(t)/dt]$.

For $\epsilon \to 0$

$$\text{Var}\left[\frac{d\Pi(t)}{dt}\right] \simeq \frac{1}{\epsilon^2}\text{Var}[\Pi(t+\epsilon) - \Pi(t)]$$

$$= \frac{1}{\epsilon^2}\text{Var}[\Pi(t+\epsilon)]$$

since the initial value of the portfolio $\Pi(t)$ is deterministic. In other words, in stochastic hedging the variance of the **final value** of the portfolio is minimized. This procedure holds even if ϵ is a finite quantity, and is the basis for finite time hedging. In summary

Instantaneous hedging $\text{Var}[\Pi(t+\epsilon)] : \text{ minimize}$

Finite time hedging $\text{Var}[\Pi(t_*)] : \text{ minimize} ; t_* \gg t$

The **minimum value** of the variance of the final value of the portfolio, for both instantaneous and finite time hedging, is called the **residual variance** and is denoted by

$$V = \text{ minimum of Var}\left[\frac{d\Pi(t)}{dt}\right] : \text{ instantaneous hedging}$$

$$V_* = \text{ minimum of Var}\left[\Pi(t_*)\right] : \text{ finite time hedging}$$

The detailed calculation for determining the hedge parameters and portfolio variance is carried out in Appendices 9.15 and 9.16 and is summarized in the Table 9.1. The result depends on the hedging matrix M_{ij} consisting of the correlation of the bonds or futures being used to hedge, and a hedging vector L_i which is the correlation of the bond being hedged with the other hedging bonds or futures.

For $N = 1$, the hedge parameter, from Eq. (9.72), reduces to

$$\Delta_1 = -\frac{L_1}{M_{11}} = -\left[\frac{\int_t^T dx \int_t^{T_1} dx' \sigma(t,x)\sigma(t,x')D(x,x';t,T_{FR})}{\int_t^{T_1} dx \int_t^{T_1} dx' \sigma(t,x)\sigma(t,x')D(x,x';t,T_{FR})}\right]\frac{P(t,T)}{P(t,T_1)}$$

$$\text{(9.37)}$$

---

[3] In stochastic models of interest rates, Gaussian models do not yield a unique solution for the hedging weights [18].

Table 9.1 *Residual variance and hedging weights for the instantaneous hedging of a zero coupon bond $P(t, T) \equiv P$ using other hedging bonds $P(t, T_i) \equiv P_i$, and futures contracts $\mathcal{F}(t, t_F, T_i) \equiv \mathcal{F}_i$ with maturity at time t_F.*
$$[\theta = x - t; \ \theta' = x' - t].$$

Portfolio $\Pi(t)$	Residual variance $V = \text{Min}\big(\text{Var}[d\Pi(t)/dt]\big)$	Weights Δ_i
P	$V_0 = P^2 \displaystyle\int_0^{T-t} d\theta \int_0^{T-t} d\theta' \sigma(\theta)\sigma(\theta')D(\theta,\theta')$	0
$P + \displaystyle\sum_{i=1}^{N}\Delta_i P_i$	$V = V_0 - L^T M^{-1} L$ $L_i = P P_i \displaystyle\int_0^{T-t} d\theta \int_0^{T_i-t} d\theta' \sigma(\theta)D(\theta,\theta')\sigma(\theta')$ $M_{ij} = P_i P_j \displaystyle\int_0^{T_i-t} d\theta \int_0^{T_j-t} d\theta' \sigma(\theta)D(\theta,\theta')\sigma(\theta')$	$-\displaystyle\sum_{j=1}^{N} L_j M_{ji}^{-1}$
$P + \displaystyle\sum_{i=1}^{N}\Delta_i \mathcal{F}_i$	$V = V_0 - \tilde{L}^T \tilde{M}^{-1} \tilde{L}$ $\tilde{L}_i = P\mathcal{F}_i \displaystyle\int_{t_F-t}^{T_i-t} d\theta \int_0^{T-t} d\theta' \sigma(\theta)D(\theta,\theta')\sigma(\theta')$ $\tilde{M}_{ij} = \mathcal{F}_i\mathcal{F}_j \displaystyle\int_{t_F-t}^{T_i-t} d\theta \int_{t_F-t}^{T_j-t} d\theta' \sigma(\theta)D(\theta,\theta')\sigma(\theta')$	$-\displaystyle\sum_{j=1}^{N} \tilde{L}_j \tilde{M}_{ji}^{-1}$

When $T_1 = T$, the hedge parameter equals minus one and is a trivial result that reduces residual variance in Eq. (9.70) to zero as $\Delta_1 = -1$, since $P = P_1$ implies $L_1 = M_{11}$. Empirical results for nontrivial hedging strategies are discussed in the next section.

On comparing Eqs. (9.33) and (9.37) for the value of the hedging parameters n_1 and Δ_1, it is seen that the strategies of stochastic delta hedging and minimization of variance give very different results. It is shown in Section 9.17 that these two expressions become equal in the HJM limit if the volatility parameter is taken to be $\propto \exp -\beta(x - t)$.

9.8 Empirical analysis of instantaneous hedging

The empirical results for hedging of a bond follows the treatment of [12]. For simplicity the analysis will be done only for the constant rigidity field theory model as well as for the empirical propagator.[4] The best-fit propagator for constant rigidity

---

[4] All the empirical analysis uses data from the Eurodollar futures market.

given in Figure 8.8 and the empirical propagator given in Figure 8.4 are used in the calculations, as well as the volatility function given in Figure 8.2.

Reduction of residual variance to zero is not feasible in practice; the best one can do is to decide the level of risk one is prepared to live with, and then include as many hedging instruments as is required to achieve this level of risk.

The hedging of a five-year zero coupon bond with other zero coupon bonds and futures contracts is the focus of this section. The current forward rates' curve is taken to be flat and equal to 5% throughout. The initial forward rates curve does not affect any of the qualitative results. N indicates the number of innstruments being used to hedge the bond $P(t, T)$.

Hedging with other Treasury Bonds

The $N = 1$ residual variance for the hedged bond portfolio is shown in Figure 9.2; as expected, the residual variance drops to zero when the same bond is used to hedge itself – since one is eliminating the original position in the process. The hedge ratio Δ_1 is shown in Figure 9.3.

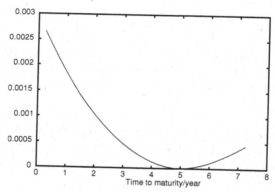

Figure 9.2 Residual variance $V = V_0 - L_1^2/M_{11}$ for a five-year bond versus the maturity of the bond being used for hedging; note the residual variance is zero for a five-year maturity hedging bond, as expected.

The parabolic nature of the residual variance is because μ is constant. A more complicated function could produce a residual variance that does not deviate monotonically as the maturity of the underlying bonds increases. The parabolic nature of the graphs, however, does appeal to ones economic intuition, which suggests that the correlation between forward rates should decrease monotonically as the distance between them increases.

The $N = 2$ case, when a Treasury Bond is hedged using two other bonds, has sharp 'ridges' in the residual variance, as shown in Figure 9.4, for both the empirical propagator as well as for the constant rigidity model. The ridges in residual

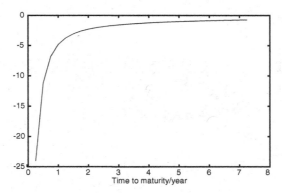

Figure 9.3 Hedging weight $\Delta_1 = -L_1/M_{11}$ for a five-year bond hedged with another bond with varying maturity.

variance become more pronounced when a short duration one-year bond is hedged with two other bonds, as shown in Figure 9.5, displaying a greater sensitivity to the fluctuations of the underlying forward rates. These ridges are typical for all such portfolios and emerge for finite time hedging as well. The existences of these 'ridges' point to potential instabilities in the values of vega for these portfolios. The causes of these ridges from the principles of finance is not clear.

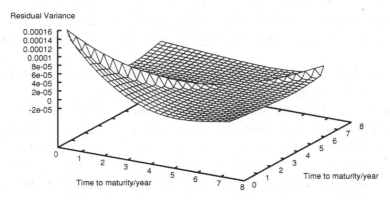

Figure 9.4 Residual variance $V = V_0 - L^T M^{-1} L$ for five-year bond versus two bond maturities being used for hedging

Hedging with futures

The residual variance is calculated for hedging a five-year zero coupon bond with futures contracts, all expiring in one year, on zero coupon bonds with various maturities. The residual variance is shown in Figure 9.6, and, as expected, is never zero.

residual variance

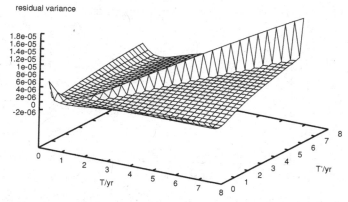

Figure 9.5 The residual variance $V = V_0 - L^T M^{-1} L$ when a one year bond is hedged with two other bonds.

The futures contracts required for the optimal hedging of a five-year bond are shown in Table 9.2. Observe from Table 9.2 that, when using only one contract, the five-year zero coupon bond is best hedged by selling a futures contracts on a 4.5-year bond; this is explained by the fact that the futures contract depends only on the variation in the forward rate curve from t_F to T but the zero coupon bond depends on the variation of the forward rate curve from t to T that is not fully covered by the fluctuations in the futures contract. Hence, a shorter underlying bond maturity is chosen for the futures contract to compensate for the forward rates curve from t to t_F.

A similar result is seen when hedging a bond with two futures contracts both expiring in one year. In this case, optimal hedging is obtained when futures contracts on the same bond as well as one on a bond with the minimum possible maturity (1.25 years) are shorted. The use of a futures contract on a short maturity bond is consistent with the high volatility of short maturity forward rates as displayed in Figure 8.2. Data show a dramatic reduction in the residual variance – by three orders of magnitude – when futures contracts are used for hedging a Treasury Bond. Table 9.2 further shows that hedging with three other instruments, be it other bonds or futures contracts, does not significantly lower the residual variance, showing that hedging with only two other instruments seems to be the best hedging strategy.

The expiry of the futures contracts is fixed to be one year from the present. This is long enough time to clearly show the effect of the expiry time as well as short enough to make practical sense (since long-term futures contracts are illiquid and unsuitable for hedging purposes).

Table 9.2 *Residual variance and hedging weights for a five-year bond hedged with one-year futures contracts on one, two or three bonds – the maturities of the bonds are indicated in the second column. Hedging with two futures contracts is seen to be the best strategy.*

Number of contracts	Futures contracts (hedge ratios) for hedging a five-year Treasury Bond $\Delta_i = -\sum_{j=1}^{N} \tilde{L}_j \tilde{M}_{ji}^{-1}$	Residual variance $V = V_0 - \tilde{L}^T \tilde{M}^{-1} \tilde{L}$
0	none	1.82×10^{-3}
1	4.5 years (-1.288)	5.29×10^{-6}
2	5 years (-0.9347), 1.25 years (-2.72497)	1.58×10^{-6}
3	5 years (-0.95875), 1.5 years (1.45535), 1.25 years (-5.35547)	1.44×10^{-6}

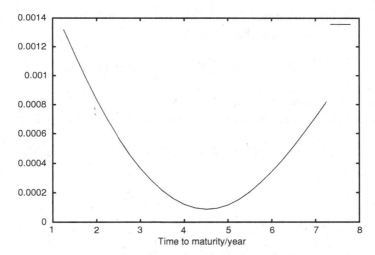

Figure 9.6 Residual variance $V = V_0 - L_1^2/M_{11}$ for a five-year bond hedged with a one-year futures contract on a bond whose maturity is plotted on the horizontal axis.

9.9 Finite time hedging

Finite time hedging of bonds with other bonds is considered. The calculations for minimizing variance can be done exactly. The hedging for finite time follows the treatment given by [10]. The hedging of bonds with futures contracts will not be considered – even though this can also be solved exactly by minimizing the variance – as it does not give any new insight for the finite time case. To see this, consider hedging with a futures contract on a zero coupon bond of duration that matures at the same as the hedging horizon. This gives exactly the same result as hedging with a bond on which the futures contract is written, since one receives a Treasury Bond on the maturity of the futures contract.

Denote initial time by t, and the hedging horizon by t_*. Consider the hedging of one bond maturing at T with N other bonds maturing at T_i, $1 \leq i \leq N$. Assume for nontrivial solutions that $T_i \neq T$ $\forall i$.

Recall from Eq. (9.35) that a hedged portfolio is given by

$$\Pi(t) = P(t, T) + \sum_{i=1}^{N} \Delta_i P(t, T_i) \tag{9.38}$$

The portfolio $\Pi(t_*)$, for $t_* > t$, is not a log normal (Gaussian) random variable; however, its **final variance** is nevertheless taken to be a suitable measure of the **risk** in the portfolio's value. The weights of the bonds $P(t, T_i)$ with maturities T_i, namely Δ_i, are chosen so that the variance of the portfolio $\Pi(t_*)$ (at the future time t_*) is a minimum. Hence the variance

$$\text{Var}[\Pi(t_*)] \equiv E[\Pi^2(t_*)] - \left\{ E[\Pi(t_*)] \right\}^2 \tag{9.39}$$

is minimized to determine the coefficients Δ_i.

The covariance between the prices $P(t_*, T_i)$ and $P(t_*, T_j)$ is given by

$$M_{ij} = E[P(t_*, T_i)P(t_*, T_j)] - E[P(t_*, T_i)]E[P(t_*, T_j)] \tag{9.40}$$

and yields [10]

$$M_{ij} = \mathcal{F}(t, t_*, T_i)\mathcal{F}(t, t_*, T_j)$$
$$\times \left\{ \exp\left(\int_t^{t_*} dt \int_{t_*}^{T_i} dx \int_{t_*}^{T_j} dx' \sigma(t, x) D(x - t, x' - t)\sigma(t, x') \right) - 1 \right\} \tag{9.41}$$

The covariance between the hedged bond of maturity T and the hedging bonds of maturity T_i is similarly given by [10]

$$L_i = \mathcal{F}(t, t_*, T)\mathcal{F}(t, t_*, T_i)$$
$$\times \left\{ \exp\left(\int_t^{t_*} dt \int_{t_*}^{T} dx \int_{t_*}^{T_i} dx' \sigma(t, x) D(x - t, x' - t)\sigma(t, x') \right) - 1 \right\} \tag{9.42}$$

Minimization of the residual variance of the hedged portfolio is straightforward, and the hedging weights are given by

$$\Delta_i = -\sum_{j=1}^{N} L_j M_{ji}^{-1} \tag{9.43}$$

The hedged portfolio is hence given by

$$\Pi(t) = P(t, T) + \sum_{i=1}^{N} \Delta_i P(t, T_i) \tag{9.44}$$

with the portfolio's minimized **residual variance** being given by

$$V_* = \text{Var}[\Pi(t_*)]\Big|_{\text{Minimum}} = \text{Var}[P(t_*, T)] - L^T M^{-1} L \tag{9.45}$$

The residual variance enables the effectiveness of the hedged portfolio to be assessed. In the next section, residual variance of hedged portfolios that include bonds of different maturities is analyzed. It can be shown that, up to rescaling by ϵ, the results of finite time hedging for the hedging weights and residual variance reduce to the instantaneous case in the limit of $t_* \rightarrow t + \epsilon$ [10].

Finite time hedging depends on the market value of the drift velocity $\alpha_M(t, x)$. The reason is that if one is not hedging continuously, then the portfolio is exposed to market risks, and therefore risk premiums encoded in $\alpha_M(t, x)$ appear in the formulae for finite time hedging.

In the calculation above, the risk-neutral drift velocity $\alpha(t, x)$ was used. However the market does not follow the risk-neutral measure, as discussed in Section 8.5, and it is better for applications to use a market estimate for the value of $\alpha_M(t, x)$. For the case of instantaneous hedging, since only short time scales are important, the stochastic term dominates making the drift term inconsequential. This, of course, is not the case for the finite time case where the importance of the drift velocity grows with an increase in the hedging time horizon.

9.10 Empirical results for finite time hedging

The calculation of L_i and M_{ij} are carried out using simple trapezoidal integration as the data itself are not very accurate. Volatility σ is assumed to be a function of only $\theta = x - t$ so that all the integrals over x are replaced by integrals over θ. The bond to be hedged is chosen to be a five-year zero coupon bond and the time horizon t_* is chosen to be one year.

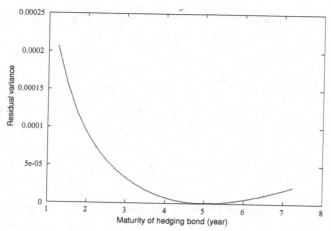

Figure 9.7 Residual variance $V_* = \text{Var}[P(t_*, T)] - L_1^2/M_{11}$ when a five-year bond is hedged with one other bond using the empirical propagator. The hedging time horizon is one year.

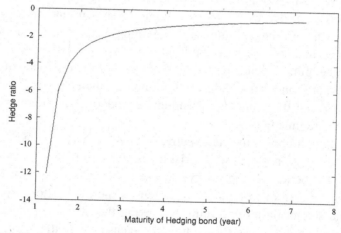

Figure 9.8 Hedging weight ratio $\Delta_1 = -L_1/M_{11}$ when a five-year bond is hedged with one other bond with the best fit for the empirical propagator. Time horizon of hedging of one year

The errors involved largely cancel themselves out, and hence the residual variances obtained are still quite accurate.

The residual variance and hedging weight for the hedged portfolio when hedging with one bond, using the empirical propagator, is shown in Figures 9.7 and 9.8. The residual variance for hedging with two bonds is shown in Figure 9.9, and has 'ridges' when the maturity of the two bonds being used for hedging

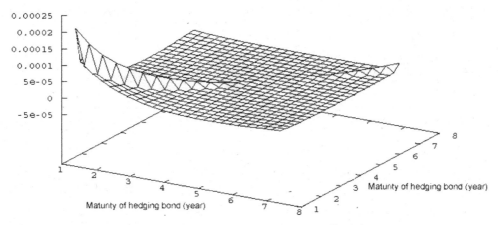

Figure 9.9 Residual variance $V_* = \mathrm{Var}[P(t_*, T)] - L^T M^{-1} L$ for a five-year bond hedged with two other bonds, using the empirical propagator. The time horizon of hedging is one year.

have nearby maturities, as has already been noted for the case of instantaneous hedging [12].

The residual variance using the field theory model for constant rigidity μ is shown in Figure 9.10, and is qualitatively the same as the result for the empirical propagator. A general feature of residual variance is that it is lower for the empirical propagator compared with the constant rigidity propagator, as can be seen by comparing Figures 9.9 and 9.10. This is to be expected since the errors in the fit of the constant rigidity model also appear in the hedging exercise.

One interesting result of finite time hedging is that the actual residual variance of the hedged portfolio, when hedging over a finite time horizon, is **less than** the value one obtains if one naively extrapolates the infinitesimal hedging result. This seems to be due to the fact that the domain of the forward rates that contributes to the variance of the bonds being used to hedge reduces as the time horizon increases. This is very clear if the maturity of the bond is close to the hedging horizon, since the volatility of bonds rapidly drop to zero as the time to maturity approaches. Apart from this reduction, the results look very similar to the infinitesimal case. This is probably due to the fact that the volatility is quite small so the nonlinear effects in the covariance matrix M_{ij} given in Eq. (9.41) are not apparent.

If very long time horizons (ten years or more) and long-term bonds are considered, the results is expected to be quite different.

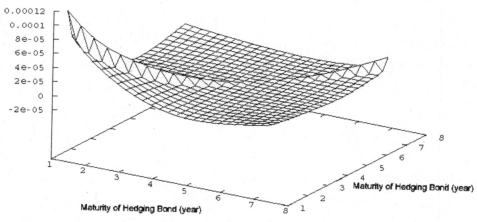

Figure 9.10 Residual variance $V_* = \text{Var}[P(t_*, T)] - L^T M^{-1} L$ when a five-year bond is hedged with two other bonds using the best-fit constant rigidity field theory model. The time horizon of hedging is one year.

9.11 Summary

The results of this chapter show that the quantum field theory of forward interest rates is quite suitable for calculating the futures and options of Treasury Bonds, and other interest rate derivatives. In particular, the formulae for the interest rate caps as well as for futures and options of Treasury Bonds, derived using Gaussian path integration, involve nontrivial correlations in the volatility of the model arising from the correlation of the forward interest rates in the maturity direction.

The concept of hedging in field theory was shown to be a natural generalization of conventional approaches, and correctly reproduces the limiting results of the HJM model. Despite the infinite-dimensional nature of the quantum field, it was shown that a low-dimensional hedge portfolio effectively hedges interest rate risk by exploiting the correlation between forward rates. Therefore, field theory models address the theoretical dilemmas of finite factor term structure models and offer a practical alternative to these models.

The study of derivatives, and the hedging of Treasury Bonds, demonstrates the far-reaching significance of the nontrivial correlation between forward interest rates with differing maturities.

9.12 Appendix: Conditional probabilities

The most important forward interest rate is the spot interest rate, namely $f(t_*, t_*)$ at some future time $t_* \in [t_0, T]$. The case of hedging a Treasury Bond against the fluctuations in the spot rate is considered.

The basic idea in forming the conditional expectation value of a bond is to **integrate out** all the forward rates **except** the rate being hedged against, in this case the spot rate $f(t_*, t_*)$. In other words, one needs to compute

$$\int Df \, \delta(f(t_*, t_*) - r) P(t_*, T) e^S \tag{9.46}$$

This naive prescription is however not completely correct. One is computing the **conditional expectation value**[5] of the Treasury Bond, given the occurrence of a specific value for the spot rate. Hence the integration over the forward interest rates has to be performed with a properly normalized probability distribution.

The conditional expectation of the Treasury bond, for spot rate $f(t_*, t_*) \equiv r(t_*)$ fixed at some value r, is denoted by $B(r; t_*, T)$. Hence

$$B(r; t_*; T) = E\big[P(t_*, T)|r\big] \tag{9.47}$$

$$= \frac{\int Df \, \delta[f(t_*, t_*) - r] P(t_*, T) e^S}{\int Df \, \delta[f(t_*, t_*) - r] e^S} \tag{9.48}$$

For hedging against the fluctuations of N-forward rates, the conditional probability is given by

$$B(f_1, f_2, \ldots, f_N; t_*, T) = \frac{\int Df \prod_{i=1}^{N} \delta[f(t_*, x_i) - f_i] P(t_*, T) e^S}{\int Df \prod_{i=1}^{N} \delta[f(t_*, x_i) - f_i] e^S} \tag{9.49}$$

The probability distribution of r is given by

$$P(r) = \frac{1}{Z} \int Df \, \delta[f(t_*, t_*) - r] e^S; \quad Z = \int Df e^S \tag{9.50}$$

Eqs. (9.48) and (9.50) yield the expected result as follows

$$E[P(t_*, T)] = \int dr \, E\big[P(t_*, T)|r\big] P(r)$$

$$= \int dr \, B(r; t_*; T) P(r)$$

$$= \frac{1}{Z} \int Df \int dr \, \delta[f(t_*, t_*) - r] P(t_*; T) e^S$$

$$= \frac{1}{Z} \int Df \, P(t_*, T) e^S$$

There are similar expressions for conditional probabilities holding N-forward rates fixed.

---

[5] Conditional probability is discussed in Appendix A.1.

9.13 Appendix: Conditional probability of Treasury Bonds

The case of the conditional probability of a Treasury Bond, given a fixed value of the interest rate at some future time, namely for fixed $r(t_*)$, is fully worked out. The derivation shows a number of important features of the Gaussian model of the forward rates, and all the detailed derivations crucially hinge on the path-integral formulation of the forward rates.

To understand the detailed nature of the conditional value of a bond $B(r; t_*, T)$ it is expressed in terms of $P(\eta|r)$, the conditional probability distribution – given r – of the zero coupon bond $P(t_*, T)$, given by

$$B(r; t_*; T) = \int_{-\infty}^{+\infty} d\eta\, e^{-\eta} P(\eta|r; t_*; T) \tag{9.51}$$

$$P(\eta|r; t_*; T) = \frac{\int Df\, \delta[f(t_*, t_*) - r]\delta[\eta - \int_{t_*}^{T} dx f(t_*, T)] e^S}{\int Df\, \delta[f(t_*, t_*) - r] e^S}$$

A straightfoward but tedious calculation gives $P(\eta|r; t_*; T)$. Using the results of the Gaussian models discussed in Chapter 7, yields the following result[6]

$$P(\eta|r; t_*; T) = \frac{1}{\sqrt{2\pi\sigma_{\eta|r}^2}} \exp\left[-\frac{1}{2\sigma_{\eta|r}^2}(\eta - \eta_0)^2\right] \tag{9.52}$$

$$\eta_0 = \int_{\mathcal{R}} \alpha(t, x) + \int_{t_*}^{T} dx f(t_0, T) + \frac{B}{A}\left[(r - f(t_0, t_*)) - \int_{t_0}^{t_*} dt\alpha(t, t_*)\right]$$

$$\sigma_{\eta|r}^2 = C - \frac{B^2}{A}$$

$$A = \int_{t_0}^{t_*} dt\sigma^2(t, t_*) D(t_*, t_*; t, T_{FR})$$

$$B = \int_{t_0}^{t_*} dt \int_{t_*}^{T} dx\sigma(t, t_*) D(t_*, x; t, T_{FR})\sigma(t, x)$$

$$C = \int_{t_0}^{t_*} dt \int_{t_*}^{T} dx dx'\sigma(t, x) D(x, x'; t, T_{FR})\sigma(t, x')$$

Note that all the integrations in the formulae above are over the rectangular domain $\mathcal{R}$ given in Figure 9.1.

The unconditional probability distribution for the zero coupon bond yields a volatility of $\sigma_\eta^2 = C$ and hence the conditional expectation reduces the volatility of the bond by B^2/A. This is to be expected since the constraint imposed by

---

[6] For Gaussian field theories, $P(\eta|r; t_*; T)$ is also Gaussian, and hence the simplest way to compute it is to find the mean and variance of the random variable η.

the requirement of conditional probability reduces the allowed fluctuations of the Treasury Bond. This reduction of variance is generalized in Eq. (9.45) to the case of hedging with N-instruments.

The largest reduction of the conditional variance $\sigma^2_{\eta|r}$ that is possible depends on the properties of the volatility function $\sigma(t, x)$, and it could be the case that hedging not against the spot rate $r(t_*) = f(t_*, t_*)$ but against some other forward rate, namely $f(t_*, x)$ $(x > t_*)$, may lead to a greater reduction in the conditional variance. The answer can only be found by empirically studying the properties of $\sigma(t, x)$. Furthermore, an N-fold constraint on the bond would clearly further reduce the variance of the conditional value of the bond.

From Eq. (9.51)

$$B(r; t_*; T) = \int_{-\infty}^{+\infty} d\eta e^{-\eta} P(\eta|r; t_*; T) = e^{-\eta_0 + \frac{1}{2}\sigma^2_{\eta|r}} \qquad (9.53)$$

Hence the final result is given by

$$B(r; t_*; T) = \frac{P(t_0, T)}{P(t_0, t_*)} \exp\left\{ -X(t_*, T)[r - f(t_0, t_*)] - a(t_*, T) \right\} \qquad (9.54)$$

with

$$X(t_*, T) = \frac{B}{A} = \frac{\int_{t_0}^{t_*} dt \int_{t_*}^{T} dx \sigma(t, t_*) D(t_*, x; t, T_{FR}) \sigma(t, x)}{\int_{t_0}^{t_*} dt \sigma^2(t, t_*) D(t_*, t_*; t, T_{FR})} \qquad (9.55)$$

and

$$a(t_*, T) = \frac{1}{2A}\left[B^2 - 2B \int_{t_0}^{t_*} dt \alpha(t, t_*) \right] - \Omega_{\mathcal{F}}(t_0, t_*, T)$$

where $\Omega_{\mathcal{F}}$ is given by Eq. (9.4).

The result obtained for $B(r; t_*; T)$ is quite different from a similar equation given in [52]. Equation (9.54) is an identity for $P(t, T)$ in the HJM model with an exponential volatility function, whereas in the field theory case, recall it is only the conditional expectation value of the Treasury Bond, given a value of $r(t_*)$ for the spot rate.

Recall the hedging parameter is given by Eq. (9.28), and using Eq. (9.54) yields

$$n_1 = -\frac{\partial B(r; t, T)}{\partial r} \bigg/ \frac{\partial B(r; t, T_1)}{\partial r}$$

$$= -\left[\frac{X(t, T)}{X(t, T_1)}\right] \frac{B(r; t, T)}{B(r; t, T_1)} \qquad (9.56)$$

Setting $t_0 = t$ and $t_* = t + \epsilon$ gives from Eq. (9.55)

$$X(t, T) = \frac{\int_t^T dx \sigma(t, t) D(t, x; t, T_{FR}) \sigma(t, x)}{\sigma(t, t) D(t, t; t, T_{FR})}$$

and hence for (instantaneous) stochastic delta hedging

$$n_1 = -\left[\frac{\int_t^T dx D(t, x; t, T_{FR}) \sigma(x, t)}{\int_t^{T_1} dx D(t, x; t, T_{FR}) \sigma(x, t)} \right] \frac{B(r; t, T)}{B(r; t, T_1)} \tag{9.57}$$

9.14 Appendix: HJM limit of hedging functions

To recover the HJM limit for the hedging functions given in [52], set the propagator $D(t_*, x; t, T_{FR}) \to 1$. Choose volatility function

$$\sigma_{hjm}(t, x) = \sigma_0 e^{-\beta(x-t)} \tag{9.58}$$

It can be shown that the results in Appendix 9.13 yield

$$A_{hjm} = \frac{\sigma_0^2}{2\beta} \left[1 - e^{-2\beta(t_*-t_0)} \right]$$

$$B_{hjm} = \frac{\sigma_0^2}{2\beta^2} \left[1 - e^{-2\beta(t_*-t_0)} \right] \left[1 - e^{-\beta(T-t_*)} \right]$$

$$C_{hjm} = \frac{\sigma_0^2}{2\beta^3} \left[1 - e^{-2\beta(t_*-t_0)} \right] \left[1 - e^{-\beta(T-t_*)} \right]^2 = \frac{B_{hjm}^2}{A_{hjm}} \tag{9.59}$$

It follows that

$$X_{hjm}(t_*, T) = \frac{B_{hjm}}{A_{hjm}} = \frac{1 - e^{-\beta(T-t_*)}}{\beta} \tag{9.60}$$

and an exact cancellation yields

$$a_{hjm}(t_*, T) = \frac{B_{hjm}^2}{2A_{hjm}}$$

$$= \frac{\sigma_0^2}{4\beta^3} \left[1 - e^{-2\beta(t_*-t_0)} \right] \left[1 - e^{-\beta(T-t_*)} \right]^2 \tag{9.61}$$

and agrees with the result given in [52].

The exponential volatility function given in Eq. (9.58) has a remarkable property that, from Eq. (9.59)

$$\sigma_{\eta|r}^2(hjm) = C_{hjm} - \frac{B_{hjm}^2}{A_{hjm}} \equiv 0$$

Hence Eqs. (9.52) and (A.10) yield

$$P_{hjm}(\eta|r; t_*; T) = \delta(\eta - \eta_0)$$

In other words, the conditional probability for the zero coupon bond is determin-istic, and once the spot rate r is fixed leads to the **identity**

$$B_{hjm}(r; t_*; T) \equiv P(t_*, T) \tag{9.62}$$

Hence, for the volatility function given by $\sigma_2 e^{-\beta(x-t)}$ the zero coupon bond for the HJM model is exactly determined by the spot interest rate. The derivation of the Treasury Bond as a function of r given in [52] for Eqs. (9.60) and (9.61) is ob-tained by a change of variables, and not by evaluating the conditional probability. This change of variables is possible because in the HJM-model there is only one random variable at every instant t driving the forward rates, and for the specific form of volatility given in Eq. (9.58) this random variable can be reduced to the spot rate $r(t)$.

9.15 Appendix: Stochastic hedging with Treasury Bonds

The hedged portfolio and its rate of change are

$$\Pi(t) = P(t, T) + \sum_{i=1}^{N} \Delta_i P(t, T_i); \quad \frac{d\Pi(t)}{dt} = \frac{dP}{dt} + \sum_{i=1}^{N} \Delta_i \frac{dP_i}{dt}$$

For notational simplicity, the bonds $P(t, T_i)$ and $P(t, T)$ are denoted P_i and P respectively. Hence,[7] one has

$$\mathrm{Var}\left[\frac{d\Pi(t)}{dt}\right] = \mathrm{Var}\left[\frac{dP}{dt}\right] + \mathrm{Var}\left[\sum_{i=1}^{N} \Delta_i \frac{dP_i}{dt}\right]$$

$$+ \sum_{i=1}^{N} \Delta_i \left[< \frac{dP}{dt}\frac{dP_i}{dt} > - < \frac{dP}{dt} >< \frac{dP_i}{dt} > \right] \tag{9.63}$$

For starters, consider the variance of an individual bond in the field theory model. The definition $P(t, T) = \exp\left(-\int_t^T dx f(t, x)\right)$ for the zero coupon bond implies

---

[7] Recall notation $< \ldots >$ stands for the expectation value of random variables.

that

$$\frac{dP}{dt} = \left[f(t,t) - \int_t^T dx \frac{\partial f(t,x)}{\partial t} \right] P$$

$$= \left[r(t) - \int_t^T dx \alpha(t,x) - \int_t^T dx \sigma(t,x) A(t,x) \right] P$$

$$= < \frac{dP}{dt} > -P \int_t^T dx \sigma(t,x) A(t,x) \tag{9.64}$$

since $P = P(t,T)$ is deterministic and $< A(t,x) > = 0$. Therefore

$$\frac{dP}{dt} - < \frac{dP}{dt} > = -P \int_t^T dx \sigma(t,x) A(t,x) \tag{9.65}$$

Squaring this expression and using the result that

$$< A(t,x) A(t,x') > = \delta(0) D(x,x';t,T_{FR}) = \frac{1}{\epsilon} D(x,x';t,T_{FR}) \tag{9.66}$$

yields the instantaneous bond price variance

$$\text{Var}\left[\frac{dP}{dt} \right] = \frac{1}{\epsilon} P^2 \int_t^T dx \int_t^T dx' \sigma(t,x) D(x,x';t,T_{FR}) \sigma(t,x') \tag{9.67}$$

For a portfolio of bonds, $\hat{\Pi}(t) = \sum_{i=1}^N \Delta_i P_i$, similar to the case of the bond, the following result holds

$$\frac{d\hat{\Pi}(t)}{dt} - < \frac{d\hat{\Pi}(t)}{dt} > = - \sum_{i=1}^N \Delta_i P_i \int_t^{T_i} dx \sigma(t,x) A(t,x) \tag{9.68}$$

and the correlation of the bonds being used to hedge yields

$$\text{Var}\left[\frac{d\hat{\Pi}(t)}{dt} \right] = \frac{1}{\epsilon} \sum_{i=1}^N \sum_{j=1}^N \Delta_i \Delta_j M_{ij}$$

with the hedging matrix given by

$$M_{ij} = P_i P_j \int_t^{T_i} dx \int_t^{T_j} dx' \sigma(t,x) \sigma(t,x') D(x,x';t,T_{FR}) \tag{9.69}$$

The cross correlation between the bond being hedged and the hedging bonds is

$$\sum_{i=1}^N \Delta_i \left[< \frac{dP}{dt} \frac{dP_i}{dt} > - < \frac{dP}{dt} > < \frac{dP_i}{dt} > \right] = \frac{2}{\epsilon} \sum_{i=1}^N \Delta_i L_i$$

with the hedging vector given by

$$L_i = P_i P \int_t^T dx \int_t^{T_i} dx' \sigma(t, x) \sigma(t, x') D(x, x'; t, T_{FR})$$

The (residual) variance of the hedged portfolio

$$\Pi(t) = P(t, T) + \sum_{i=1}^N \Delta_i P(t, T_i)$$

may now be computed in a straightforward manner. From equation (9.63) the instantaneous change in the variance of the hedged portfolio equals the following (ignoring the overall factor of $1/\epsilon$)

$$P^2 \int_t^T dx \int_t^T dx' \sigma(t, x) \sigma(t, x') D(x, x'; t, T_{FR})$$

$$+ 2 \sum_{i=1}^N \Delta_i L_i + \sum_{i=1}^N \sum_{j=1}^N \Delta_i \Delta_j M_{ij} \qquad (9.70)$$

The hedge parameters Δ_i that minimize the residual variance in equation (9.70) are obtained by differentiating equation (9.70) with respect to Δ_i, and setting the result to zero; this is subsequently solved for Δ_i and yields

$$\Delta_i = - \sum_{j=1}^N M_{ij}^{-1} L_j \qquad (9.71)$$

and represents the optimal amounts of $P(t, T_i)$ to include in the hedge portfolio when hedging $P(t, T)$.

For $N = 1$, the hedge parameter in Eq. (9.71) reduces to

$$\Delta_1 = - \frac{L_1}{M_{11}} = - \left(\frac{\int_t^T dx \int_t^{T_1} dx' \sigma(t, x) \sigma(t, x') D(x, x'; t, T_{FR})}{\int_t^{T_1} dx \int_t^{T_1} dx' \sigma(t, x) \sigma(t, x') D(x, x'; t, T_{FR})} \right) \frac{P}{P_1} \qquad (9.72)$$

Substituting the value of Δ_i given in Eq. (9.71) into Eq. (9.70) yields the variance of the hedged portfolio (ignoring the overall factor of $1/\epsilon$)

$$V = P^2 \int_t^T dx \int_t^T dx' \sigma(t, x) \sigma(t, x') D(x, x'; t, T_{FR}) - \sum_{i=1}^N \sum_{j=1}^N L_i M_{ij}^{-1} L_j$$

$$(9.73)$$

and which declines monotonically as N increases.

The residual variance in Eq. (9.73) enables the effectiveness of the hedge portfolio to be evaluated. Therefore, Eq. (9.73) is the basis for studying the effect of including different bonds in the hedge portfolio.

9.16 Appendix: Stochastic hedging with futures contracts

The material in (previous) Appendix 9.15 allows the hedging properties of futures contracts on bonds to be studied. Proceeding as before, the appropriate hedge parameters for futures contracts expiring at time t_F (taken to be one year in the empirical study) are computed. The futures price $\mathcal{F}(t, t_F, T)$, from Eq. (9.1), is given by

$$\mathcal{F}(t, t_F, T) = F(t, t_F, T)e^{\Omega_\mathcal{F}} = e^{-\int_{t_F}^T dx f(t,x)} e^{\Omega_\mathcal{F}}$$

where $F(t, t_F, T)$ is the forward price and $\Omega_\mathcal{F}(t, t_F, T)$ is a deterministic quantity. The dynamics of the futures price $\mathcal{F}(t, t_F, T)$ is given by

$$\frac{d\mathcal{F}(t, t_F, T)}{dt} = \left[\frac{d\Omega_\mathcal{F}(t, t_F, T)}{dt} - \int_{t_F}^T dx \frac{\partial f(t, x)}{\partial t} \right] \mathcal{F}(t, t_F, T) \qquad (9.74)$$

which implies

$$\frac{d\mathcal{F}(t, t_F, T)}{dt} - < \frac{d\mathcal{F}(t, t_F, T)}{dt} > = -\mathcal{F}(t, t_F, T) \int_{t_F}^T dx \sigma(t, x) A(t, x) \qquad (9.75)$$

since $< A(t, x) > = 0$. Squaring both sides leads, from Eq. (9.66), to the instantaneous variance of the futures price

$$\text{Var}\left[\frac{d\mathcal{F}(t, t_F, T)}{dt} \right] = \frac{1}{\epsilon} \mathcal{F}^2(t, t_F, T) \int_{t_F}^T dx \int_{t_F}^T dx' \sigma(t, x) D(x, x) \sigma(t, x') \qquad (9.76)$$

The following definitions are the futures contracts analogs of the results of the hedging by bonds. Let $\mathcal{F}_i$ denote the futures price $\mathcal{F}(t, t_F, T_i)$ of a contract expiring at time t_F on a zero coupon bond maturing at time T_i. The hedged portfolio in terms of the futures contract is given by

$$\Pi(t) = P + \sum_{i=1}^N \Delta_i \mathcal{F}_i \qquad (9.77)$$

where $\mathcal{F}_i$ represent observed market prices. For notational simplicity, define the following terms

$$\tilde{L}_i = P\mathcal{F}_i \int_{t_F}^{T_i} dx \int_t^T dx' \sigma(t, x) D(x, x'; t, T_{FR}) \sigma(t, x')$$

$$\tilde{M}_{ij} = \mathcal{F}_i \mathcal{F}_j \int_{t_F}^{T_i} dx \int_{t_F}^{T_j} dx' \sigma(t, x) D(x, x'; t_F, T_{FR}) \sigma(t, x')$$

The hedge parameters and the residual variance, when futures contracts are used as the underlying hedging instruments, have identical expressions to those in Eqs. (9.71) and (9.73) but are based on Eq. (9.77).

Hedge parameters, similar to the case of hedging with bonds, are given by [12]

$$\Delta_i = -\sum_{j=1}^{N} \tilde{M}_{ij}^{-1} \tilde{L}_j$$

while the residual variance of the hedged portfolio equals [12]

$$V = P^2 \int_t^T dx \int_t^T dx' \sigma(t, x) \sigma(t, x') D(x, x'; t, T_{FR}) - \sum_{i=1}^{N} \sum_{j=1}^{N} \tilde{L}_i \tilde{M}_{ij}^{-1} \tilde{L}_j$$

for $\tilde{L}_i$ and $\tilde{M}_{ij}$ given in Eq. (9.78).

The proof follows directly from repeating the derivations for the hedging with Treasury Bonds.

9.17 Appendix: HJM limit of the hedge parameters

The HJM limit of the hedge parameters Δ_1 and n_1 is analyzed for the specific exponential volatility function that is considered by Jarrow and Turnbull [52].

The HJM limit is taken by $D(x, x'; t, T_{FR}) \to 1$. Hence, Eq. (9.57) gives n_1 as

$$n_1 \to - \left[\frac{\int_t^T dx \sigma(x, t)}{\int_t^{T_1} dx \sigma(x, t)} \right] \frac{B(r; t, T)}{B(r; t, T_1)}$$

For the hedge parameter Δ_1, Eq. (9.72) reduces to

$$\Delta_1 \to - \left[\frac{\int_t^T dx \int_t^{T_1} dx' \sigma(t, x) \sigma(t, x')}{\left(\int_t^{T_1} dx \sigma(t, x) \right)^2} \right] \frac{P}{P_1} = - \left[\frac{\int_t^T dx \sigma(t, x)}{\int_t^{T_1} dx \sigma(t, x)} \right] \frac{P}{P_1} \quad (9.78)$$

Δ_1 is similar to n_1 except that the ratio P/P_1 appears in Δ_1 instead of B/B_1. For the exponential volatility function $\sigma(t, x) = \sigma_2 e^{-\beta(x-t)}$ Eq. (9.62) leads to $B(r; t, T) = P(t, T)$, and hence $\Delta_1 = n_1$.

The reason that stochastic delta hedging and minimization of residual variance give the same result for the exponential volatility HJM model is because for this choice of volatility the variance of the conditional probability is zero; hence the minimum for the residual volatility is automatically satisfied by stochastic delta hedging.

The following explicit calculation shows how the field theory result encoded in Δ_1 is equivalent to stochastic delta hedging in the HJM model. Under the assumption of exponential volatility, Eq. (9.78) becomes

$$\Delta_1 = - \left(\frac{1 - e^{-\beta(T-t)}}{1 - e^{-\beta(T_1-t)}} \right) \frac{P}{P_1}$$

In terms of the function $X_{hjm}(t, T) = (1 - e^{-\beta(T-t)})/\beta$ given in Eq. (9.60), the following holds

$$\Delta_1 = - \left(\frac{X_{hjm}(t, T)}{X_{hjm}(t, T_1)} \right) \frac{P(t, T)}{P(t, T_1)} \tag{9.79}$$

For emphasis, note that the following delta hedging equation holds in a one factor HJM model

$$\frac{\partial [P(t, T) + \Delta_1 P(t, T_1)]}{\partial r(t)} = 0 \tag{9.80}$$

The equation above can be verified using equation Eqs. (9.79) and (9.54) (recall for the HJM model with exponential volatility $B(r; t, T) = P(t, T)$ as shown in Eq. (9.62)) since

$$\frac{\partial [P(t, T) + \Delta_1 P(t, T_1)]}{\partial r(t)} = -P(t, T)X_{hjm}(t, T) - \Delta_1 P(t, T_1)X_{hjm}(t, T_1)$$

$$= -P(t, T)X_{hjm}(t, T) + P(t, T)X_{hjm}(t, T) = 0$$

10

Field theory Hamiltonian of forward interest rates

The Hamiltonian formulation of quantum field theory is equivalent to, and independent of, its formulation based on the Feynman path integral and the Lagrangian. There are many advantages of having multiple formulations of the same theory, since for some calculations the Hamiltonian formulation may be more transparent and calculable than the Lagrangian formulation.

The path-integral formulation of the forward interest rates, discussed in some detail Chapter 7, is useful for calculating the expectation values of the quantum fields. To study questions related to the time evolution of quantities of interest, one needs to derive the Hamiltonian for the system from its Lagrangian. This route is the opposite to the one taken in Chapter 5 where the Lagrangian for option pricing was derived starting from the Hamiltonian formulation [5].

Many of the derivations in Chapter 7 that are feasible for Gaussian field theories cannot be replicated for nonlinear field theories, but which, in some cases, are nevertheless tractable in the Hamiltonian formulation. In particular, the risk-neutral martingale measure for the linear theory of the forward rates was derived by performing a Gaussian path integral, and this derivation is no longer tractable for nonlinear forward rates.

A rather remarkable result for the theory of nonlinear forward rates is that the martingale condition can be solved by generalizing the infinitesimal formulation of the condition for the martingale measure that was discussed for the case of a single security in Section 4.7.

The generator of infinitesimal time evolution of the forward interest rates, namely the Hamiltonian, is obtained for both the linear and nonlinear forward interest rates, as well as for the case of stochastic volatility. To obtain the Hamiltonian it is necessary to first define the underlying state space of the forward rates' quantum field. The state space of the forward rates is shown to be time dependent – moving with a constant 'velocity' – and this in turn yields a time-dependent Hamiltonian for the forward interest rates.

The Hamiltonian formulation of the martingale condition for the forward interest rates yields an exact solution of the risk-neutral measure for the case of nonlinear forward rates, and for the case of forward rates with stochastic volatility.

The change of numeraire for nonlinear forward rates is solved using the forward rates' Hamiltonian. The pricing kernel for the forward rates as well as the European bond option price is derived to illustrate the utility of the Hamiltonian.

10.1 Forward interest rates' Hamiltonian

The following are a number of general features of the Hamiltonian's derivation from a Lagrangian that were discussed in Appendix 6.11, and that are also valid for the derivation of the forward rates' Hamiltonian.[1]

- Implicit in going from Lagrangian to the Hamiltonian for the Black–Karasinski model, as seen in Eqs. (6.82) and (6.83), is the use of the completeness equation for the ϕ degree of freedom, namely that $\int d\phi |\phi><\phi| = \mathcal{I}$. This feature will turn out to be more complicated for the forward rates.
- The representation of $e^{\epsilon L(n)}$ by means of a Gaussian integration is a necessary step for the derivation, but will be more complicated for the forward rates.
- The integration measure term $e^{-\nu \phi_n}$ in Eq. (7.59) is included in the definition of the nonlinear forward rates' path integral, similar to Eq. (6.83), and is necessary for obtaining a well-defined Hamiltonian.
- The Hamiltonian and the state space of the forward rates are two independent ingredients of the quantum theory, as was discussed in general terms in Section 4.1; taken together they reproduce the various forward rates' linear and nonlinear actions discussed in Chapter 7.

One would like to represent the forward rates' path integral given in Eq. (7.50), in analogy with Eq. (6.81), as the matrix element of the Hamiltonian in the following manner

$$ Z = \int Df e^{S[f]} =? =< f_{\text{initial}}|e^{-(T_f - T_i)\mathcal{H}}|f_{\text{final}} > \quad : \text{ incorrect} \quad (10.1) $$

where

$$ S[f] = \int_{\mathcal{P}} \mathcal{L}[f] $$

$$ \int Df \equiv \prod_{(t,x)\epsilon\mathcal{P}} \int_{-\infty}^{+\infty} df(t,x) $$

Recall the domain $\mathcal{P}$ of the forward rates is given in Figure 7.3.

---

[1] It would be useful for a reader unfamiliar with the derivation of a Hamiltonian from a Lagrangian to review Appendix 6.11.

The expression for $\mathcal{H}$ in Eq. (10.1) is valid only for Hamiltonians that do not explicitly depend on the parameter of time t. This condition of time independence does **not** hold for the forward rates' Hamiltonian due to the parallelogram shape of the domain $\mathcal{P}$ on which the theory is defined; the shape of the domain $\mathcal{P}$ imposes the condition that for every instant of time t the state space changes; there is a **different** state space for each instant t, and a different (time-dependent) Hamiltonian $\mathcal{H}(t)$ acting on this time-dependent state space.

Hence one cannot mechanically follow the procedure used in deriving the Black–Karasinski Hamiltonian, since in that case both the Hamiltonian $\mathcal{H}_{BK}$ and the completeness equation are time–independent. Instead, in order to derive the forward rates Hamiltonian $\mathcal{H}(t)$ from the action $S[f]$ one first needs to examine the state space of the forward rates before one can proceed any further.

10.2 State space for the forward interest rates

The Lagrangian for the forward rates given in Eq. (7.60) has derivatives in time of the form $(\partial f(t, x)/\partial t)^2$, and hence an infinitesimal generator, namely the Hamiltonian $\mathcal{H}$ exists for it. As discussed in Section 10.1, in order to obtain the Hamiltonian for the forward rates, one needs to specify the state space on which it acts, and the rest of this section will be spent in deriving this state space.

This section is rather technical, but the final result is quite intuitive and simple.

For notational brevity, the forward rates quantum field $f(t, x)$ is taken to represent both the forward rates' quantum fields $f(t, x)$ and the stochastic volatility quantum field $h(t, x)$.

The state space of a field theory, similar to all quantum systems as discussed in Section 4.1, is a linear vector space – denoted by $\mathcal{V}$ – that consists of functionals of the field configurations at some fixed time t. From Section 4.1 the dual space of $\mathcal{V}$ – denoted by $\mathcal{V}_{Dual}$ – consists of all linear mappings from $\mathcal{V}$ to the complex numbers, and is also a linear vector space. Recall that the Hamiltonian $\mathcal{H}$ is an operator – the quantum analog of energy – that is an element of the tensor product space $\mathcal{V} \otimes \mathcal{V}_{dual}$.

Obtaining the Hamiltonian for the forward rates will turn out to be a complicated exercise due to the non-trivial structure of the underlying domain $\mathcal{P}$ given, as in Figure 7.4, with $x \geq t$. In particular, it will be seen that the forward rates' quantum field has a distinct state space $\mathcal{V}_t$ for every instant t.

For greater clarity, discretize both time and maturity time into a finite lattice, with lattice spacing in both directions taken to be ϵ. [2] On the lattice the minimum time for futures contract is time ϵ; for most applications $\epsilon =$ one day. The points comprising the discrete domain $\tilde{\mathcal{P}}$ are shown in Figure 10.1.

---

[2] For a string moving with 'velocity' v, the maturity lattice would have spacing of $v\epsilon$; for the forward rates $v = 1$.

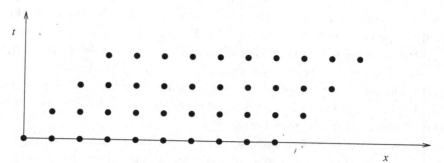

Figure 10.1 Lattice in time and maturity directions

The (discrete) lattice domain $\tilde{\mathcal{P}}$ has been discussed Appendix 7.18, and is given by

$$(t, x) \rightarrow \epsilon(n, l) \; ; \; n, l : \text{ integers}$$
$$(T_i, T_f, T_{FR}) \rightarrow \epsilon(N_i, N_f, N_{FR})$$
$$\text{Lattice } \tilde{\mathcal{P}} = \{(n, l)|N_i \leq n \leq N_f \; ; \; n \leq l \leq (n + N_{FR})\} \quad (10.2)$$
$$f(t, x) \rightarrow f_{n,l}$$
$$\frac{\partial f(t, x)}{\partial t} \simeq \frac{f_{n+1,l} - f_{n,l}}{\epsilon} \; ; \quad \frac{\partial f(t, x)}{\partial x} \simeq \frac{f_{n,l+1} - f_{n,l}}{\epsilon} \quad (10.3)$$

The partition function is now given by a finite multiple integral, namely

$$Z = \prod_{(n,l)\in\tilde{\mathcal{P}}} \int df_{n,l} e^{S[f]} \quad (10.4)$$

$$S[f] = \sum_n S(n) \quad (10.5)$$

Consider two adjacent time slices labelled by n and $n+1$, as shown in Figure 10.2. $S(n)$ is the action connecting the forward rates of these two time slices.

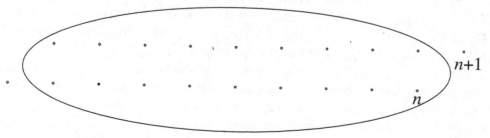

Figure 10.2 Two consecutive time slices for $t = n\epsilon$ and $t = (n + 1)\epsilon$

As can be seen from Figure 10.2, for the two time slices there is a mismatch of the two-lattice sites on the edges, namely, lattice sites (n, n) at time n and $(n+1, n+1+N_{FR})$ at time $n+1$ are not in common. Isolating the un-matched variables yields the following

Variables at time n :

$\{f_{n,n}, F_n\}$; $F_n \equiv \{f_{n,l}|n+1 \leq l \leq n+N_{FR}\}$

Variables at time $(n+1)$:

$\{F_{n+1}, f_{n+1,n+1+N_{FR}}\}$; $F_{n+1} \equiv \{f_{n+1,l}|n+1 \leq l \leq n+N_{FR}\}$ (10.6)

The variables F_n refer to time n, and variables F_{n+1} refer to later time $n + 1$. From Figure 10.2 it can be seen that both **sets of variables** F_{n+1} and F_n cover the **same** lattice sites in the maturity direction, namely $n + 1 \leq l \leq n + N_{FR}$, and hence have the same number of forward rates, namely $N_{FR} - 1$. It will be shown that the Hamiltonian is expressible solely in terms of these variables.

From the discretized time derivatives defined in Eq. (10.3) the discretized action $S(n)$ contains terms that couple only the common points in the lattice for the two time slices, namely the variables belonging to the sets F_n; F_{n+1}. Hence the action is given by the following Langrangian density $\mathcal{L}_n$

$$S(n) = \epsilon \sum_{\{l\}} \mathcal{L}_n[f_{n,l}, f_{n+1,l}] \qquad (10.7)$$

$$= \epsilon \sum_{\{l\}} \mathcal{L}_n[F_n; F_{n+1}] \qquad (10.8)$$

As shown in Figure 10.3, the action for the entire domain $\tilde{\mathcal{P}}$ shown in Figure 10.1 can be constructed by repeating the construction given in Figure 10.2 and summing over the action $S(n)$ over all time $N_i \leq n \leq N_f$.

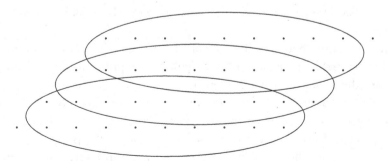

Figure 10.3 Reconstructing the lattice from the two time slices

The Hamiltonian of the forward rates is an operator that acts on the space of states of the forward rates; one hence needs to determine the co-ordinates of the state space.

Consider again the two consecutive time slices n and $n+1$ given in Figure 10.2. The forward rates for two adjacent instants, namely $\{f_{n,n}, F_n\}$ and $\{F_{n+1}, f_{n+1,n+1+N_{FR}}\}$ given in Eq. (10.6) are interpreted as the co-ordinates of the state spaces $\mathcal{V}_n$ and $\mathcal{V}_{n+1}$ respectively.

For every instant of time n there is a distinct state space $\mathcal{V}_n$, and its dual $\mathcal{V}_{\text{Dual},n}$. The co-ordinates of the state spaces $\mathcal{V}_n$ and $\mathcal{V}_{n+1}$ are given by the tensor product of the space of state for every maturity point l, namely

$$< f_n| = \bigotimes_{n \leq l \leq n+N_{FR}} < f_{n,l}| \equiv < f_{n,n}| < F_n|$$

: co-ordinate state of $\mathcal{V}_{\text{dual},n}$

$$|f_{n+1}> = \bigotimes_{(n+1) \leq l \leq n+1+N_{FR}} |f_{n+1,l} > \equiv |F_{n+1} > |f_{n+1,n+1+N_{FR}} >$$

: co-ordinate state of $\mathcal{V}_{n+1}$

$$(10.9)$$

The state space $\mathcal{V}_n$ consists of all possible functions of N_{FR} forward rates $\{f_{n,n}, F_n\}$. The state spaces $\mathcal{V}_n$ differ for different n by the fact that a different set of forward rates comprise their set of independent variables.

Although the state spaces $\mathcal{V}_n$ and $\mathcal{V}_{n+1}$ are not the identical, there is an intersection of these two spaces, namely $\mathcal{V}_n \cap \mathcal{V}_{n+1}$ that covers the same interval in the maturity direction, and is coupled by the action $S(n)$. The intersection yields the **physical state space**, namely $\mathcal{F}_n$, on which the Hamiltonian evolution of the forward rates takes place.

The choice of what constitutes the state space, and what is its dual, will determine the direction of propagation. In the discussion (in a simpler setting) for the Fokker–Planck Hamiltonian in Appendix 6.8, it was pointed out that in Dirac's notation the ket vector represents $|$starting state $>$ and the bra (dual) vector represents $<$ ending state$|$.

The choice that is made for the forward rates' state space and Hamiltonian is dictated by the fact that in finance time evolution is primarily for the purpose of **discounting** the value of an instrument from the future to its present day value. Hence evolution should go backwards; in keeping with this interpretation of the Hamiltonian, the later time state space $\mathcal{V}_{n+1}$ is chosen to contain $|$starting state $>$ and the earlier time state space is chosen to be the dual $\mathcal{V}_{\text{Dual},n}$ containing the $<$ ending state$|$. This choice of the state space yields an interest rates **backward Hamiltonian** propagating the system backwards in time, from the future into the past.[3] In symbols

$$\mathcal{V}_{n+1} = \mathcal{F}_{n+1} \otimes |f_{n+1,n+1+N_{FR}} >$$
$$\mathcal{V}_{\text{Dual},n} = < f_{n,n}| \otimes \mathcal{F}_{\text{dual},n}$$
$$\mathcal{H}_n : \mathcal{F}_{n+1} \rightarrow \mathcal{F}_{n+1} \Rightarrow \mathcal{H}_n \in \mathcal{F}_{n+1} \otimes \mathcal{F}_{\text{dual},n+1}$$

The Hamiltonian $\mathcal{H}_n$ is an element on the tensor product space spanned by the operators $|F_{n+1} >< F_{n+1}|$, namely the space of operators given by $\mathcal{F}_{n+1} \otimes \mathcal{F}_{\text{Dual},n+1}$. The state spaces $\mathcal{F}_n$ and $\mathcal{F}_{n+1}$ are isomorphic (identical), and the index n is required to track where in the domain $\tilde{\mathcal{P}}$ they are located. Hence both states $|F_{n+1} >$ and $< F_n|$ belong to the same state space $\mathcal{F}_n$ and its dual. Note as one proceeds to different values of n the state space changes, and hence $\mathcal{F}_n$ is time-dependent.

The vector spaces $\mathcal{V}_n$ and the Hamiltonian $\mathcal{H}_n$ acting on these spaces is shown in Figure 10.2.

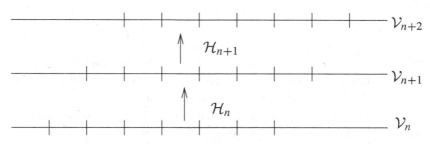

Figure 10.4 Hamiltonians $\mathcal{H}_n$ propagating the space of forward rates $\mathcal{V}_n$

As one scans through all possible values for the forward rates $\{f_{nl}|n \leq l \leq n + N_{FR}\}$, one obtains a complete basis for the state space $\mathcal{V}_n$. In particular, the resolution of the identity operator for $\mathcal{V}_n$ – denoted by $\mathcal{I}_n$ – is a reflection that the basis states are complete, and is given by [3]

$$\mathcal{I}_n = \prod_{n \leq l \leq n+N_{FR}} \int df_{n,l} |f_n >< f_n|$$

$$\equiv \int df_{n,n} \, DF_n \, |f_{n,n}; F_n >< f_{n,n}; F_n| \tag{10.10}$$

The Hamiltonian of the system $\mathcal{H}_n$ is defined by the Feynman formula (up to a normalization), from Eq. (10.7), as

$$\rho_n e^{\epsilon \sum_{\{l\}} \mathcal{L}_n[f_{n,l}, f_{n+1,l}]} = < f_{n,n}, F_n |e^{-\epsilon \mathcal{H}_n}|F_{n+1}, f_{n+1,n+1+N_{FR}} > \tag{10.11}$$

---

[3] Making the other choice for the state space and its dual will lead to a Hamiltonian that is the Hermitian conjugate of what is obtained, and in effect would reverse the sign of the drift velocity.

where in general ρ_n is a field-dependent measure term. In the discrete action given in Eq. (10.8) that connects the two time slices, as shown in Figure 10.2, the unmatched forward rates $f_{n,n}$ and $f_{n+1,n+1+N_{FR}}$ are decoupled from the action, and therefore

$$\rho_n e^{\epsilon \sum_{\{l\}} \mathcal{L}_n[F_n, F_{n+1}]} = \; < f_{n,n}, F_n | e^{-\epsilon \mathcal{H}_n} | F_{n+1}, f_{n+1,n+1+N_{FR}} > \qquad (10.12)$$

$$= \; < F_n | e^{-\epsilon \mathcal{H}_n} | F_{n+1} > \qquad (10.13)$$

Equation (10.13) is the main result of this section.

It is worth emphasizing that Eq. (10.13) follows from Eq. (10.12) because the action $S(n)$ connecting time slices n and $n+1$ does **not** contain the variables $f_{n,n}$ and $f_{n+1,n+1+N_{FR}}$ respectively. This leads to the result that the Hamiltonian $\mathcal{H}_n$ consequently does not depend on these variables.

The interpretation of Eq. (10.13) is that the Hamiltonian $\mathcal{H}_n$ propagates the 'initial' state $|F_{n+1}>$ backward through time interval ϵ to the 'final' state $< F_n|$. The relation

$$< f_{n,n}, F_n | e^{-\epsilon \mathcal{H}_n} | F_{n+1}, f_{n+1,n+1+N_{FR}} > = < F_n | e^{-\epsilon \mathcal{H}_n} | F_{n+1} > \qquad (10.14)$$

shows that there is an asymmetry in the time direction, with the Hamiltonian being **independent** of the **earliest** forward rate $f_{n,n}$ and of the **latest** forward rate $f_{n+1,n+1+N_{FR}}$. It is this **asymmetry** in the propagation of the forward rates which yields the parallelogram domain $\mathcal{P}$ given in Figure 10.1, and reflects the asymmetry that the forward rates $f(t, x)$ exist only for $x > t$.

Putting in the co-ordinates of stochastic volatility, Eq. (10.13) gives the following

$$\rho_n e^{S(n)} = \rho_n e^{\epsilon \sum_{\{l\}} \mathcal{L}_n[H_{n+1}, H_n; F_{n+1}, F_n]} \qquad (10.15)$$

$$= \; < F_n; H_n | e^{-\epsilon \mathcal{H}_n} | F_{n+1}; H_{n+1} > \qquad (10.16)$$

where the volatility quantum field $h(t, x)$ has been explicitly included in the equation above.

A general expression for the Hamiltonian has been obtained in terms of the action S as given in Eq. (10.13), and this formula needs to be applied to the case of a specific Lagrangian of the forward interest rates to obtain the expression for its Hamiltonian $\mathcal{H}$.

Continuum notation

For notational simplicity, consider the maturity direction x to be continuous, and let the time direction be discrete. In the continuum notation, the subtleties of the variables at time t and $t + \epsilon$ are accounted for by carefully analyzing the variables appearing on the **boundaries** of the interval $[t \leq x \leq t + T_{FR}]$. The action $S(n)$

for time $t = n\epsilon$, is given by the following

$$S(n) = \epsilon \int_x \mathcal{L}_n(t, x) \tag{10.17}$$

$$\int_x \equiv \int_t^{t+T_{FR}} dx \tag{10.18}$$

In continuum notation the state space is labelled by $\mathcal{V}_t$, and state vector by $|f_t >$. The elements of the state space of the forward rates $\mathcal{V}_t$ includes **all** possible financial instruments that are traded in the market at time t. In continuum notation, from Eq. (10.9)

$$|f_t > = \dot{\bigotimes_{t \leq x \leq t+T_{FR}}} |f(t, x) >$$

$$|F_t > = \bigotimes_{t \leq x < t+T_{FR}} |f(t, x) > \tag{10.19}$$

The only difference between state vectors $|f_t >$ and $|F_t >$ is that, in Eq. (10.19), the point $x = t + T_{FR}$ is excluded in the continuous tensor product for $|F_t >$.

The partition function Z given in Eq. (10.4) can be reconstructed from the Hamiltonian and state space by recursively applying the procedure discussed for the two time slices. In continuum notation, this yields

$$Z = \int Df e^{S[f]}$$

$$= < f_{\text{initial}} | \mathcal{T} \left\{ \exp - \int_{T_i}^{T_f} \mathcal{H}(t) \, dt \right\} | f_{\text{final}} > \tag{10.20}$$

where the symbol $\mathcal{T}$ in the equation above stands for time ordering the time-dependent state space and Hamiltonian operator in the argument, with the earliest time being placed to the left.

To recapitulate: the time-dependent Hamiltonian $\mathcal{H}(t)$ propagates the interest rates **backwards** in time, taking the final state $|f_{\text{final}} >$ given at time T_f backwards to an initial state $< f_{\text{initial}}|$ at the earlier time T_i. The Hamiltonian propagates the system backwards in time due to the choice made for the forward interest rates' state space and its dual.

Although Eq. (10.20) looks superficially similar to Eq. (10.1), the two are very different. Eq. (10.1) is an appropriate expression for a Hamiltonian that is defined on a time-independent state space that corresponds to a rectangular domain $\mathcal{R}$, given in Figure 9.1, for the corresponding path integral, whereas Eq. (10.20) generates the trapezoidal domain $\mathcal{T}$, given in Figure 7.3, as the domain of its path integral.

10.3 Treasury Bond state vectors

The coupon and zero coupon bond are important state vectors in the theory of forward interest rates.

Consider a risk-free zero coupon Treasury Bond that matures at time T with a payoff of \$1. Recall from Eq. (2.8) that the price of a zero coupon bond at time $t < T$ is given by

$$P(t, T) = e^{-\int_t^T f(t,x)dx} \equiv P[f; t, T]$$

The ket state vector $|P(t, T) >$ is an element of the state space $\mathcal{V}_t$. The bond state vector is written as follows

$$P(t, T) \equiv < f_t | P(t, T) >= e^{-\int_t^T f(t,x)dx} \qquad (10.21)$$

The coupon bond $|\mathcal{B} >$ is a state vector, with fixed payoffs of amount c_l at time T_l, and with a final payoff of L at time T, and is represented by a linear superposition of the zero coupon bonds given by

$$|\mathcal{B}(t) >= \sum_l c_l | P(t, T_l) > +L | P(t, T) >$$

10.4 Hamiltonian for linear and nonlinear forward rates

To derive the Hamiltonian for the linear (Gaussian) forward rates one needs to repeat all the steps discussed in Appendix 6.11. For simplicity, the case of constant rigidity is analyzed. From Eq. (7.6) the Lagrangian density $\mathcal{L}[f]$ is given by

$$\mathcal{L}[f] = \mathcal{L}_{kinetic}[f] + \mathcal{L}_{rigidity}[f] \qquad (10.22)$$

$$= -\frac{1}{2}\left[\left\{\frac{\frac{\partial f(t,x)}{\partial t} - \alpha(t, x)}{\sigma(t, x)}\right\}^2 + \frac{1}{\mu^2}\left\{\frac{\partial}{\partial x}\left(\frac{\frac{\partial f(t,x)}{\partial t} - \alpha(t, x)}{\sigma(t, x)}\right)\right\}^2\right]$$

$$-\infty \le f(t, x) \le +\infty \qquad (10.23)$$

On discretizing the Lagrangian, and using the Neumann boundary conditions given in Eq. (7.10) yields

$$S(n) = \epsilon \int_x \mathcal{L}_n = -\frac{1}{2\epsilon}\int_x A\left(1 - \frac{1}{\mu^2}\frac{\partial^2}{\partial x^2}\right) A \qquad (10.24)$$

$$A = \sigma^{-1}(t, x)(f_{t+\epsilon} - f_t - \epsilon\alpha_t)(x) \qquad (10.25)$$

where $f(t, x) \equiv f_t(x)$ has been written to emphasize that time t is a parameter in the Hamiltonian formulation of the forward rates.

Eq. (10.24) is re-written using Gaussian integration and (ignoring henceforth irrelevant constants), using notation $\prod_x \int dp(x) \equiv \int Dp$, yields

$$e^{S(n)} = \int Dp e^{-\frac{\epsilon}{2}\int_{x,x'} p(x)D(x,x';t)p(x')+i\int_x p(x)A(x)} \tag{10.26}$$

$$\left(1 - \frac{1}{\mu^2}\frac{\partial^2}{\partial x^2}\right)D(x,x';t) = \delta(x-x') \; ; \; \text{Neumann B.C.'s}$$

with the propagator $D(x,x';t)$ given in Eq. (7.27). Re-scaling the variable $p(x)$ by $p(x) \to \sigma(t,x)p(x)$ gives (up to a constant)

$$e^{S(n)} = \int Dp e^{i\int_x p(x)(f_{t+\epsilon}-f_t-\epsilon\alpha_t)(x)-\frac{\epsilon}{2}\int_{x,x'}\sigma(t,x)p(x)D(x,x';t)\sigma(t,x')p(x')} \tag{10.27}$$

Recall from Eq. (10.13) that the Hamiltonian is defined by

$$\rho_n e^{S(n)} = < f_t|e^{-\epsilon\mathcal{H}_f}|f_{t+\epsilon}> \tag{10.28}$$

$$= e^{-\epsilon\mathcal{H}_n(\delta/\delta f_t)} \int Dp e^{i\int_x p(f_t-f_{t+\epsilon})} \tag{10.29}$$

where for linear forward rates $\rho_n = 1$. The Hamiltonian is written as a (functional) derivative in the co-ordinates of the **dual** state space variables f_t for reasons discussed in the derivation of Eq. (6.86). Since there are infinitely many independent forward rates (degrees of freedom) for each instant of time, represented by the collection of variables $f_t(x), x \in [t, t + T_{FR}]$ one needs to use functional derivatives, discussed in Eq. (A.15), to represent the Hamiltonian as a differential operator.

The degrees of freedom $f(x)$ (or equivalently $f(\theta)$) refer to time t only through the domain on which the Hamiltonian is defined. Unlike the action $S[f]$ that spans all instants of time from the initial to the final time, the Hamiltonian is an infinitesimal generator in time, and refers only to the instant of time at which it acts on the state space. This is the reason that in the Hamiltonian the time index t can be dropped for the variables $f_t(x)$, with $f(x)$, $t \leq x \leq t + T_{FR}$. Hence, the Hamiltonian for the linear forward rates, similar to Eq. (6.86), is given by

$$\mathcal{H}_f(t) = -\frac{1}{2}\int_t^{t+T_{FR}} dxdx'\sigma(t,x)D(x,x';t)\sigma(t,x')\frac{\delta^2}{\delta f(x)\delta f(x')}$$

$$-\int_t^{t+T_{FR}} dx\alpha(t,x)\frac{\delta}{\delta f(x)} \tag{10.30}$$

The Hamiltonian is non-Hermitian, as is typical for the the case of finance. The European bond option and pricing kernel are derived in Appendix 10.13 using this Hamiltonian.

Psychological future time

For psychological future time given by nonlinear maturity variable $z = z(\theta)$; $\theta = x - t = \theta(z)$, the Hamiltonian is given by

$$
\mathcal{H}_{f,z}(t) = -\frac{1}{2}\int_{z(0)}^{z(T_{FR})} dz dz' \sigma(t,z)G(z,z')\sigma(t,z')\frac{\delta^2}{\delta f(\theta(z))\delta f(\theta(z'))}
$$

$$
- \int_{z(0)}^{z(T_{FR})} dz\alpha(t,z)\frac{\delta}{\delta f(\theta(z))} \qquad (10.31)
$$

It can be shown that the Hamiltonian above corresponds to the action given in Eq. (7.86). The $\theta = \theta(z)$ variable continues to be the label of the forward rates functional derivative $\delta/\delta f(\theta)$, but is otherwise replaced everywhere in the action by the nonlinear variable $z(\theta)$. These features of the Hamiltonian are a reflection of the defining equation (7.85) for psychological future time.

Hamiltonian for nonlinear forward rates

For the case of positive forward rates $f(t,x) = f_0 e^{\phi(t,x)}$, the derivation proceeds exactly in the same manner as the linear case discussed in this section. The only differences are the following.

- One starts from Eq. (7.57) with the action for the nonlinear forward rates given by

$$
\mathcal{L}[\phi] = -\frac{1}{2}\left[\left\{\frac{f_0\frac{\partial\phi(t,x)}{\partial t} - \alpha(t,x)}{\sigma_0(t,x)e^{\nu\phi(t,x)}}\right\}^2 + \frac{1}{\mu^2}\left\{\frac{\partial}{\partial x}\left(\frac{f_0\frac{\partial\phi(t,x)}{\partial t} - \alpha(t,x)}{\sigma_0(t,x)e^{\nu\phi(t,x)}}\right)\right\}^2\right]
$$

- The quantity A in Eq. (10.25) is now defined by

$$
A_\phi = f_0\sigma_0^{-1}e^{-\nu\phi}(\phi_{t+\epsilon} - \phi_t - \epsilon f_0^{-1}\alpha)
$$

- The measure term is defined as

$$
\rho_n = \prod_x e^{-\nu\phi}(n\epsilon,x)
$$

- The re-scaling of $p(x)$ is as follows

$$
p(x) \to f_0^{-1}\sigma_0 e^{\nu\phi}p(x)
$$

and cancels the measure term on the right-hand side of Eq. (10.28),

The Hamiltonian for nonlinear forward rates with stochastic volatility, similar to Eq. (10.30), is given by

$$\mathcal{H}_\phi(t) = -\frac{1}{2f_0^2} \int dx dx' \sigma_0 e^{\nu\phi}(x) D(x, x'; t) \sigma_0 e^{\nu\phi}(x') \frac{\delta^2}{\delta\phi(x)\delta\phi(x')}$$
$$-\frac{1}{f_0} \int dx \alpha(t, x) \frac{\delta}{\delta\phi(x)} \qquad (10.32)$$

The case of nonlinear forward rates for deterministic volatility σ_0 is given by setting $\nu = 0$, and gives

$$\mathcal{H}_\phi(t)\Big|_{\nu=0} = -\frac{1}{2f_0^2} \int dx dx' \sigma_0(t, x) D(x, x'; t) \sigma_0(t, x') \frac{\delta^2}{\delta\phi(x)\delta\phi(x')}$$
$$-\frac{1}{f_0} \int dx \alpha \frac{\delta}{\delta\phi(x)} \qquad (10.33)$$

The Hamiltonian given above for nonlinear forward interest rates with deterministic volatility ($\nu = 0$) is a generalization of the Black–Karasinski Hamiltonian for the spot interest rate that was obtained in Eq. (6.87).

10.5 Hamiltonian for forward rates with stochastic volatility

Consider the case when both the forward interest rates and its volatility fluctuate independently, and are represented by separate quantum fields. To obtain the Hamiltonian the following Lagrangian given in Eq. (7.60) needs to be examined

$$\mathcal{L}(t, x) = -\frac{1}{2(1 - \rho^2)} \left(\frac{\frac{\partial f}{\partial t} - \alpha}{\sigma} - \rho \frac{\frac{\partial h}{\partial t} - \beta}{\xi} \right)^2 - \frac{1}{2} \left(\frac{\frac{\partial h}{\partial t} - \beta}{\xi} \right)^2$$
$$-\frac{1}{2\mu^2} \left(\frac{\partial}{\partial x} \left(\frac{\frac{\partial f}{\partial t} - \alpha}{\sigma} \right) \right)^2 - \frac{1}{2\kappa^2} \left(\frac{\partial}{\partial x} \left(\frac{\frac{\partial h}{\partial t} - \beta}{\xi} \right) \right)^2 \qquad (10.34)$$

where recall

$$\sigma(t, x) = \sigma_0 e^{h(t,x)}; \quad -\infty \le f(t, x), h(t, x) \le +\infty$$

Discretizing time, and for notational simplicity suppressing the time (and sometimes) the maturity labels, the Lagrangian $\mathcal{L}$ in matrix notation is as follows

$$S(n) = -\frac{1}{2\epsilon} \int_{x,x'} [\sigma^{-1}A \ \xi^{-1}B](x) \mathcal{M}(x, x'; t) \begin{bmatrix} \sigma^{-1}A \\ \xi^{-1}B \end{bmatrix}(x') \qquad (10.35)$$

where

$$M(x, x'; t) = \begin{bmatrix} \frac{1}{1-\rho^2} - \frac{1}{\mu^2}\frac{\partial^2}{\partial x^2} & -\frac{\rho}{1-\rho^2} \\ -\frac{\rho}{1-\rho^2} & \frac{1}{1-\rho^2} - \frac{1}{\kappa^2}\frac{\partial^2}{\partial x^2} \end{bmatrix} \delta(x - x') \qquad (10.36)$$

and

$$A \equiv \tilde{f}_{t+\epsilon} - f_t - \epsilon\alpha$$
$$B \equiv \tilde{h}_{t+\epsilon} - h_t - \epsilon\beta$$

In obtaining Eq. (10.35) for $S(n)$ the Neumann boundary conditions on the fields given in Eqs. (7.10) and (7.61) have been used.

Eq. (10.35) is re-written using Gaussian integration and yields (ignoring irrelevant constants)

$$e^{S(n)} = \int DpDq e^{-\frac{\epsilon}{2}\int_{x,x'}\begin{bmatrix} p & q \end{bmatrix}\mathcal{M}^{-1}\begin{bmatrix} p \\ q \end{bmatrix} + i\int_x\begin{bmatrix} p & q \end{bmatrix}\begin{bmatrix} \sigma^{-1}A \\ \xi^{-1}B \end{bmatrix}} \qquad (10.37)$$

Define the measure term by

$$\rho_n \equiv \prod_x e^{-h(x)} \qquad (10.38)$$

and rescale the p and q variables in Eq. (10.37) for each x as

$$p \rightarrow \sigma p$$
$$q \rightarrow \xi q$$

Eq. (10.37) yields

$$\rho_n e^{S(n)} = \int DpDq e^{-\frac{\epsilon}{2}\int_{x,x'}\begin{bmatrix} \sigma p & \xi q \end{bmatrix}\mathcal{M}^{-1}\begin{bmatrix} \sigma p \\ \xi q \end{bmatrix} + i\int_x\begin{bmatrix} p & q \end{bmatrix}\begin{bmatrix} \tilde{f} - f - \epsilon\alpha \\ \tilde{h} - h - \epsilon\beta \end{bmatrix}}$$

showing that the measure term cancels out. Hence from above

$$\rho_n e^{S(n)} = < f; h|e^{-\epsilon\mathcal{H}_n}|\tilde{f}; \tilde{h} > \qquad (10.39)$$
$$= e^{-\epsilon\mathcal{H}_n(\delta/\delta f, \delta/\delta h)} \int DpDq e^{i\int_x[p(f-\tilde{f})+q(h-\tilde{h})]}$$

and which yields the following Hamiltonian for forward rates and volatility quantum fields

$$\mathcal{H}(t) = \frac{1}{2}\int_{x,x'}\begin{bmatrix} \sigma\frac{\delta}{i\delta f} & \xi\frac{\delta}{i\delta h} \end{bmatrix}\mathcal{M}^{-1}\begin{bmatrix} \sigma\frac{\delta}{i\delta f} \\ \xi\frac{\delta}{i\delta h} \end{bmatrix} - \int_x\left\{\alpha\frac{\delta}{\delta f} + \beta\frac{\delta}{\delta h}\right\} \qquad (10.40)$$

It is shown in Eq. (10.73) that the inverse of matrix in Eq. (10.36) is given by

$$\mathcal{M}^{-1}(x, x'; t) =$$

$$C \begin{bmatrix} D_- - D_+ + \frac{1-\rho^2}{\kappa^2}(r_+D_+ - r_-D_-) & \rho(D_- - D_+) \\ \rho(D_- - D_+) & D_- - D_+ + \frac{1-\rho^2}{\mu^2}(r_+D_+ - r_-D_-) \end{bmatrix}$$

$$(10.41)$$

where constants $C, r_\pm$ and functions $D_\pm = D_\pm(x, x'; t)$ are given in Appendix 10.11.

For obtaining the forward interest rates' martingale measure the following matrix element will be needed

$$E(x; x', t) \equiv \mathcal{M}_{11}^{-1}(x, x'; t) \tag{10.42}$$

$$= \frac{\mu^2}{\sqrt{(\kappa^2 - \mu^2)^2 + 4\rho^2\mu^2\kappa^2}}[\kappa^2(D_- - D_+) + (1 - \rho^2)(r_+D_+ - r_-D_-)]$$

10.6 Hamiltonian formulation of the martingale condition

The existence of a risk-neutral measure is central to the theory of arbitrage-free pricing of financial instruments, and a path-integral formulation of this principle has been discussed in Section 7.6. The Gaussian Lagrangians discussed in Chapter 7 are quadratic in the fields, and hence the martingale condition could be solved exactly as in Section 7.6 by performing a Gaussian path integration.

For the cases of nonlinear forward interest rates and their stochastic volatility, the Lagrangian is nonlinear and hence the evaluation of the risk-neutral measure cannot be done explicitly using the path integral; for this reason the derivation of the risk-neutral measure is reformulated using the Hamiltonian. The Hamiltonian formulation provides an exact solution for the martingale measure of the nonlinear theory of the forward rates with stochastic volatility.

The result given in Eq. (4.43) shows that the existence of a martingale measure is equivalent to a risk-free Hamiltonian that annihilates the underlying security S. It will be seen that a similar condition holds for the Hamiltonian of the forward rates, but with a number of complications arising from the nontrivial domain of the forward rates, and the fact that the spot interest rate $r(t) = f(t, t)$ is itself a stochastic quantity.

The martingale condition given in Eq. (2.10) states that the price of the bond $P(t_*, T)$ at some future time $T > t_* > t$ is equal to the price of the bond at time t_*, discounted by the risk-free interest rate $r(t) = f(t, t)$. In other words

$$P(t, T) = E_{[t, t_*]}[e^{-\int_t^{t_*} r(t)dt} P(t_*, T)] \tag{10.43}$$

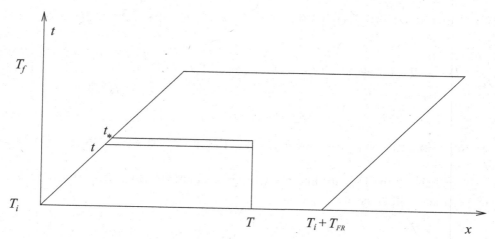

Figure 10.5 Domains for risk-neutral measure based on Treasury Bonds

where, as before, $E_{[t,t_*]}[X]$ denotes the average value of X over all the stochastic variables in the time interval (t, t_*).

In terms of the Feynman path integral, Eq. (10.43) yields (for measure ρ)

$$P(t, T) = \int Df \rho[f] e^{-\int_t^{t_*} r(t)dt} P(t_*, T) e^{S[f]} \qquad (10.44)$$

There are two domains involved involved in the path integral given in Eq. (10.44), namely the domain for the Treasury Bonds that is nested inside the domain of the forward rates. These domains are shown in Figure 10.5.

Although written in an integral form, the martingale condition given in Eq. (10.44), similar to the case of a single security – is clearly a differential condition since it holds for any value of t_*. Hence take $t_* = t + \epsilon$. The reason that one needs to consider an infinitesimal change in the forward rates is due to the time-dependent nature of the state space $\mathcal{V}_t$. For an infinitesimal evolution in time, the functional integral in Eq. (10.44) collapses to an integration over the final time variables $\tilde{f}_{t+\epsilon}$ defined on time slice $t_* = t + \epsilon$. Hence

$$P(t, T) = \int D\tilde{f}_{t+\epsilon} \rho_{t+\epsilon} \, e^{-\epsilon f(t,t)} e^{\epsilon \int \mathcal{L}[f,\tilde{f}]} P[\tilde{f}; t + \epsilon, T] \qquad (10.45)$$

The above equation is re-written in the language of state vectors, namely that

$$< f_t | P(t, T) > = \int D\tilde{f}_{t+\epsilon} < f_t | e^{-\epsilon f(t,t)} e^{-\epsilon \mathcal{H}} | \tilde{f}_{t+\epsilon} > < \tilde{f}_{t+\epsilon} | P(t + \epsilon, T) >$$

$$(10.46)$$

The completeness equation given in Eq. (10.10) is

$$\mathcal{I}_{t+\epsilon} = \int D\tilde{f}_{t+\epsilon} |\tilde{f}_{t+\epsilon}><\tilde{f}_{t+\epsilon}| \tag{10.47}$$

Hence from Eq. (10.46)

$$< f_t|P(t,T)> = < f_t|e^{-\epsilon f(t,t)}e^{-\epsilon\mathcal{H}(t)}|P(t+\epsilon,T)> \tag{10.48}$$
$$\Rightarrow |P(t,T)> = e^{-\epsilon f(t,t)}e^{-\epsilon\mathcal{H}(t)}|P(t+\epsilon,T)> \tag{10.49}$$

It can be verified, using the explicit representation of the zero coupon bond given in Eq. (10.21), that

$$e^{+\epsilon f(t,t)}|P(t,T)> = |P(t+\epsilon,T)> \tag{10.50}$$

Hence

$$|P(t+\epsilon,T)> = e^{-\epsilon\mathcal{H}(t)}|P(t+\epsilon,T)> \tag{10.51}$$
$$\Rightarrow \mathcal{H}(t)|P(t+\epsilon,T)> = 0 \tag{10.52}$$

Since there is nothing special about the bond that is being considered, one arrives at the differential formulation of the martingale measure, namely that all zero coupon bonds – and consequently all coupon bonds – are eigenfunctions of the Hamiltonian $\mathcal{H}$ that are annihilated by $\mathcal{H}$, that is, have zero eigenvalue[4]

$$\mathcal{H}(t)|P(t,T)> = 0 \text{ for all } t,\ T \tag{10.53}$$

or more explicitly

$$< f_t|\mathcal{H}(t)|P(t,T)> = \mathcal{H}(t)e^{-\int_t^T dx f(t,x)} = 0 \tag{10.54}$$

The above equation is the field theory generalization of the case of a single security given in Eq. (4.43).

The role of the discounting factor is very different for a security that it is for a bond. The spot rate r is a constant for the case of a security, and is introduced into the Hamiltonian 'by hand' using hedging arguments. In contrast, for the case of the forward rates the spot interest rate is part of the forward rates and is essential in obtaining the martingale condition. In particular, due to the difference in the domain for the state space at two different instants, the discounting by the spot rate, namely $e^{-\epsilon f(t,t)}$, is precisely the factor required to transform the Treasury Bond at a later time $P(t+\epsilon,T)$ to a bond at an infinitesimally earlier time, namely $P(t,T)$.

The transformation from the initial bond state to the later time final bond state, as in Eq. (10.50), is essential in obtaining the equivalence of the martingale condition with the (eigenfunction) condition that the forward rates Hamiltonian annihilates

---

[4] Similar to the case of the stock price in Section 4.7, the bonds are not normalizable state vectors of the forward rates' state space.

the Treasury Bond; the fact that the initial and final bond state vectors are defined on unequal maturity time intervals is compensated by the discounting factor.

In a Hamiltonian derivation of the change of numeraire, discussed in Section 10.9, discounting by a Treasury Bond $P(t, t_*)$ replaces discounting by the money market account $\exp\{-\int_t^T dt'r(t')\}$; it is seen in this case as well that the discounting factor makes the initial and final bond state vectors equivalent.

10.7 Martingale condition: linear and nonlinear forward rates

The martingale condition obtained in Eq. (10.53) is applied to evaluate the martingale evolution equation for the linear and nonlinear forward rates.

Linear Hamiltonian

Recall the Hamiltonian for the linear (Gaussian) forward interest rates is given is Eq. (10.30) by

$$
\mathcal{H}_f(t) = -\frac{1}{2} \int_t^{t+T_{FR}} dx dx' \sigma(t, x) D(x, x'; t) \sigma(t, x') \frac{\delta^2}{\delta f(x) \delta f(x')}
$$

$$
- \int_t^{t+T_{FR}} dx \alpha(t, x) \frac{\delta}{\delta f(x)}
$$

From Eq. (10.21) the zero coupon bond is given by

$$
P(t, T) = \exp\left(-\int_t^T dx f(t, x)\right) \tag{10.55}
$$

and which yields

$$
\frac{\delta^n}{\delta f(t, x)^n} P(t, T) = \begin{cases} (-1)^n P(t, T), & t < x < T \\ 0, & x > T \end{cases}
$$

$$
= (-1)^n P(t, T) \Theta(T - x) \tag{10.56}
$$

The martingale condition requires that

$$
\mathcal{H}_f(t) | P_t(T) >= 0 \tag{10.57}
$$

and hence, from above equation and Eqs. (10.30) and (10.56)

$$
\left[-\frac{1}{2} \int_t^T dx dx' \sigma(t, x) D(x, x'; t) \sigma(t, x') + \int_t^T dx \alpha(t, x)\right] | P_t(T) >= 0
$$

$$
\Rightarrow \alpha(t, x) = \sigma(t, x) \int_t^x dx' D(x, x'; t) \sigma(t, x') \tag{10.58}
$$

: condition for martingale measure

The expression for the drift velocity given in Eq. (7.38), that was obtained using Gaussian path integration, has been recovered from the Hamiltonian approach.

The effective Hermitian Hamiltonian for the linear forward interest rates is derived in Appendix 10.12.

Linear Hamiltonian: psychological future time

From Eq. (10.21) the zero coupon bond is given by

$$P(t, T) = \exp \left(- \int_0^{T-t} d\theta f(t, \theta) \right) \tag{10.59}$$

Hence, similar to Eq. (10.56), for psychological future time

$$\frac{\delta^n}{\delta f(t, \theta(z))^n} P(t, T) = (-1)^n P(t, T) \Theta(\theta - T + t) \tag{10.60}$$

Imposing the martingale condition for the psychological future time Hamiltonian given in Eq. (10.31) leads to

$$
\mathcal{H}_{f,z}(t) P(t, T) = 0
$$
$$
= \left[-\frac{1}{2} \int_{z(0)}^{z(T-t)} dz dz' \sigma(t, z) G(z, z') \sigma(t, z') + \int_{z(0)}^{z(T-t)} dz \alpha(t, z) \right] P(t, T)
$$
$$
\Rightarrow \alpha(t, z) = \sigma(t, z) \int_{z(0)}^{z} dz' G(z, z') \sigma(t, z') \tag{10.61}
$$

Nonlinear Hamiltonian

For the nonlinear forward rates, the zero coupon bond is given by

$$P(t, T) = \exp \left(-f_0 \int_t^T dx e^{\phi(x)} \right) \tag{10.62}$$

and which yields

$$
\frac{\delta}{\delta \phi(x)} P(t, T) = -f_0 e^{\phi(x)} P(t, T) \Theta(T - x)
$$
$$
\frac{\delta^2}{\delta \phi(x) \phi(x')} P(t, T) = \left[-f_0 \delta(x - x') e^{\phi(x)} \Theta(T - x) \right.
$$
$$
\left. + f_0^2 e^{\phi(x)} e^{\phi(x')} \Theta(T - x) \Theta(T - x') \right] P(t, T) \tag{10.63}
$$

The Hamiltonian for nonlinear forward rates given by Eq. (10.32)

$$\mathcal{H}_\phi(t) = -\frac{1}{2f_0^2} \int dx dx' \sigma_0 e^{\nu\phi}(x) D(x, x'; t) \sigma_0 e^{\nu\phi}(x') \frac{\delta^2}{\delta\phi(x)\delta\phi(x')}$$

$$-\frac{1}{f_0} \int dx \alpha(t, x) \frac{\delta}{\delta\phi(x)}$$

and the martingale condition $\mathcal{H}_\phi(t)P(t, T) = 0$ yields from Eq. (10.63)

$$\left[-\frac{1}{2} \int_t^T dx dx' \sigma_0 e^{(\nu+1)\phi}(x) D(x, x'; t) \sigma_0 e^{(\nu+1)\phi}(x') \right.$$
$$\left. + \frac{1}{2f_0} \int_t^T dx D(x, x; t) \sigma_0^2 e^{(2\nu+1)\phi}(x) + \int_t^T dx \alpha(t, x) e^{\phi(x)} \right] P(t, T) = 0$$

Hence the drift velocity is obtained as

$$\alpha(t, x) = -\frac{\sigma_0^2 e^{2\nu\phi(t, x)}}{2f_0} D(x, x; t)$$
$$+ \sigma_0 e^{\nu\phi(t, x)} \int_t^x dx' D(x, x'; t) \sigma_0 e^{\nu\phi(t, x')} e^{\phi(t, x')} \quad (10.64)$$

where the time index for the field, namely $\phi(t, x)$ has been restored in the expression for α.[5] For the case of nonlinear fields with deterministic volatility $\nu = 0$. The simplest nonlinear theory of the forward rates has $\sigma_0 =$ constant, and the drift velocity is given by

$$\alpha_{\nu=0}(t, x) = -\frac{\sigma_0^2}{2f_0} D(x, x; t) + \sigma_0^2 \int_t^x dx' D(x, x'; t) e^{\phi(t, x')} \quad (10.65)$$

For the nonlinear forward rates, the exponential variable $e^{\phi(t, x)}$ appears in the action only through the drift term $\alpha_{\nu=0}(t, x)$. All studies of the nonlinearities of the forward rates have a natural starting point in studying this simplest and irreducible nonlinear model.

The martingale measure for the nonlinear forward interest rates obtained above is not contained in the class of solutions for which the drift velocity is a quadratic function of the volatility functions (fields) [2], as is the case for linear field theories as well as the HJM model. The appearance of the forward rates $f(t, x)$ directly in the drift velocity emerges naturally in the field-theoretic formulation, and is a reflection of the kinetic term $(\partial f/\partial t)^2$ in the Lagrangian for the case of $f \in [-\infty, +\infty]$, being replaced for the case of positive forward rates $f \geq 0$ by $(\partial \ln(f/f_0)/\partial t)^2$.

---

[5] The first term in Eq. (10.64) was inadvertently omitted in [3].

The derivation of the martingale condition for the linear theory of the forward interest rates can be carried out for any of the Gaussian propagators, and the result is obtained by replacing the propagator $D(x, x'; t)$ by the appropriate one; in particular, the result for the stiff Lagrangian is obtained by replacing $D(x, x'; t)$ with $G(x, x'; t)$.

10.7.1 General form of the action

The following general result for the action $S[\phi]$ is proven for the case of when (stochastic) volatility is a function of the forward rates. A general form for the Lagrangian can be written as

$$\mathcal{L}_{general} = \mathcal{L}\left[\left(\frac{f_0 \frac{\partial \phi(t,x)}{\partial t} - \alpha(t,x)}{\sigma_0(t,x)e^{v\phi(t,x)}}\right)\right] + \int U(t,x)\frac{\partial \phi}{\partial t} + \int W(t,x)$$

U, W : arbitrary local functions of $f(t, x)$

It can be shown that the martingale condition yields [3]

$$U(t, x) = W(t, x) = 0$$

In particular, a string tension term in the Lagrangian of the form

$$W(t, x) \propto \left(\frac{\partial f}{\partial x}\right)^2$$

is excluded by requiring a martingale measure.

10.8 Martingale condition: forward rates with stochastic volatility

From the Hamiltonian given in Eq. (10.40) it can be seen that $\delta/\delta h$ yields zero in Eq. (10.54) since the zero coupon bond does not depend explicitly on the volatility field. Hence, from Eqs. (10.54), (10.40) and (10.63)

$$\left[-\frac{1}{2}\int_t^T dx dx' \sigma(t, x)E(x, x'; t)\sigma(t, x') + \int_t^T dx\alpha(t, x)\right]P(t, T) = 0$$

$$\Rightarrow \alpha(t, x) = \sigma(t, x)\int_t^x dx' E(x, x'; t)\sigma(t, x') \qquad (10.66)$$

$$: \text{condition for martingale measure}$$

Incorporating the expression for $\alpha(t, x)$ given in Eq. (10.66) into the Lagrangian yields the final result. For notational convenience define the following non-local function of the volatility field

$$v(t, x) = \int_t^x dx' E(x, x'; t)\sigma(t, x') \qquad (10.67)$$

One the obtain the Lagrangian given below that is a complete description of the theory of linear forward rates with an independent stochastic volatility quantum field

$$
\mathcal{L}(t,x) = -\frac{1}{2(1-\rho^2)} \left(\sigma^{-1}\frac{\partial f}{\partial t} - v - \rho \frac{\frac{\partial h}{\partial t} - \beta}{\xi} \right)^2 - \frac{1}{2} \left(\frac{\frac{\partial h}{\partial t} - \beta}{\xi} \right)^2
$$
$$
- \frac{1}{2\mu^2} \left(\frac{\partial}{\partial x} \left(\sigma^{-1}\frac{\partial f}{\partial t} - v \right) \right)^2 - \frac{1}{2\kappa^2} \left(\frac{\partial}{\partial x} \left(\frac{\frac{\partial h}{\partial t} - \beta}{\xi} \right) \right)^2
$$

$$(10.68)$$

All the parameters in the theory, namely the function $\beta(t,x)$ and the parameters μ, κ, ξ and ρ need to be determined from market data. Due to the presence of the field $v(t,x)$ the Lagrangian is non-local, with the function $E(x,x';t)$ containing all the information regarding the risk-neutral martingale measure.

Since at present there is no instrument in the financial markets that trades in volatility of the forward rates, one cannot apply the condition of martingale to the volatility field, and, in particular, one cannot fix the drift velocity of the volatility field, namely $\beta(t,x)$, to be a function of the other fields and parameters of the theory. For this reason β has to be determined empirically from the market. To obtain the limit of deterministic volatility, the limit of ξ, ρ and $\kappa \to 0$ has to be taken, and this yields

$$
\xi, \rho, \kappa \to 0
$$
$$
r_+ \to \mu
$$
$$
r_- \to 0
$$
$$
E(x,x';t) \to D(x,x';t)
$$

with propagator $D(x,x';t)$ given by Eq. (7.27).

There are two further generalizations that can made for the Lagrangian obtained in Eq. (10.68), namely (a) the propagator $E(x,x';t)$ can include more complex effects such as stiffness, psychological future time $z = z(x-t)$ etc., and (b) the forward rate can be made nonlinear and positive, that is, $f > 0$.

10.9 Nonlinear change of numeraire

The martingale condition that was obtained in Section 10.6 is based on discounting the Treasury Bond $P(t,T)$ with the money market account given by $B(t_0, t) = \exp \int_{t_0}^{t} r(t')dt'$, where $r(t)$ is the spot interest rate. The requirement that the discounted instrument $P(t,T)/B(t_0,t)$ is a martingale led to the equation $\mathcal{H}(t)P(t,T) = 0$, which in turn fixes the drift velocity α.

As discussed in Section 7.7 one can choose to discount a Treasury Bond by any positive-valued instrument, and then choose the new drift velocity α_* to make the discounted Treasury Bond a martingale. In particular, one can discount the Treasury Bond $P(t, T)$ by another Treasury Bond $P(t, t_*)$ so that $P(t, T)/P(t, t_*)$ is a martingale.

The drift velocity α_* is fixed by demanding that the Hamiltonian $\mathcal{H}(t)$ should annihilate the instrument $P(t, T)/P(t, t_*)$. In other words, analogous to Eq. (10.54), by repeating the derivation given in Section 10.6, one derives the martingale condition as

$$
\mathcal{H}(t) \left[\frac{P(t, T)}{P(t, t_*)} \right] = 0
$$

$$
\Rightarrow\; < f_t | \mathcal{H}(t) | \frac{P(t, T)}{P(t, t_*)} > = \mathcal{H}(t) \exp\left\{ -\int_{t_*}^{T} dx f(t, x) \right\} = 0 \quad (10.69)
$$

The only effect of the discounting factor $P(t, t_*)$ is to change, on the right-hand side of Eq. (10.69), the lower limit of integration of the forward rates $f(t, x)$ from t to t_*. This in turn yields, for example similar to Eq. (10.56), that

$$
\frac{\delta^n}{\delta f(t, x)^n} \left[\frac{P(t, T)}{P(t, t_*)} \right] = (-1)^n \frac{P(t, T)}{P(t, t_*)} \Theta(T - x) \Theta(x - t_*) \quad (10.70)
$$

In all the derivations for the drift velocity α when discounting by the spot rate, the Hamiltonian $\mathcal{H}(t)$ was defined for all future time $x > t$, and forward time x for the drift velocity was restricted to the range of $[t, T]$, due to martingale condition satisfied by the Treasury Bond $P(t, T)$. However, with the change of numeraire, due to Eq. (10.70), in the derivations of drift velocity α_*, forward time x is now restricted to lie in the range $[t_*, T]$. Hence, to obtain α_* from α, one needs to only change the lower limit of integration for x from t to t_*, which yields the following modified drift velocities.

- Linear forward rates. Analogous to Eq. (10.58)

$$
\alpha_*(t, x) = \sigma(t, x) \int_{t_*}^{x} dx' D(x, x'; t) \sigma(t, x') \quad (10.71)
$$

- Nonlinear forward rates. From Eq. (10.64)

$$
\alpha_*(t, x) = -\frac{\sigma_0^2 e^{2v\phi}(t, x)}{2 f_0} D(x, x; t)
$$
$$
+ \sigma_0 e^{v\phi}(t, x) \int_{t_*}^{x} dx' D(x, x'; t) \sigma_0 e^{(v+1)\phi}(t, x')
$$

- For stochastic volatility. From Eq. (10.66)

$$\alpha_*(t,x) = \sigma(t,x) \int_{t_*}^{x} dx' E(x,x';t)\sigma(t,x')$$

10.10 Summary

The Hamiltonian and state space of the forward interest rates provide an independent formulation of the quantum field theory of the forward rates, and can lead to new insights on the behaviour of the forward rates.

The Hamiltonian of the forward rates was derived for both the linear and nonlinear theory. The Hamiltonian is a time-dependent differential operator in infinitely many independent degrees of freedom that requires the machinery of quantum field theory for its mathematical description. A key step in obtaining the Hamiltonian was to first derive the underlying time-dependent state space of the forward rates – and which turned to be quite an involved derivation due to the trapezoidal structure of the underlying manifold.

The Hamiltonian generates infinitesimal translations in time, and due to the choice made for the state space, the forward rates' Hamiltonian propagates the interest rates **backwards** in time. Since the Hamiltonian is non-Hermitian, the forward and backward Hamiltonians – similar to the case for the Fokker–Planck Hamiltonian – are inequivalent.

One can model the forward rates starting with a Hamiltonian, and obtain the Lagrangian following a procedure similar to the followed in the earlier studies of the Black–Scholes equation and of spot interest rates. This route looks deceptively simple, since to obtain the Lagrangian one also needs to know the state space on which the Hamiltonian acts.

It was shown that many crucial properties of the time-dependent state space are encoded in the action $S[f]$ – and a considerable amount of effort was spent in decoding it. Without knowledge of the full action, it would be difficult to obtain the state space starting only from the Hamiltonian, since the state space of the forward rates is an **independent** quantity that needs to be specified in addition to the Hamiltonian.

The non-Hermiticity of the Hamiltonian always arises due to the drift term that is required for having a martingale time evolution, and a similarity transformation for the linear case makes the forward rates Hamiltonian equivalent to a Hermitian Hamiltonian.

The model for nonlinear forward rates, and the case of linear forward rates with stochastic volatility could be fully analyzed for its martingale measure due to the Hamiltonian formulation of the martingale condition.

There are a number of free parameters in the action for nonlinear forward interest rates and stochastic volatility that need to be determined from the market. Hence the empirical procedures followed in Chapter 8 need to be extended to the case of nonlinear forward rates with stochastic volatility.

The pricing kernel and European call option for a Treasury Bond were derived for the linear Hamiltonian to illustrate a few concrete calculations using the field theory Hamiltonian.

10.11 Appendix: Propagator for stochastic volatility

The inverse of the matrix operator given in Eq. (10.36) is derived in this appendix.

$$
\mathcal{M}(x, x'; t) = \begin{bmatrix} \frac{1}{1-\rho^2} - \frac{1}{\mu^2}\frac{\partial^2}{\partial x^2} & -\frac{\rho}{1-\rho^2} \\ -\frac{\rho}{1-\rho^2} & \frac{1}{1-\rho^2} - \frac{1}{\kappa^2}\frac{\partial^2}{\partial x^2} \end{bmatrix} \delta(x - x') \qquad (10.72)
$$

All the elements of the matrix operator $\mathcal{M}(x, x'; t)$ commute, and hence it can be inverted as an ordinary 2×2 matrix leading to the following result

$$
\mathcal{M}^{-1}(x, x'; t) = \begin{bmatrix} \frac{1}{1-\rho^2} - \frac{1}{\kappa^2}\frac{\partial^2}{\partial x^2} & \frac{\rho}{1-\rho^2} \\ \frac{\rho}{1-\rho^2} & \frac{1}{1-\rho^2} - \frac{1}{\mu^2}\frac{\partial^2}{\partial x^2} \end{bmatrix} \frac{1}{\det \mathcal{M}} \delta(x - x') \ .
$$

The inverse of the determinant, using notation $\partial^2/\partial x^2 = \partial^2$, can be written as

$$
\frac{1}{\det \mathcal{M}} \delta(x - x') = (1 - \rho^2)C \left[\frac{1}{-\partial^2 + r_-} - \frac{1}{-\partial^2 + r_+} \right] \delta(x - x')
$$
$$
= (1 - \rho^2)C \left[D_-(x, x'; t) - D_+(x, x'; t) \right]
$$

where

$$
C = \frac{\mu^2 \kappa^2}{\sqrt{(\kappa^2 - \mu^2)^2 + 4\rho^2\mu^2\kappa^2}}
$$
$$
r_\pm = \frac{1}{2(1 - \rho^2)} \left[\mu^2 + \kappa^2 \pm \sqrt{(\kappa^2 - \mu^2)^2 + 4\rho^2\mu^2\kappa^2} \right]
$$
$$
\left(-\frac{\partial^2}{\partial x^2} + r_\pm \right) D_\pm(x, x'; t) = \delta(x - x') \ : \ \text{Neumann B.C.'s}
$$

Collecting results derived above yields

$$\mathcal{M}^{-1}(x, x'; t) =$$

$$C \begin{bmatrix} D_- - D_+ + \frac{1-\rho^2}{\kappa^2}(r_+ D_+ - r_- D_-) & \rho(D_- - D_+) \\ \rho(D_- - D_+) & D_- - D_+ + \frac{1-\rho^2}{\mu^2}(r_+ D_+ - r_- D_-) \end{bmatrix}$$

$$(10.73)$$

which is the result quoted in Eq. (10.41). The identity

$$-\partial^2 \big[D_-(x, x'; t) - D_+(x, x'; t) \big] = -r_- D_-(x, x'; t) + r_+ D_+(x, x'; t)$$

simplifies the diagonal terms of $\mathcal{M}^{-1}(x, x'; t)$ in Eq. (10.73).

10.12 Appendix: Effective linear Hamiltonian

Recall that the Hamiltonian for the linear forward rates is given by Eq. (10.30) by

$$\mathcal{H}_f(t) = -\frac{1}{2} \int_t^{t+T_{FR}} dx dx' \sigma(t, x) D(x, x'; t) \sigma(t, x') \frac{\delta^2}{\delta f(x) \delta f(x')}$$

$$- \int_t^{t+T_{FR}} dx \alpha(t, x) \frac{\delta}{\delta f(x)}$$

The drift velocity, from Eq. (10.58), is given by

$$\alpha(t, x) = \sigma(t, x) \int_t^x dx' D(x, x'; t) \sigma(t, x')$$

Similar to the transformation carried out for option pricing Hamiltonians in Eq. (4.46), one can write the field theory Hamiltonian as

$$\mathcal{H}_f(t) = e^{-\mathcal{O}(t)} \mathcal{H}_{\text{eff}}(t) e^{\mathcal{O}(t)} \qquad (10.74)$$

Using the notation that $\delta_x \equiv \delta/\delta f(x)$ gives

$$e^{-\mathcal{O}} \delta_x \delta_{x'} e^{\mathcal{O}} = \delta_x \mathcal{O} \delta_{x'} \mathcal{O} + \delta_x \delta_{x'} \mathcal{O} + \delta_x \mathcal{O} \delta_{x'} + \delta_{x'} \mathcal{O} \delta_x + \delta_x \delta_{x'}$$

Using the notation $\int_t^{t+T_{FR}} dx \equiv \int_x$, choose the following transformation

$$\mathcal{O}(t) = \int_x \beta(t, x) f(x) \; ; \quad \delta_x \mathcal{O} = \beta(t, x)$$

$$\beta(t, x) = \int_{y, z} \sigma(t, y) D(y, z; t) \theta(z - y) \theta(y - t) D^{-1}(z, x; t) \sigma^{-1}(t, x)$$

To obtain $\mathcal{H}_{\text{eff}}(t)$ the coefficient β has been chosen to exactly cancel the drift term in the Hamiltonian. [6] Hence

$$\int_{x,x'} \sigma(t,x)D(x,x';t)\sigma(t,x')\delta_{x'}\mathcal{O}\delta_x = \int_{x,x'} \sigma(t,x)D(x,x';t)\sigma(t,x')\beta(t,x')\delta_x$$

$$= \int_x \sigma(t,x)\int_y \sigma(t,y)D(x,y;t)\theta(x-y)\theta(y-t)\delta_x = \int_t^{t+T_{FR}} dx\alpha(t,x)\frac{\delta}{\delta f(x)}$$

The effective Hermitian Hamiltonian is hence given by

$$\mathcal{H}_{\text{eff}}(t) = \mathcal{H}_0(t) + \Gamma(t) = \mathcal{H}_{\text{eff}}^{\dagger}(t)$$

$$\mathcal{H}_0(t) = -\frac{1}{2}\int_{x,x'} \sigma(t,x)D(x,x';t)\sigma(t,x')\frac{\delta^2}{\delta f(x)\delta f(x')}$$

$$\Gamma(t) = \frac{1}{2}\int_{x,x'} \sigma(t,x)\beta(t,x)D(x,x';t)\sigma(t,x')\beta(t,x')$$

The effective Hamiltonian can be used to price barrier and other options on Treasury Bonds.

10.13 Appendix: Hamiltonian derivation of European bond option

The main focus in the analysis of the Hamiltonian has been on obtaining the martingale measure for nonlinear forward rates. To illustrate other useful features of the Hamiltonian, the price of a European bond option $C(t_0, t_*, T, K)$ given in (6.53) is derived using the field theory Hamiltonian.

For calculating the price of options, as discussed in Eqs. (7.47) and (7.48), it is more convenient to discount the future price of the option by the discount factor $P(t, t_*)$ that yields, for $t_0 < t_*$, the price of option

$$C(t_0, t_*, T, K) = P(t_0, t_*)E_*[(P(t_*, T) - K)_+] \tag{10.75}$$

For numeraire given by $P(t, t_*)$, from Eq. (10.71), the drift velocity is

$$\alpha_*(t,x) = \sigma(t,x)\int_{t_*}^x dx'D(x,x';t)\sigma(t,x')$$

---

[6] In Section 10.9 a change of the discounting factor (numeraire) is considered. For the new drift velocity given by α_*, the β_* for obtaining the effective Hamiltonian is given by $\beta_*(t,x) = \int_{y,z} \sigma(t,y)D(y,z;t)\theta(z-y)\theta(y-t_*)D^{-1}(z,x;t)\sigma^{-1}(t,x)$.

The Hamiltonian for the linear forward rates, from Eq. (10.30), is

$$\mathcal{H}_f^*(t) = -\frac{1}{2} \int_t^{t+T_{FR}} dx dx' \sigma(t, x) D(x, x'; t) \sigma(t, x') \frac{\delta^2}{\delta f(x) \delta f(x')}$$
$$- \int_t^{t+T_{FR}} dx \alpha_*(t, x) \frac{\delta}{\delta f(x)}$$

To make the content of the payoff function $g[f_*] \equiv (P(t_*, T) - K)_+$ more explicit, consider the dual basis states

$$< f_*| \equiv \prod_{t_* \leq x \leq t_* + T_{FR}} < f(x)| \quad ; \quad < f_0| \equiv \prod_{t_0 \leq x \leq t_0 + T_{FR}} < f(x)| \quad (10.76)$$

Hence

$$< f_*|g > = \begin{cases} (P(t_*, T) - K)_+, & t_* \leq x \leq T \\ 0, & x > T \end{cases} \quad (10.77a)$$
$$\equiv g[f_*] \quad (10.77b)$$

From above, it can be seen that the payoff function $|g >$ has non-zero components in the future direction x **only** in the interval $t_* \leq x \leq T$.

The call option price at time t_0 is given by propagating, **backwards** in time, the payoff function $|g >$ that matures at time $t_* > t_0$ to present time t_0, as given in Eq. (10.20), discounted by the numeraire $P(t_0, t_*)$. Hence

$$C(t_0, t_*, T, K) = P(t_0, t_*) E_*[(P(t_*, T) - K)_+]$$
$$= P(t_0, t_*) < f_0|T\left\{\exp - \int_{t_0}^{t_*} dt \mathcal{H}_f^*(t)\right\}|g > \quad (10.78)$$

Using the completeness equation $\int Df_* |f_* >< f_*| = \mathcal{I}$

$$< f_0|T\left\{\exp - \int_{t_0}^{t_*} dt \mathcal{H}_f^*(t)\right\}|g >$$
$$= \int Df_* < f_0|T\left\{\exp - \int_{t_0}^{t_*} dt \mathcal{H}_f^*(t)\right\}|f_* >< f_*|g > \quad (10.79)$$

In the matrix element of the Hamiltonian above, due to the nontrivial nature of the time-ordering symbol T, the bra vector $< f_0|$ is an element of state space of functions on the interval $[t_0, t_0 + T_{FR}]$ and the ket vector $< f_*|$ is an element of the state space of functions on the interval $[t_*, t_* + T_{FR}]$. Eq. (10.79) can be simplified further. The overlap of $< f_*|$ with $|g >$, that is $< f_*|g >$ is non-zero only in the interval $x \in [t_*, T]$; therefore the non-zero overlap of the the basis state $< f_0|$ with $|g >$ is also only in the interval $x \in [t_*, T]$.

Hence the domain $\mathcal{R}$ that is involved in computing the matrix element in Eq. (10.79) is the one given in Figure 9.1. The domain $\mathcal{R}$ has the important feature that the subspace of the state space V_t that enter the matrix element in Eq. (10.79) are all fixed in time and identical for all $t \in [t_0, t_*]$, being spanned by variables $f(x)$, $x \in [t_*, T]$; moreover, on domain $\mathcal{R}$, the Hamiltonian is also time independent and hence $[\mathcal{H}_f^*(t), \mathcal{H}_f^*(t')] = 0$.

For these reasons the time ordering $\mathcal{T}$ in Eq. (10.79) is no longer necessary, and can be ignored.[7] Hence restrict $\mathcal{H}_f^*(t)$ to the domain $\mathcal{R}$ by limiting the range of $x \in [t_*, T]$. Defining $\int_{t_*}^T dx \equiv \int_{*x}$ yields

$$
\mathcal{H}_f^*(t)|_{\mathcal{R}} = -\frac{1}{2} \int_{*x, x'} \sigma(t, x) D(x, x'; t) \sigma(t, x') \frac{\delta^2}{\delta f(x) \delta f(x')}
$$
$$
- \int_{*x} \alpha_*(t, x) \frac{\delta}{\delta f(x)}
$$

Since there is no longer any time ordering for the Hamiltonian, the operator driving the option price is given by

$$
W \equiv \int_{t_0}^{t_*} dt \mathcal{H}_f^*(t)|_{\mathcal{R}}
$$
$$
= -\frac{1}{2} \int_{*x, x'} q^2(x, x') \frac{\delta^2}{\delta f(x) \delta f(x')} - \int_{*x} j(x) \frac{\delta}{\delta f(x)}
$$
$$
\Rightarrow q^2(x, x') = \int_{t_0}^{t_*} dt \sigma(t, x) D(x, x'; t) \sigma(t, x') \; ; \; j(x) = \int_{t_0}^{t_*} dt \alpha_*(t, x)
$$

$$(10.80)$$

The matrix element given in Eq. (10.79) is hence $(g[f_*] \equiv \, < f_* | g >)$

$$
< f_0| \mathcal{T} \left\{ \exp - \int_{t_0}^{t_*} dt \mathcal{H}_f^*(t) \right\} | f_* > < f_* | g > = < f_0 | e^{-W} | f_* > \Big|_{\mathcal{R}} g[f_*]
$$

$$(10.81)$$

Recall from Eq. (6.55) that a representation of the payoff function is given by

$$
(P(t_*, T) - K)_+ = \int_{-\infty}^{+\infty} dG \frac{d\xi}{2\pi} e^{i\xi(G + \int_{*x} dx f(t_*, x))} (e^G - K)_+
$$

---

[7] If the discounting factor is $\exp(-\int_{t_0}^{t_*} dt r(t))$, the domain for evaluating the matrix element in Eq. (10.79) is the trapezoidal domain given in Figure 6.4. This is because the discounting factor extends the non-zero overlap of the basis state $< f_0|$ with $\exp(-\int_{t_0}^{t_*} dt r(t))|g >$ to the interval $x \in [t, T]$; furthermore, the time-ordering symbol $\mathcal{T}$ cannot be ignored, since the underlying state space and Hamiltonian are now time dependent, and lead to a separate calculation for each $t \in [t_0, t_*]$.

The matrix elements of e^{-W} are evaluated in the domain $\mathcal{R}$, and hence from Eqs. (10.78), (10.79) and (10.81)

$$
C(t_0, t_*, T, K) = P(t_0, t_*) \int Df_* < f_0|e^{-W}|f_* > \Big|_{\mathcal{R}} g[f_*]
$$

$$
= P(t_0, t_*) \int_{-\infty}^{+\infty} dG \frac{d\xi}{2\pi} (e^G - K)_+ e^{i\xi G} \int Df_* < f_0|e^{-W}|f_* > \Big|_{\mathcal{R}} e^{i\xi \int_{*x} f(t_*,x)}
$$

To further simplify the matrix element, the field theory generalization of the momentum basis given in Eq. (4.34) is useful. Hence, similar to the Black–Scholes case given in Eq. (4.38), it follows that[8]

$$
Z \equiv \int Df_* < f_0|e^{-W}|f_* > \Big|_{\mathcal{R}} e^{i\xi \int_{*x} f(t_*,x)}
$$

$$
= \int Df_* Dp < f_0|e^{-W}|p > \Big|_{\mathcal{R}} < p|f_* > e^{i\xi \int_{*x} f(t_*,x)}
$$

$$
= \int Df_* Dp \; e^{-\frac{1}{2} \int_{*x,x'} q^2(x,x') p(x) p(x')} \times
$$

$$
e^{i \int_{*x} p(x) j(x)} e^{i \int_x p(x)[f(t_0,x) - f(t_*,x)]} e^{i\xi \int_{*x} f(t_*,x)} \tag{10.82}
$$

Performing the final forward rates integrations $\int Df_*$, where $f_*(x) \equiv f(t_*, x)$, yields a product of delta functions that constrains all the momentum integrations. Hence

$$
Z = \int Dp e^{-\frac{1}{2} \int_{*x,x'} q^2(x,x') p(x) p(x')} e^{i \int_{*x} p(x)[f(t_0,x) + j(x)]} \prod_{t_* \leq x \leq T} \delta(p(x) - \xi)
$$

$$
= e^{-\frac{q^2}{2}\xi^2} e^{i\xi \int_x [f(t_0,x) + j(x)]} \quad \text{with} \quad q^2 = \int_{*x,x'} q^2(x,x')
$$

and from Eq. (9.9)

$$
\int_{*x} j(x) = \int_{t_*}^T dx \int_{t_0}^{t_*} dt \alpha_*(t, x) = \int_{\mathcal{R}} \alpha_*(t, x) = \frac{q^2}{2}
$$

Collecting all the results gives

$$
C(t_0, t_*, T, K) = P(t_0, t_*) \int_{-\infty}^{+\infty} dG \frac{d\xi}{2\pi} (e^G - K)_+ e^{-\frac{q^2}{2}\xi^2} e^{i\xi(G + \int_{t_*}^T dx f(t_0,x) + \frac{q^2}{2})}
$$

which, on doing the Gaussian integration on ξ, yields Eq. (9.10) as expected.

---

[8] Recall that the derivatives in Hamiltonian, namely $\delta/\delta f(x)$, act on the dual basis and are, consequently, inside the matrix element, equal to $\delta/\delta f_0(x)$.

10.13.1 Forward interest rates' pricing kernel

The pricing kernel, similar to the case for stocks as in Eq. (4.39), can be evaluated for the forward rates and is sufficient for pricing any path-independent option if the discounting is done with a bond $P(t_0, t_*)$. It is defined, from Eq. (10.78) for remaining time $\tau = t_* - t_0$ as follows

$$p[f_0, f_*; \tau] = \, < f_0 | T \left\{ \exp - \int_{t_0}^{t_*} dt \mathcal{H}_f^*(t) \right\} | f_* >$$

$$= \, < f_0 | e^{-W} | f_* > \Big|_{\mathcal{R}pk}$$

The non-lap overlap of $< f_0|$ with $|f_* >$ is for values of the maturity variable x in the interval $[t_*, t_0 + T_{FR}]$, and the path integral for the pricing kernel is over the rectangular domain $\mathcal{R}pk : (t, x) \in [t_0, t_*] \times [t_*, t_0 + T_{FR}]$. A calculation similar to the one carried out in Eq. (10.82) yields the result ($\int_{pk:x} \equiv \int_{t_*}^{t_0 + T_{FR}} dx$)

$p[f_0, f_*; \tau]$

$$= \int Dp \; e^{-\frac{1}{2} \int_{pk:x,x'} q^2(x,x')p(x)p(x')} e^{i \int_{pk:x} p(x)j(x)} e^{i \int_{pk:x} p(x)[f(t_0,x) - f(t_*,x)]}$$

$$= \mathcal{N} \exp \left\{ -\frac{1}{2} \int_{pk:x,x'} [f(t_0, x) - f(t_*, x) + j(x)] q^{-2}(x, x') \right.$$

$$\left. [f(t_0, x') - f(t_*, x') + j(x')] \right\}$$

where $\mathcal{N}$ is a normalization constant. $q^2(x, x')$ is given by Eq. (10.80) and

$$\int_{pk:x'} q^2(x, x') q^{-2}(x', y) = \delta(x - y)$$

with

$$j(x) = \int_{t_0}^{t_*} dt \alpha_*(t, x) = \int_{t_0}^{t_*} dt \sigma(t, x) \int_{t_*}^{x} D(x, x'; t) \sigma(t, x')$$

11

Conclusions

A wide range of topics have been covered, from stock options to forward interest rates. The definition, valuation and hedging of financial instruments have been formulated in the mathematics of quantum theory. Most of the models used in mathematical finance are based on stochastic calculus, and have been shown to belong to the class of problems that require only a finite number of degrees of freedom. Path integrals for systems with a finite number of degrees of freedom are completely finite, and can be used to compute all quantities of interest. In particular it was shown that for stock options the pricing kernel (conditional probability) can be efficiently computed using path integrals.

The Hamiltonian formulation of random systems yields many new results and insights. Barrier options, as well as path-dependent stock options, can be modelled using the concept of potentials and eigenfunctions, and the pricing kernel can be evaluated using concepts such as the completeness equation. The Hamiltonian provides a formulation of the concept of a martingale that is independent of stochastic calculus. The field theory Hamiltonian and state space were essential in obtaining the exact martingale measure for nonlinear forward interest rates.

New financial instruments, in particular path-dependent stock options, can be designed and hedged based on the Hamiltonian approach to option pricing.

Quantum field theory is required for studying systems with infinitely many degrees of freedom. In a completely general theory of the forward interest rates, at a given instant t, the forward rate $f(t, x)$ for each x is an independent random variable, yielding a system with infinitely many degrees of freedom. The forward interest rates, considered as a randomly fluctuating curve, is consequently a two-dimensional quantum field. The field theory of the forward rates yields many results on the fundamental properties of the forward interest rates, some examples of which are the following:

- Gaussian field theory models can accurately describe the empirical behaviour of the forward rates
- Nonlinear forward rates with stochastic volatility can be mathematically modelled using quantum field theory
- The hedging of Treasury Bonds can be carried out effectively for field theory models of the interest rates
- The martingale condition and change of numeraire for nonlinear forward rates can be solved exactly using the techniques of quantum field theory

The topics covered convincingly show that quantum field theory provides a suitable mathematical structure for studying the forward interest rates.

All calculations, as well as empirical analysis, of forward interest rates were carried out for linear (Gaussian) theories. Market data show an excellent agreement of the Gaussian 'stiff' field theory results with the observed behaviour of the forward rates.

The encouraging results obtained for linear theory provide strong motivation for studying the much more complicated and relatively intractable nonlinear theories of forward interest rates. The infinitely many degrees of freedom of a quantum field lead to potential divergences for nonlinear field theories. The procedure of renormalization that has been mentioned earlier is the manner in which these divergences are studied. Only renormalizable theories are consistent, and the main challenge in the study of nonlinear forward interest rates is to devise (analytical and numerical) computational techniques to determine whether the nonlinear theories are renormalizable. In case the nonlinear theories are renormalizable, one needs to develop empirical methods for testing and applying such nonlinear forward interest rates theories to the debt market.

Appendix A

Mathematical background

The subjects discussed in this appendix are all primarily of a mathematical nature, and constitute background material for the main text. The topics are included for the readers easy reference.

A.1 Probability distribution

The notation for probability theory is discussed, and in particular the definition of conditional probability. The concept of a martingale is discussed as this has important applications in finance.

Random variables are real- or discrete-valued variables that take values in some pre-specified range determined by their probability distributions. Random variables are designated by either upper case such as X or lower case such as r. A stochastic process refers to a collection of random variables. The stochastic process can be (a) a continuous process with an independent random variable $r(t)$ for every t, with the continuous label t in some range $t \in [t_0, t_*]$, or can be (b) a discrete process with a collection of random variables Z_n with n: integer. Degrees of freedom, in the terminology of physics, refer to the number of independent random variables at a given instant, and hence each degree of freedom corresponds to a independent stochastic process.

Consider a collection of $N + 1$ random variables Z_i; $1 \leq i \leq N + 1$, with joint probability distribution functions given by $p(z_1, z_2, \ldots, z_{N+1})$. The expectation value of some arbitrary function of the random variables $f(z_1, z_2, \ldots, z_{N+1})$ is given by

$$E[f] = \int dz_1 dz_2, \ldots, dz_{N+1} f(z_1, z_2, \ldots, z_{N+1}) p(z_1, z_2, \ldots, z_{N+1})$$

$$E[1] = 1 = \int dz_1 dz_2, \ldots, dz_{N+1} p(z_1, z_2, \ldots, z_{N+1})$$

The joint probability distribution functions given by $p(z_1, z_2, \ldots, z_{N+1})$ generates the so-called marginal probability distributions

$$p^{(N+1)}(z_1, z_2, \ldots, z_N) = \int dz_{N+1} p(z_1, z_2, \ldots, z_{N+1})$$

$$\cdots \cdots$$

$$p^{(N+1, N, \ldots, 2)}(z_1) = \int dz_2, \ldots, dz_N dz_{N+1} p(z_1, z_2, \ldots, z_{N+1})$$

The conditional expectation for the outcome of some function $f(z_{N+1})$, given that the

fixed values z_i occur for the random variable Z_i; $1 \leq i \leq N$, is given by

$$E\big[f(z_{N+1})|z_1, z_2, z_3, \ldots, z_N\big] = \int dz_{N+1} \, f(z_{N+1})p(z_1, z_2, \ldots, z_{N+1}) \quad \text{(A.1)}$$

A.1.1 Martingale

A martingale refers to a special category of stochastic processes. An arbitrary discrete stochastic process X_i is a martingale if it satisfies

$$E\big[X_{n+1}|x_1, x_2, \ldots, x_n\big] = x_n \quad : \quad \text{Martingale} \quad \text{(A.2)}$$

In other words, the expected value of the random variable X_{n+1}, conditioned on the occurrence of $x_1, x_2, \ldots, x_n$ for random variables $X_1, X_2, \ldots, X_n$, is simply x_n itself.

One can think of the martingale as describing a gambling game; the given condition x_n is the amount of money that the gambler has on the conclusion of the nth game, and the random variable X_{n+1} represents the various possible outcomes of the $n + 1$th game. The martingale condition states that the **expected value** of the gamblers money at the end of the $n + 1$th game is equal only to the money with which he enters the $n + 1$-th game, namely x_n, and not on his history of wins or loses. A martingale is a mathematical representation of a fair game. Using Eq. (A.2) one can prove the following result.

$$
\begin{aligned}
E\big[X_{n+1}\big] &= \int dx_1 dx_2, \ldots, dx_n dx_{n+1} E\big[X_{n+1}|x_1, x_2, \ldots, x_n\big] p(x_1, x_2, \ldots, x_{n+1}) \\
&= \int dx_1 dx_2, \ldots, dx_n dx_{n+1} \, x_n \, p(x_1, x_2, \ldots, x_{n+1}) \\
&= E[X_n] \\
\Rightarrow E\big[X_{n+1}\big] &= E\big[X_n\big] = E\big[X_{n-1}\big] = \ldots = E\big[X_1\big] \\
\Rightarrow E\big[X_n\big] &= E\big[X_1\big]
\end{aligned}
\quad \text{(A.3)}
$$

From above it is seen that in a (fair) game obeying the martingale condition, the gambler, on average, neither loses or wins, and leaves with the money he comes in with.

Random stopping time: Wald's Equation

An important application of the concept of a martingale is in the stopping time problem. Consider a discrete stochastic process denoted by $X_1, X_2, \ldots, X_N$, where all the random variables are independent and identically distributed. In other words, the stochastic process starts, in appropriate units, at time 1 and stops at time N. The stopping N itself is a discrete random variable and hence the stopping time is random.

One can think of the above process as representing the position of a particle doing a random walk, with X_i its position after i steps. One can ask the question: what is the average position of the particle when it stops – given by the expectation of $\sum_{i=1}^{N} X_i$?

The concept of a martingale provides the following solution.

Suppose $E[X] = \mu$; then the stochastic process defined by

$$Z_n = \sum_{i=1}^{n} (X_i - \mu)$$

is a martingale. The following is a proof.

$$E[Z_{n+1}|z_1, z_2, \ldots, z_n] = E\left[\sum_{i=1}^{n+1}(X_i - \mu)|z_1, z_2, \ldots, z_n\right]$$

$$= E\left[\left\{X_{n+1}' - \mu + \sum_{i=1}^{n}(X_i - \mu)\right\}\bigg|z_1, z_2, \ldots, z_n\right]$$

$$= E[(X_{n+1} - \mu)|z_1, z_2, \ldots, z_n] + E[Z_n|z_1, z_2, \ldots, z_n]$$

$$= E(X_{n+1} - \mu) + E[Z_n|z_1, z_2, \ldots, z_n]$$

$$= z_n : \text{ martingale}$$

For random stopping time N

$$E[Z_N] = E\left[\sum_{i=1}^{N}(X_i - \mu)\right] = E\left[\sum_{i=1}^{N}X_i - N\mu\right]$$

$$= E\left[\sum_{i=1}^{N}X_i\right] - E[N]\mu \tag{A.4}$$

But from Eq. (A.3)

$$E[Z_N] = E[Z_1] = E(X_1 - \mu) = 0$$

Hence from Eq. (A.4) and above, the average final position of the particle doing a random walk – with random stopping time given by random integer N– is given by

$$E\left[\sum_{i=1}^{N}X_i\right] = E[N]\mu : \text{ Walds equation}$$

A.2 Dirac Delta function

The Dirac Delta function is useful in the study of continuous spaces, and some of its properties are reviewed. Dirac Delta functions are not ordinary Lebesgue measureable functions since they have support on a set that has zero measure; rather they are generalized functions also called distributions. The Dirac Delta function is the continuum generalization of the discrete Kronecker delta function.

Consider a continuous line labelled by co-ordinate x such that $-\infty \leq x \leq +\infty$, and let $f(x)$ be an infinitely differentiable function. The Dirac Delta function, denoted by $\delta(x - a)$, is defined by

$$\delta(x - a) = \delta(a - x) : \text{ even function}$$

$$\delta(c(x - a)) = \frac{1}{|c|}\delta(x - a)$$

$$\int_{-\infty}^{+\infty} dx f(x)\delta(x - a) = f(a) \tag{A.5}$$

$$\int_{-\infty}^{+\infty} dx f(x)\frac{d^n}{dx^n}\delta(x - a) = (-1)^n \frac{d^n}{dx^n}f(x)|_{x=a}$$

$$\tag{A.6}$$

The Heaviside step function $\Theta(t)$ is defined by

$$\Theta(t) = \begin{cases} 1 & t > 0 \\ \frac{1}{2} & t = 0 \\ 0 & t < 0 \end{cases} \tag{A.7}$$

From its definition $\Theta(t) + \Theta(-t) = 1$. The following is a representation of the δ function.

$$\int_{-\infty}^{b} \delta(x - a) = \Theta(b - a) \tag{A.8}$$

$$\Rightarrow \int_{-\infty}^{a} \delta(x - a) = \Theta(0) = \frac{1}{2} \tag{A.9}$$

where last equation is due to the Dirac Delta function being an even function. From Eq. (A.8)

$$\frac{d}{db}\Theta(b - a) = \delta(b - a)$$

A representation of the δ-function based on the Gaussian distribution is

$$\delta(x - a) = \lim_{\sigma \to 0} \frac{1}{\sqrt{2\pi\sigma^2}} \exp\left\{-\frac{1}{2\sigma^2}(x - a)^2\right\} \tag{A.10}$$

Moreover

$$\delta(x - a) = \lim_{\mu \to \infty} \frac{1}{2}\mu \exp\left\{-\mu|x - a|\right\}$$

From the definition of Fourier transforms

$$\delta(x - a) = \int_{-\infty}^{+\infty} \frac{dp}{2\pi} e^{ip(x-a)} \tag{A.11}$$

To see the relation of the Dirac Delta function with the discrete Kronecker Delta, recall for n, m integers

$$\delta_{n-m} = \begin{cases} 0 & n \neq m \\ 1 & n = m \end{cases}$$

Discretize continuous variable x into a lattice of discrete points $x = n\epsilon$, and let $a = m\epsilon$; then $f(x) \to f_n$. Discretizing Eq. (A.5) gives

$$\int_{-\infty}^{+\infty} dx f(x)\delta(x - a) \to \epsilon \sum_{-\infty}^{+\infty} f_n\delta(x - a) = f_m$$

$$\Rightarrow \delta(x - a) \to \frac{1}{\epsilon}\delta_{n-m} \tag{A.13}$$

Hence, taking the limit of $\epsilon \to 0$ in the equation above

$$\delta(x - a) = \lim_{\epsilon \to 0} \frac{1}{\epsilon}\delta_{n-m} = \begin{cases} 0 & x \neq a \\ \infty & x = a \end{cases}$$

A.2.1 Functional derivatives

The concept of partial derivatives has a natural generalization for the case of infinitely many independent variables. Consider n independent variables X_n, with the index running over all integers, namely $n = 0, \pm 1, \pm 2, \pm 3, \ldots, \pm\infty$. Hence

$$\frac{\partial X_m}{\partial X_n} = \delta_{n-m}$$

Suppose $t = n\epsilon$, $t' = m\epsilon$, and consider the limit of $\epsilon \to 0$; this yields an infinite collection of independent variables $X(t)$; the functional derivative $\delta/\delta X(t)$ is defined by

$$\frac{\delta X(t')}{\delta X(t)} \equiv \lim_{\epsilon \to 0} \frac{1}{\epsilon} \frac{\partial X_m}{\partial X_n}$$

$$= \lim_{\epsilon \to 0} \frac{1}{\epsilon} \delta_{n-m}$$

$$\Rightarrow \frac{\delta X(t')}{\delta X(t)} = \delta(t - t') \tag{A.15}$$

For example

$$\frac{\delta}{\delta X(t)} \int_{-\infty}^{+\infty} ds \ f(s) X(s) = f(t)$$

A.3 Gaussian integration

Gaussian integration permeates all of theoretical finance, as well as forming one of the foundations of quantum theory. One-dimensional and multi-dimensional Gaussian integration are briefly reviewed. Gaussian integrals have the remarkable property that they can be generalized to infinite dimensions, and is briefly discussed.

A.3.1 One-dimensional Gaussian integral

Consider the one-dimensional Gaussian integral

$$Z[j] = \mathcal{N} \int_{-\infty}^{+\infty} e^{-\frac{1}{2}\lambda x^2 + jx} dx$$

All the moments of x can be obtained by

$$E[x^n] = \frac{d^n Z[j]}{dj^n}\bigg|_{j=0}$$

and hence $Z[j]$ is called the moment generating function for the Gaussian distribution.

The normalization constant $\mathcal{N}$ is chosen so that $Z(0) = 1$. Squaring $Z[0]$, and converting to polar coordinates, gives

$$Z^2[0] = \mathcal{N}^2 \int_{-\infty}^{+\infty} \int_{-\infty}^{\infty} e^{-\frac{1}{2}\lambda(x^2 + y^2)} dx dy = \mathcal{N}^2 \int_0^{\infty} \int_0^{2\pi} r e^{-\frac{1}{2}\lambda r^2} dr d\theta$$

$$1 = \mathcal{N}^2 2\pi \int_0^{\infty} d\xi e^{-\lambda\xi} = \mathcal{N}^2 \frac{2\pi}{\lambda} \Rightarrow \mathcal{N} = \sqrt{\frac{\lambda}{2\pi}}$$

Shifting $x \to x - \frac{j}{\lambda}$ leaves the integration measure invariant, and hence yields the final result

$$Z[j] = e^{\frac{1}{2\lambda}j^2} \mathcal{N} \int_{-\infty}^{+\infty} e^{-\frac{1}{2}\lambda x^2} dx$$

$$= e^{\frac{1}{2\lambda}j^2} \tag{A.16}$$

A.3.2 Higher-dimensional Gaussian integral

The general n-dimensional Gaussian integral, with variables $x_1, x_2, \ldots, x_n$, can be written as

$$Z[J] = \mathcal{N} \int_{-\infty}^{+\infty} e^S dx_1 dx_1 dx_2 \ldots dx_n$$

$$S = -\frac{1}{2} \sum_{i,j=1}^{n} x_i A_{ij} x_j + \sum_{i=1}^{n} J_i x_i$$

with normalization constant chosen so that $Z(0) = 1$. In quantum theory, S is called the action.

Let A be a $n \times n$ symmetric matrix which can be diagonalized by an orthogonal matrix M and yields

$$A = M^T \text{diag}(\lambda_1, \lambda_2, \ldots, \lambda_n) M$$

$$MM^T = \mathcal{J}_{n \times n}$$

where $\mathcal{J}_{n \times n}$ is a $n \times n$ unit matrix, and M^T is the transpose of M.

Only matrices A with positive eigenvalues $\lambda_i \geq 0$ are considered. A change variables gives

$$x_i = \sum_{j=1}^{n} M_{ij} z_j$$

$$\prod_{i=1}^{n} dx_i = \det(M) \prod_{i=1}^{n} dz_i = \prod_{i=1}^{n} dz_i$$

and hence the n-dimensional Gaussian integral is given by

$$Z[J] = \mathcal{N} \prod_{i=1}^{n} \left[\int_{-\infty}^{+\infty} dz_i e^{-\frac{1}{2}\lambda_i z_i^2 + \tilde{J}_i z_i} \right]$$

$$\tilde{J}_i \equiv \sum_{j=1}^{n} J_j M_{ji}^T$$

The n-dimensional Gaussian integral has completely factorized into a product of one-dimensional Gaussian integrals, all of which can be evaluated by the result given in

Eq. (A.16). Hence

$$Z[J] = \mathcal{N} \prod_{i=1}^{n} \left[\sqrt{\frac{2\pi}{\lambda_i}} \, e^{\frac{1}{2\lambda_i} \tilde{J}_i^2} \right] \tag{A.17}$$

In matrix notation

$$\mathcal{N} \prod_{i=1}^{n} \sqrt{\frac{2\pi}{\lambda_i}} = \mathcal{N}(2\pi)^{n/2} \frac{1}{\sqrt{\det A}} = 1$$

$$\sum_{i=1}^{n} \frac{1}{\lambda_i} \tilde{J}_i^2 = J \frac{1}{A} J \equiv J A^{-1} J$$

Hence, the final result can be written as

$$Z[J] = \exp\left(\frac{1}{2} J A^{-1} J\right) \tag{A.18}$$

A.3.3 Infinite-dimensional Gaussian integration

Consider a continuum number of integration variables $x(t)$, with $-\infty \leq t \leq +\infty$, and with the 'action' given by

$$S = -\frac{1}{2} \int_{-\infty}^{+\infty} dt\, dt'\, x(t) D^{-1}(t, t') x(t') + \int_{-\infty}^{+\infty} dt\, J(t) x(t) \tag{A.19}$$

By discretizing the variable t, following the steps taken in the derivation of $n \times n$ case, and then taking the limit of $n \to \infty$ yields

$$Z[J] = \mathcal{N} \prod_{t=-\infty}^{+\infty} \int_{-\infty}^{+\infty} dx(t) e^S = \exp\left\{ \frac{1}{2} \int_{-\infty}^{+\infty} dt\, dt'\, J(t) D(t, t') J(t') \right\} \tag{A.20}$$

$$\int_{-\infty}^{+\infty} ds\, D^{-1}(t, s) D(s, t') = \delta(t - t')$$

The normalization $\mathcal{N}$ is now a divergent quantity, and ensures the usual normalization $Z(0) = 1$. In discussions on quantum theory, Eq. (A.20) plays a central role.

The fundamental reason why Gaussian integration generalizes to infinite dimensions is because the measure is invariant under translations, that is under $x(t) \to x(t) + \xi(t)$; one can easily verify that this symmetry of the measure yields the result obtained in A.20.

Example

Consider the infinite time action of the 'harmonic oscillator' given by

$$S = -\frac{m}{2} \int_{-\infty}^{+\infty} dt \left[\left(\frac{dx(t)}{dt}\right)^2 + \omega^2 x^2(t) \right]$$

$$= -\frac{m}{2} \int_{-\infty}^{+\infty} dt\, x(t) \left(-\frac{d^2}{dt^2} + \omega^2\right) x(t)$$

$$\Rightarrow D^{-1}(t, t') = m \left(-\frac{d^2}{dt^2} + \omega^2\right) \delta(t - t')$$

where an integration by parts was done, discarding boundary terms at $\pm\infty$. The propagator $D(t, t')$ is given by

$$D(t, t') = \frac{1}{2\pi m} \int_{-\infty}^{+\infty} dp \frac{e^{ip(t-t')}}{p^2 + \omega^2}$$

$$= \frac{1}{2m|\omega|} e^{-|\omega||t-t'|}$$

The result above can be verified by using Eq. (A.11).

A.3.4 Normal random variable

The Normal, or Gaussian, random variable – denoted by $N(\mu, \sigma)$ – is a variable x that has a probability distribution given by

$$P(x) = \frac{1}{\sqrt{2\pi\sigma^2}} \exp\left\{-\frac{1}{2\sigma^2}(x - \mu)^2\right\}$$

From Eq. (A.16)

$$E[x] \equiv \int_{-\infty}^{+\infty} x P(x) = \mu \quad : \text{mean}$$

$$E[(x - \mu)^2] \equiv \int_{-\infty}^{+\infty} (x - \mu)^2 P(x) = \sigma^2 \quad : \text{variance}$$

Any normal random variables is equivalent to the $N(0, 1)$ random variable via the following linear transformation

$$X = N(\mu, \sigma) \; ; \; Z = N(0, 1)$$
$$\Rightarrow X = \mu + \sigma Z$$

All the moments of the random variable $Z = N(0, 1)$ can be determined by the generating function given in Eq. (A.16); namely

$$E[z^n] = \frac{d^n}{dj^n} Z[j]|_{j=0} \; ; \; Z[j] = e^{\mu j + \frac{1}{2}\sigma^2 j^2} \tag{A.21}$$

The cumulative distribution for the normal random variable $N(x)$ is defined by

$$\text{Prob}(-\infty \le z \le x) = N(x) = \frac{1}{\sqrt{2\pi}} \int_{-\infty}^{x} e^{-\frac{1}{2}z^2} dz \tag{A.22}$$

A sum of normal random variables is also another normal random variable

$$X_1 = N(\mu_1, \sigma_1); \quad X_2 = N(\mu_2, \sigma_2); \quad \ldots, X_n = N(\mu_n, \sigma_n)$$
$$\Rightarrow X = \sum_{i=1}^{n} \alpha_i X_i = N(\mu, \sigma) \text{ with } \mu = \sum_{i=1}^{n} \alpha_i \mu_i; \quad \sigma^2 = \sum_{i=1}^{n} \alpha_i^2 \sigma_i^2$$

The result above can be proven using the generating function given in Eq. (A.21).

A.4 White noise

The fundamental properties of Gaussian white noise are that

$$< R(t) >= 0; \quad < R(t)R(t') >= \delta(t - t')$$

Discretize time, namely $t = n\epsilon$, with $R(t) \to R_n$. The probability distribution function of white noise is given by

$$P(R_n) = \sqrt{\frac{\epsilon}{2\pi}} e^{-\frac{\epsilon}{2} R_n^2} \tag{A.23}$$

Hence, R_n is a Gaussian random variable with zero mean and $1/\sqrt{\epsilon}$ variance denoted by $N(0, 1/\sqrt{\epsilon})$. The following result is essential in deriving the rules of Ito calculus

$$R_n^2 = \frac{1}{\epsilon} + \quad \text{random terms of } 0(1) \tag{A.24}$$

To prove result stated in Eq. (A.24), it is shown that the generating function of R_n^2 can be derived from a R_n^2 that is deterministic. All the moments of R_n^2 can be determined from its generating function, namely

$$E\left[(R_n^2)^k\right] = \frac{d^k}{dt^k} E\left[e^{t R_n^2}\right]\Big|_{t=0}$$

Note one needs to evaluate the generating function $E\left[e^{t R_n^2}\right]$ only in the limit of $t \to 0$. Hence, for ϵ small but fixed

$$\lim_{t \to 0} E\left[e^{t R_n^2}\right] = \int_{-\infty}^{+\infty} dR_n e^{t R_n^2} \sqrt{\frac{\epsilon}{2\pi}} e^{-\frac{\epsilon}{2} R_n^2}$$

$$= \frac{1}{\sqrt{1 - \frac{2t}{\epsilon}}}$$

$$\sim \exp\left(\frac{t}{\epsilon}\right) + O(1)$$

The probability distribution function for R_n^2 which gives the above generating function is given by

$$P(R_n^2) = \delta\left(R_n^2 - \frac{1}{\epsilon}\right)$$

In other words, although R_n is a random variable, the quantity R_n^2, to leading order in ϵ, is **not** a random variable, but is instead fixed at the value of $1/\epsilon$.

To write the probability measure for $R(t)$, with $t_1 \leq t \leq t_2$ discretize $t \to n\epsilon$, with $n = 1, 2, \ldots, N$, and with $R(t) \to R_n$. White noise $R(t)$ has the probability distribution given in Eq. (A.23). The probability measure for the white noise random variables in the interval

$t_1 \le t \le t_2$ is given by

$$\mathcal{P}[R] = \prod_{n=1}^{N} P(R_n) = \prod_{n=1}^{N} e^{-\frac{\epsilon}{2}R_n^2} \tag{A.25}$$

$$\int dR = \prod_{n=1}^{N} \sqrt{\frac{\epsilon}{2\pi}} \int_{-\infty}^{+\infty} dR_n$$

For notational simplicity take the limit of $\epsilon \to 0$; for purposes of rigor, the continuum notation is simply a short-hand for taking the continuum limit of the discrete multiple integrals given above. For $t_1 < t < t_2$

$$\mathcal{P}[R, t_1, t_2] \to e^S \tag{A.26}$$

$$S = -\frac{1}{2} \int_{t_1}^{t_2} dt\, R^2(t) \tag{A.27}$$

$$\int dR \to \int DR \tag{A.28}$$

The 'action functional' S is ultra-local with all the variables being decoupled.

A.4.1 Integral of white noise

Consider the following integral of white noise

$$I = \int_{t}^{T} dt'\, R(t') \sim \epsilon \sum_{n=0}^{M} R_n \; ; \; M = \left[\frac{T-t}{\epsilon} \right]$$

where ϵ is an infinitesimal. For Gaussian white noise

$$R_n = N\left(0, \frac{1}{\sqrt{\epsilon}}\right) \quad \Rightarrow \epsilon R_n = N(0, \sqrt{\epsilon})$$

The integral of white noise is a sum of normal random variables and hence, from Eq. (A.23) and above, is also a Gaussian random variable given by

$$I \sim N(0, \sqrt{\epsilon M}) \to N(0, \sqrt{T-t}) \tag{A.29}$$

In general, for

$$Z = \int_{t}^{T} dt'\, a(t') R(t') \quad \Rightarrow \quad Z = N(0, \sigma^2)\,; \; \sigma^2 = \int_{t}^{T} dt'\, a^2(t')$$

A.5 The Langevin equation

A physical formulation of stochastic processes is given by the Langevin equation. Langevin considered the equation of motion of a particle in a fluid which is classically given by

$$M\frac{dv}{dt} + \gamma v + \Phi(v) = 0$$

where γ is the coefficient of friction, and $\Phi(v)$ is a potential that can be an arbitrary function of the velocity v.

For large objects the trajectory followed is very smooth, and hence the equation is adequate. However, for tiny particles the size of a grain of pollen, it is known that they undergo more than 10^9 collisions per second, and so it is not very meaningful to talk of a smooth trajectory. To describe the motion of a tiny speck of pollen, Langevin proposed that the medium the tiny particle is moving in be replaced by a random force that continuously acts on the particle, randomly changing the magnitude and direction of its velocity. The random force in effect makes the particle's velocity random, and thus making the particle's path random.

Langevin generalized the deterministic equation of motion to the following stochastic differential equation[1]

$$M\frac{dv}{dt} + \gamma v + \Phi(v) = F(t) \tag{A.30}$$

where $F(t)$ is a random (stochastic) force with zero mean and covariance given by

$$E[F(t)] = 0; \; E[F(t)F(s)] = 2D\delta(t-s)$$

Equation (A.30), for $\Phi(v) = 0$, can be written as

$$\frac{dv}{dt} = -\frac{\gamma v}{M} + \frac{\sqrt{2D}}{M}R$$

where R is white noise.

The formal solution for the Langevin equation gives

$$v = v_0 e^{-\gamma t/M} + \frac{1}{M}\int_0^t e^{-\frac{\gamma}{M}(t-\tau)}F(\tau)d\tau$$

so that

$$E[v(t)] = v_0 e^{-\gamma t/M} + \frac{1}{M}\int_0^t e^{\frac{\gamma}{M}(t-\tau)}E[F(\tau)]d\tau = v_0 e^{-\gamma t/M}$$

and hence

$$E\left[\left(v(t) - v_0 e^{-\gamma t/M}\right)^2\right] = \left(\frac{1}{M}\right)^2 \int_0^t \int_0^t e^{-\frac{\gamma}{M}(t-\eta)}e^{-\frac{\gamma}{M}(t-\tau)}E[F(\eta)F(\tau)]d\eta\, d\tau$$

$$= \frac{D}{\gamma M}\left(1 - e^{-2\gamma t/M}\right)$$

A similar derivation, for $v_0 = 0$, gives the unequal and equal time velocity correlators as

$$E[v(t)v(t')] = \frac{D}{\gamma M}\left[e^{-\frac{\gamma}{M}(t-t')} - e^{-\frac{\gamma}{M}(t+t')}\right]; \; t > t'$$

$$\Rightarrow \lim_{t\to\infty} E[v^2(t)] = \frac{D}{\gamma M}$$

---

[1] A differential equation is called stochastic if it includes a random function; for the case of the Langevin equation, the random force $F(t)$ is an arbitrary function of time; viewed as a deterministic differential equation, a stochastic differential equation has infinitely many solutions, one solution for each of the random configurations that is possible for $F(t)$.

As $t \to \infty$, the particle attains equilibrium with its surroundings. Hence, the velocity distribution should be Maxwellian

$$P(v) = \left(\frac{M}{2\pi kT}\right)^{1/2} \exp\left(-\frac{Mv^2}{2kT}\right) \quad \Rightarrow \quad E[v^2(t)] = \frac{kT}{M}$$

which when compared with the solution above yields $D = kT\gamma$, which is the Einstein relation.

Let the position of the randomly evolving particle be $x(t)$ so that $v = dx/dt$. Integrating the Langevin equation, and ignoring the effects of the initial condition on the long time evolution of the particle gives

$$E[x(t)x(t')] = \frac{2D}{\gamma^2}t' + \frac{DM}{\gamma^3}\left[-2 + 2e^{-\frac{\gamma}{M}t} + 2e^{-\frac{\gamma}{M}t'} - e^{-\frac{\gamma}{M}|t-t'|} - e^{-\frac{\gamma}{M}(t+t')}\right]; \ t > t'$$

which yields the equal time correlator

$$E[x^2(t)] = \frac{2D}{\gamma^2}t + \frac{DM}{\gamma^3}\left[-3 + 4e^{-\frac{\gamma}{M}t} - e^{-\frac{\gamma}{M}(t+t')}\right]$$

$$\lim_{t\to\infty}\sqrt{E[x^2(t)]} \to \sqrt{\frac{2D}{\gamma^2}} \times \sqrt{t}$$

This is an important result: for a particle undergoing random motion, the average dispersal from its starting point, namely $\sqrt{E[x^2(t)]}$ is proportional to the $\sqrt{t}$. Recall that the average (expected) value of the particle's position is always zero since $E[x(t)] = 0$; what the result states is that the distance travelled away from the origin is proportional to the square root of time. This fact is encountered in a more abstract setting in Ito calculus.

Consider the stochastic differential equation for the logarithm of the stock price, given by Eq. (3.12) as

$$\frac{dx}{dt} = r - \frac{\sigma^2}{2} + \sigma R(t)$$

Changing variables to $v = x - (r - \sigma^2/2)t$ yields the Langevin equation with $\gamma = 0$ and white noise $R(t)$.

The main importance of the Langevin equation is that it gives a different formulation of stochastic processes, and can sometimes lead to different methods for solving the stochastic differential equation.

A.5.1 *Martingale condition*

The martingale condition for a continuous stochastic processes $v(t)$ states that, given the occurrence of the value $v(t)$ at time t, the conditional expectation at the next instant is the following: $E[v(t + \epsilon)|v(t)] = v(t)$.

The martingale condition can be discussed using the Langevin equation. A stochastic process $v(t)$ for which $\gamma v + \Phi(v) = 0$ has

$$M\frac{dv}{dt} = F(t) \quad \Rightarrow \quad E\left[\frac{dv}{dt}\right] = 0 \ : \ \text{no drift} \tag{A.31}$$

The conditional expectation value can be evaluated by discretizing time, and yields

$$\frac{M}{\epsilon} E[v(t+\epsilon) - v(t)|v(t)] = E[F(t)] = 0$$

$$\Rightarrow E[v(t+\epsilon)|v(t)] = v(t) : \text{Martingale}$$

In other words, any stochastic process that does not have a drift term, or more generally a potential term, is a martingale.

A.6 Fundamental theorem of finance

Consider a financial market that is (a) a complete market and (b) in which the condition of no arbitrage holds. The fundamental theorem of finance states that conditions (a) and (b) are equivalent to the evolution of financial instruments obeying the martingale condition. Furthermore, once the discounting factor is fixed, for such a financial market there exists a **unique** risk-free, or risk-neutral, probability measure for which the evolution of all the financial instruments obey the martingale condition [40].

Suppose one has m securities, namely $S_i(t)$, with drift velocities $\mu_i(t)$ and that are driven by n white noise $R_j(t)$ and volatilities $\sigma_j^i(t)$, with $i = 1, 2, \ldots, m$ and $j = 1, 2, \ldots, n$. One obtains the following m coupled Langevin equations

$$\frac{dS_i(t)}{dt} = \mu_i(t)S_i(t) + S_i(t)\sum_{j=1}^{n}\sigma_j^i(t)R_j(t) \tag{A.32}$$

$$E[R_j(t)R_k(t')] = \delta_{j-k}\delta(t-t') ; \quad \Rightarrow R_j(t)R_k(t') \sim \frac{1}{\epsilon}\delta_{j-k}$$

(Unless necessary, all dependence on time t will henceforth suppressed.) Completeness of the market implies that the volatility functions $\sigma_j^i(t)$ are linearly independent for each i.

Consider a portfolio designed to cancel all the fluctuations in the securities $S_i(t)$, namely

$$\Pi = \sum_{i=1}^{m}\theta_i S_i$$

such that

$$\sum_{i=1}^{m}\theta_i S_i\sigma_j^i = 0 \text{ for each } j$$

Since all terms with white noise are cancelled out in $d\Pi/dt$, the principle of no arbitrage gives the following result

$$\frac{d\Pi}{dt} = \sum_{i=1}^{m}\theta_i\mu_i(t)S_i(t) : \text{deterministic}$$

$$\text{no arbitrage} \Rightarrow \frac{d\Pi}{dt} = r\Pi = r\sum_{i=1}^{m}\theta_i S_i \tag{A.33}$$

Hence, from Eqs. (A.32) and (A.33)

$$\mu_i = r + \sum_{j=1}^{n}\lambda_j\sigma_j^i \tag{A.34}$$

where λ_j has the important property that it is independent of all the securities S_i.

λ_j is the market premium for obtaining a profit above the risk-free return r that the market offers to risk takers, and is valid for all securities. The fact that all the λ_j's are independent of the securities reflects the condition of no arbitrage, which requires that the market premium must be the same for all securities.

From Eqs. (A.32) and (A.34)

$$\frac{dS_i(t)}{dt} = r S_i + S_i \sum_{j=1}^{n} \sigma_j^i [R_j + \lambda_j] \tag{A.35}$$

Define the money market account by

$$B(t) = e^{\int_0^t r(s)ds}$$

and define the discounted security by

$$X_i(t) \equiv e^{-\int_0^t r(s)ds} S_i(t)$$

Hence, from Eq. (A.35)

$$\frac{dX_i(t)}{dt} = X_i \sum_{j=1}^{n} \sigma_j^i [R_j + \lambda_j] \tag{A.36}$$

The λ_i term above is a drift term, and hence $X_i(t)$ is not a martingale; to obtain a martingale process introduce the random variable

$$\rho(t) = \exp\left[-\sum_{j=1}^{n} \int_0^t \lambda_j(s)R_j(s)ds - \frac{1}{2}\sum_{j=1}^{n} \int_0^t \lambda_j^2(s)ds \right] \tag{A.37}$$

By its construction $\rho(t)$ satisfies the martingale condition

$$E_{(s,t)}[\rho(t)|\rho(s)] = \rho(s) \; ; \; s < t \tag{A.38}$$

as can be seen by performing the integrations over white noise using Eq. (A.27).

Using Ito's chain rule given in Eq. (3.9), and from Eqs. (A.36) and (A.37) [42]

$$\frac{d(\rho X_i)}{dt} = \rho X_i \sum_{j=1}^{n} [\sigma_j^i - \lambda_j] R_j \quad : \quad \text{martingale} \tag{A.39}$$

Since Eq. (A.39) has no drift term, it follows from Appendix A.5.1 that ρX_i is a martingale condition, namely

$$E_{(s,t)}[\rho(t)X(t)|\rho(s)X(s)] = \rho(s)X(s) \; ; \; s < t \; : \; \text{martingale}$$

One can redefine the probability measure for white noise $R(t)$ given in Eq. (A.27) to incorporate the exponential martingale $\rho(t)$, and the new measure is called the 'risk-neutral' martingale measure, and with respect to which the instrument $X(t) = e^{-\int_0^t r(s)ds} S_i(t)$ obeys the martingale condition. In other words, denoting the expectation with respect to the martingale measure by $E^M[\ldots]$, yields the following

$$E_{(t_*,t)}^M[e^{-\int_0^t r(s)ds} S_i(t)|X(t_*)] = e^{-\int_0^{t_*} r(s)ds} S_i(t_*) \; ; \; t_* < t \; : \; \text{martingale}$$

In particular, for $t_* = 0$

$$S(0) = E^M_{(0,t)}\left[e^{-\int_0^t r(t')dt'} S(t) | S(0)\right] \tag{A.40}$$

It can further be shown that the risk-neutral martingale measure is a unique measure for a given complete and arbitrage-free financial market.

A.7 Evaluation of the propagator

Given the importance of the propagator, it is evaluated using two different methods. The eigenfunction expansion is essentially the same derivation given in Section 5.5, whereas the Greens function approach is a different and powerful method for solving for the propagator with a more complex structure.

A.7.1 Eigenfunction expansion

The propagator $D(x, x'; t, T_{FR})$ is evaluated using the eigenfunctions. From Eq. (5.30)

$$\left[1 - \frac{1}{\mu^2}\frac{\partial^2}{\partial x^2}\right] D(x, x'; t, T_{FR}) = \delta(x - x') \; ; \; t \leq x \leq t + T_{FR} \tag{A.41}$$

The Neumann boundary conditions yield

$$\frac{\partial}{\partial x} D(x, x'; t, T_{FR})\Big|_{x=t} = 0 = \frac{\partial}{\partial x} D(x, x'; t, T_{FR})\Big|_{x=t+T_{FR}} \tag{A.42}$$

The normalized eigenfunctions on the interval $[t, t + T_{FR}]$ that satisfy the Neumann condition of vanishing derivatives at $x = t$ and $x = t + T_{FR}$, from Eq. (5.31), are given by

$$\psi_0(x) = \frac{1}{\sqrt{T_{FR}}} \; ; \; \psi_m(x) = \frac{2}{\sqrt{T_{FR}}} \cos\left\{\frac{m\pi(x - t)}{T_{FR}}\right\} \quad m = 1, 2, 3, \ldots, \infty$$

that satisfy the eigenvalue equation

$$\left(-\frac{1}{\mu^2}\frac{\partial^2}{\partial x^2} + 1\right)\psi_m(x) = \left[\left(\frac{m\pi}{\mu T_{FR}}\right)^2 + 1\right]\psi_m(x) \equiv \lambda_m \psi_m(x) \; ; \; m = 0, 1, 2, 3, \ldots, \infty$$

It can be shown that any arbitrary function $f(O)$ of an operator O can be expressed in terms of the eigenvalues and eigenfunctions of operator O as follows

$$< x|f(O)|x' > = \sum_{m=0}^{\infty} \psi_m(x)\psi_m(x')f(\lambda_m)$$

Hence, since the propagator is the inverse of Hermitian differential operator as given in Eq. (A.41), one has the following expansion

$$D(x, x'; t, T_{FR}) = \sum_{m=0}^{\infty} \frac{\psi_m(x)\psi_m(x')}{\lambda_m}$$

The summation above has been performed in Eqs. (5.34)–(5.37) and yields

$$D(x, x'; t, T_{FR}) = \frac{\mu}{2 \sinh \mu T_{FR}} [\cosh\{\mu T_{FR} - \mu|x - x'|)$$
$$+ \cosh(\mu T_{FR} - \mu(x + x' - 2t)\}] \qquad (A.43)$$

A.7.2 Greens function

The method of Greens function is used to evaluate the propagator. From Eqs. (A.41), for $x \neq x'$, the propagator satisfies the equation

$$\left[1 - \frac{1}{\mu^2} \frac{\partial^2}{\partial x^2} \right] D(x, x'; t, T_{FR}) = 0 \; ; \; x \neq x' \qquad (A.44)$$

Consider the associated equation

$$\frac{\partial^2}{\partial x^2} u(x) = \mu^2 u(x)$$

which has the general solution

$$u(x) = a e^{\mu x} + b e^{-\mu x}$$

Let $u_<(x), u_>(x)$ be solutions that satisfy the boundary condition at $x = t$ and $t + T_{FR}$ respectively.

The Greens function (propagator) is then given by

$$D(x, x'; t, T_{FR}) = u_<(x) u_>(x') \Theta(x - x') + u_<(x') u_<(x) \Theta(x' - x)$$
$$= D(x, x'; t, T_{FR}) \; : \; \text{symmetric in } x, x'$$

To produce the correct coefficient for the $\delta(x - x')$ in Eq. (A.41), the (Wronskian of the) solution for $D(x, x')$ must obey

$$\frac{du_<(x)}{dx} u_>(x) - u_<(x) \frac{du_>(x)}{dx} = \mu^2$$

Neumann and Dirichlet Boundary Conditions

Suppose the propagator satisfies the Neumann boundary conditions at the two boundaries as given in Eq. (A.42), namely at $x = t$ and at $x = t + T_{FR}$. Choose solutions that satisfy the Neumann conditions, as in Eq. (A.42), at only one of the boundaries, namely

$$u_<(x) = A \cosh \mu(x - t)$$
$$u_>(x) = B \cosh \mu(T_{FR} + t - x)$$

Hence

$$\frac{du_<(x)}{dx} u_>(x) - u_<(x) \frac{du_>(x)}{dx} = \mu^2$$
$$\Rightarrow AB = \frac{\mu}{\sinh \mu T_{FR}}$$

The propagator is hence given by

$$D(x, x'; t, T_{FR}) = \mu \frac{\cosh \mu(x - t) \cosh \mu(T_{FR} + t - x')}{\sinh \mu T_{FR}} \Theta(x' - x)$$
$$+ \mu \frac{\cosh \mu(x' - t) \cosh \mu(T_{FR} + t - x)}{\sinh \mu T_{FR}} \Theta(x - x')$$

and can be shown to be equal to the result obtained in Eq. (A.43).

The mixed Dirichlet–Neumann boundary conditions are given by

$$D(x, x'; t, T_{FR})\Big|_{x=t} = 0 = \frac{\partial}{\partial x} D(x, x^*; t, T_{FR})\Big|_{x=t+T_{FR}} \tag{A.45}$$

and the Dirichlet–Dirichlet boundary conditions are given by

$$D(x, x'; t, T_{FR})\Big|_{x=t} = 0 = D(x, x'; t, T_{FR})\Big|_{x=t+T_{FR}}$$

The propagator for the various boundary conditions can be obtained by the same procedure as used for the Neumann–Neumann boundary conditions. The results are summarized in Table A.7.2, with the full propagator being obtained by using its symmetry under x, x'.

Table A.1 *Summary of the propagators for the different boundary conditions*

Boundary conditions	Propagator $D(x, x'; t, T_{FR}) : x > x'$
Neumann–Neumann	$\mu \dfrac{\cosh \mu(x-t) \cosh \mu(T_{FR}+t-x')}{\sinh \mu T_{FR}}$
Dirichlet–Dirichlet	$\mu \dfrac{\sinh \mu(x-t) \sinh \mu(T_{FR}+t-x')}{\sinh \mu T_{FR}}$
Dirichlet–Neumann	$\mu \dfrac{\sinh \mu(x-t) \cosh \mu(T_{FR}+t-x')}{\cosh \mu T_{FR}}$

Taking the limit of $\mu \to 0$ for the propagator with Dirichlet–Dirichlet boundary conditions yields

$$D_{DD}(x, x'; t, T_{FR}) = \mu \left[(x - t)\Theta(x' - x) + (x' - t)\Theta(x - x') \right.$$
$$\left. - \frac{(x - t)(x' - t)}{T_{FR}} \right] \tag{A.46}$$

where $\Theta(x) + \Theta(-x) = 1$ has been used.

Brief Glossary of Financial Terms

Arbitrage. Gaining a risk-free profit above the spot interest rate by simultaneously entering into two or more financial transactions.

Bond. An instrument of debt.

Capital. Economic value of society's real assets.

Capital market. A market that trades in the primary forms of financial instruments, namely in instruments of equity, debt and derivatives.

Coupon bond. A financial instrument of debt that promises a pre-determined series of cash flows.

Derivative securities. Financial assets that are derived from other financial assets, including other derivatives. The main forms of derivatives are forward, futures and option contracts.

Discounting. The process that yields the factor relating the future value of money to its present value.

Equity. A share in the ownership of a real asset, like a company.

Efficient market hypothesis. For a financial market in equilibrium, changes in the prices of all securities are random.

Financial assets. Pieces of paper that entitle its holder to the ownership of (a fraction of) real assets, and to the income (if any) that is generated by the underlying real assets.

Financial market. Market where trade in financial assets and instruments is conducted.

Financial instrument. A specific form of a financial asset – be it a stock or a bond.

Fixed income securities. Instruments of debt issued by corporations and governments that promise either a single fixed payment or a stream of fixed payments. Also called corporate and sovereign bonds, respectively.

Forward contract. A contract between a buyer and a seller, in which the seller agrees to provide the commodity or financial instrument at some future time for a price fixed at present time, with only a single cash flow when the contract matures.

Forward interest rates. The forward interest rate, also called the forward rate, $f(t, x)$ is the agreed upon, at time $t < x$, future interest rate for an instantaneous loan at a **future time** x.

Futures contract. A contract similar to a forward contract; a major difference is that, unlike a forward contract, for a futures contract there is a series of cash flows for the duration of the contract.

Hedging. A general term for the procedure of **reducing** the random fluctuations in the value of a financial instrument, and hence reducing the risk to its future value.

Interest rates. The cost of borrowing money. See spot and forward interest rates.

Ito–Wiener process. A continuous stochastic process, usually indexed by the time variable t.

Market equilibrium. For a market in equilibrium, all information has been assimilated leading to all securities having their fair price. Theoretically, it is expected that all trading ceases for a market in equilibrium.

Martingale process. A mathematical formulation of a fair game. A martingale is a stochastic process such that it's expectation value at present, conditioned on the occurrence of $x_1, x_2, \ldots, x_n$ values of the stochastic process at n previous steps, is equal to it's value at the nth step. The mathematical definition of a martingale process is

$$E\left[X_{n+1}|x_1, x_2, \ldots, x_n\right] = x_n \quad : \quad \text{Martingale}$$

Money market. Market for trading in money market instruments such as short-term debt, cash and cash equivalents, foreign currency transactions and so on.

Numeraire. The discounting factor used in computing the present value of a financial instrument from its (pre-fixed) future value.

Option. A contract with a fixed maturity, and in which the buyer has the right to – but is not obliged to – either buy a security from, or sell a security to, the seller of the option at some pre-determined (but not necessarily fixed) strike price. Options are written on underlying financial instruments such as stocks, bonds and derivatives.

Pricing kernel. The conditional probability for the occurrence of the final value of a financial instrument, given its present value.

Principle of no arbitrage. No risk-free financial instrument can yield a rate of return above that of the (risk-free) spot interest rate.

Random variables. A numerical variable that has no fixed value, and instead takes values in a whole range, with the probability of its various outcomes given by a probability distribution.

Real assets. Capital goods, skilled management and labour force, and so on, that are necessary for producing goods and services in the real economy.

Return. The profit obtained from an investment.

Risk. The uncertainty is obtaining a return on investment.

Security. A financial instrument.

Spot interest rate. The spot rate $r(t)$ is the interest rate for an instantaneous loan at time t.

Stochastic process. A (time ordered) collection of random variables with outcomes governed by a joint probability distribution. The collection of random variables is given by X_s, where s is a continuous (discrete) index for a continuous (discrete) stochastic process.

Stocks and shares. Financial instruments representing (part) ownership of equity.

Treasury Bond. A zero coupon Treasury Bond is an instrument of debt that has no risk of default, also called a risk-free bond. Coupon Treasury Bonds are similarly risk-free coupon bonds.

Volatility. The standard deviation of any random variable (including financial instruments).

Zero coupon bond. A financial instrument of debt that promises a pre-determined single cash-flow consisting of say a $1 payoff at some future time T.

Brief Glossary of Physics Terms

Action. The time integral of the Lagrangian.

Bra and ket vectors. The 'bra' vector $< d|$ represents an element of the dual state space and the 'ket' vector $|v >$ representing a vector from the state space, and with the bracket $< d|v >$ being a complex number.

Completeness equation. A statement that the basis states for the state space form a complete basis by linearly spanning the entire state space.

Dual state space. A space associated with a vector space, consisting of all mappings of the state space into the complex numbers.

Degree of freedoms. The number of independent random variables that the system has at a given instant of time. Each degree of freedom corresponds to a unique stochastic process.

Eigenfunctions. Special state vectors that are associated with an operator such that under the action of the operator, the eigenfunctions are only changed up to a multiplicative constant, called the **eigenvalues** of the respective eigenfunctions.

Feynman path integral. See path integral.

Field. A function of two or more variables, denoted by say $f(t, x, y, \ldots)$; the field is different from a path in that in addition to dependence on time t, it also depends in general, on other variables $x, y, \ldots$

Fluctuation. One possible configuration of the quantum field, or one possible path for a randomly evolving particle.

Functional. A quantity that depends on a complete function. For example, the integral of a function is a functional of the integrand.

Gaussian distribution. Generic term for probability distributions that are given by an exponential of the quadratic function of the random variables. The normal distribution is the simplest example.

Generating functional. A functional from which all the moments of a collection of (infinitely many) random variables can be produced by differentiation.

Hamiltonian. A differential operator that evolves the system in time. In finance the Hamiltonian's primary role is to discount the future value of a financial instrument to yield its present value by evolving the system backwards in time.

Hermitian conjugation. The transposition and complex conjugation of the elements of an operator.

Lagrangian. A functional that assigns probabilities for the occurrence in, the path integral, of the various random paths or random field configurations.

Linear field theory. Theories such that the Lagrangian is a quadratic function of the fields. Linear theories are also called free, or Gaussian, field theories.

Nonlinear field theory. Theories with Lagrangian that have terms that are cubic or higher powers of the quantum fields. Also called interacting, or non-Gaussian, field theories.

Operators. The infinite-dimensional generalization of matrices that act on the elements of a state space. Empirically observable quantities are represented by operators, the most important being the energy, position and momentum operators for a quantum particle.

Path. One possible trajectory followed by a particle evolving in time, denoted by $x(t)$, where t is usually time.

Path integral. A functional integral, also called the Feynman path integral; over all the possible random paths taken by a random (stochastic) process. For a quantum field that takes random values on say a plane, the path integral is an integral over all possible functions on the plane.

Partition function. The functional integral of the exponential of the action over all possible configurations for the system in question.

Quantum mechanics. The mechanics of quantum particles that evolve randomly in time, with physically measurable quantities being represented by state vectors and operators.

Quantum field. The collection of all possible configurations of a field. For a two-dimensional plane, a quantum field is the collection of all possible functions on the plane.

Quantum field theory. The theory of modelling (empirical) systems with quantum fields. Empirically observed quantities are realized by correlation functions of the quantum field, obtained by averaging over all possible configurations of the quantum field, weighted by the (appropriately normalized) exponential of the action.

Rigidity. A term in the action that yields a propagator for the forward interest rates $f(t, x)$ with a second-order derivative in the forward direction.

State space. A linear vector space, the generalization of a finite-dimensional vector space, that is used to describe the state of a quantum (random) system.

Stiffness. An action for the forward rates that yields a propagator with up to fourth-order derivatives in the forward future direction.

White Noise. A continuous set of random variables that at every instant has a probability distribution given by the normal distribution.

Main symbols

Only new symbols introduced in a chapter are listed. A consistent system of notation has been used as far as possible.

Chapter 2 Introduction to finance

R	return on portfolio
$\equiv$	definition; identity
$E[X]$	expectation value of random variable X
σ	volatility of a stock or of a portfolio
$E[X_{n+1}\|x_1, \ldots, x_n]$	expectation of X_{n+1} conditioned on $x_1, \ldots, x_n$
$D(S)$	value of a derivative of S
$\Pi(S, t)$	portfolio for stock price S
$r(t)$	spot interest rate
$P(t, T)$	value of a Treasury Bond, at time t, maturing at T
$\mathcal{B}(t, T)$	value of a coupon Treasury Bond
$f(t, x)$	forward interest rates, at time t, for an instantaneous loan at future time x

Chapter 3 Derivative securities

$F(t, T)$	value of a forward contract at time t, maturing at T
$\mathcal{F}(t, T)$	value of a futures contract at time t, maturing at T
$C(t, S(t))$	value of a call option at time t
$P(t)$	value of a put option at time t
$S(t)$	stock price at time t
σ	volatility of the stock
$R(t)$	white noise
$< X >$	expectation value of random variable X
$P(r_n)$	probability distribution for discretized white noise
ϵ	infinitesimal quantity
$x(t)$	logarithm of the stock price $S(t)$
$\tau = T - t$	time remaining till maturity at time T
$P_m(x)$	martingale probability distribution for Black–Scholes
$N(d_\pm)$	cumulative distribution for the normal random variable
$V = \sigma^2$	variance of the stock price

ρ	correlation of white noises
λ, μ, ξ	parameters for stochastic volatility

Chapter 4 Hamiltonians and stock options

$\mathcal{V}$	linear vector space
$\mathcal{V}_{\text{dual}}$	dual linear vector space
$\lvert g >$	ket vector belonging to $\mathcal{V}$
$< p \rvert$	bra vector belonging to $\mathcal{V}_{\text{dual}}$
$\lvert x >$	ket basis vector for the real line
$\psi(x)$	components of ket vector $\lvert \psi >$ equal to $< x \lvert \psi >$
H	Hamiltonian
$\lvert \psi_E >$	eigenfunction of H
E	eigenvalue of $\lvert \psi_E >$
H_{BS}	Black–Scholes Hamiltonian
H_{MG}	Merton–Garman Hamiltonian
$p(x, y, T - t, x', y')$	pricing kernel for stock price with stochastic volatility
$\lvert C, t >$	option price ket vector
$C(t, x)$	components of option price ket vector $< x \lvert C, t >$
$p_{BS}(x, T - t, x')$	Black–Scholes pricing kernel
$\lvert p >$	momentum ket basis vector
H_{eff}	effective option pricing Hamiltonian
$V(x)$	potential representing path-dependent option
H_V	Hamiltonian with potential
H_{DO}, H_{DB}	barrier option Hamiltonians
$< x \lvert \phi_n >$	eigenfunction for double barrier option

Chapter 5 Path integrals and stock options

L	Lagrangian
S	action
$\int DX, \int DY$	path-integration measure
L_{BS}	Black–Scholes Lagrangian
$g[x, K]$	path-dependent payoff function
$Z(t_i, t_f; j)$	generating functional for harmonic oscillator
$D(t, t'; t_i, t_f)$	propagator
L_{MG}	Merton–Garman Lagrangian
S_0, S_1	stochastic volatility action
$Z(i, y, p)$	generating function for moments of stock and volatility
F, F'	solution for generating function $Z(i, y, p)$

Chapter 6 Stochastic interest rates' Hamiltonians and path integrals

$a(r, t), \sigma(r, t)$	drift and volatility of spot rate
$P_F(r, t : r_0)$	forward conditional probability for r
H_F, H_B	forward and backward Hamiltonians
L_F, S_F	forward Lagrangian and action
Z_B, S_B	backward partition function and action
S_V	Vasicek action
x	future time

$\alpha(t, x)$	drift velocity for forward rates
$\sigma_i(t, x)$	volatility functions for forward rates
$Z[j, t_1, t_2]$	generating functional for HJM model
t_0	present time
t_*	future time $> t_0$, usually maturity time for options
$E_{[t_0, t_*]}[X]$	expectation value of X over all random variables in $[t_0, t_*]$
$\mathcal{T}, \Delta_0, \mathcal{R}$	domains of integration for forward interest rates
$\mathcal{F}(t_0, t_*, T)$	futures price on a Treasury Bond
$\Omega, \Omega_{\mathcal{F}}$	functions in the bond futures price
$C(t_0, t_*, T, K)$	call option (maturity t_*) on a Treasury Bond (maturity T)
q^2	volatility in bond call option
$\Psi(G, t_*, T)$	function appearing in price of bond option
S_F^{bk}, S_B^{bk}	Black–Karasinski forward and backward actions
L_B^{bk}, H_B^{bk}	Black–Karasinski backward Lagrangian and Hamiltonian

Chapter 7 Quantum field theory of forward interest rates

S_{String}	string action
Z	forward rates' partition function
T_{FR}	maximum value for future time
$S[f], \mathcal{L}[f]$	forward rates' action and Lagrangian
t	time variable
x	future time $> t$
$\sigma(t, x)$	forward rates' volatility
$\int Df$	functional integration over all forward rates $f(t, x)$
$A(t, x)$	forward rates' velocity quantum field
$Z[J]$	generating function of velocity field
$\delta/\delta J(t, x)$	functional derivative with respect to $J(t, x)$
μ	rigidity parameter in the forward rates' Lagrangian
$D(x, x'; t)$	constant rigidity forward rates' propagator
θ	future time co-ordinate $x - t$
$\theta_{\pm}$	future time co-ordinates $\theta \pm \theta'$
m	slope of curve orthogonal to propagator's diagonal axis
$B(t_0, t_*)$	numeraire for money market account
S_*	forward rates' action with numeraire $P(t, T_*)$
$\alpha_*(t, x)$	drift of forward rates for numeraire $P(t, T_*)$
$\phi(t, x)$	(nonlinear) logarithm $\ln(f(t, x)/f_0)$ of forward rates
$\psi[t, x; A]$	$\phi(t, x)$ as a function of the $A(t, x)$ field
$\mathcal{J}(t, x; t'x'; A)$	matrix for the change of variables from $\phi(t, x)$ to $A(t, x)$
$c^{\dagger}(t, x), c(t, x)$	fermion quantum fields
ν	exponent for forward rates' stochastic volatility
ρ, ξ, κ	parameters for forward rates' stochastic volatility
$D_C(\theta, \theta')$	constrained forward rates' propagator
$D_M(\theta, \theta')$	non-constant rigidity forward rates' propagator
λ	stiffness parameter in forward rates' action
$G(\theta, \theta')$	forward rates' stiff propagator
r_{ϱ}	curvature orthogonal to propagator's diagonal axis
$z(\theta)$	psychological future time
$Z_f[J]$	forward rates' generating functional
f_{mn}, A_{mn}	lattice forward rates and velocity quantum fields

Chapter 8 Empirical forward rates and field theory models

$L(t,T)$	Libor simple interest rate
τ	time interval of 90 days
$<\delta f(t,\theta)>$	forward rates average rate of change
$<\delta f(t,\theta)\delta f(t,\theta')>_c$	forward rates' velocity correlation functions
$\kappa(t,\theta)$	kurtosis of forward rates
$\mathcal{C}(\theta,\theta')$	normalized correlation function
$<f(t,x)>$	expectation value of forward rates
σ_r	volatility of spot interest rate
$\mathcal{C}_R(\theta,\theta')$	correlation function for constant rigidity
$\mathcal{C}_C(\theta,\theta')$	correlation function for constrained spot rate
$\mathcal{C}_M(\theta,\theta')$	correlation function for non-constant rigidity
$\mathcal{C}_Q(\theta,\theta')$	correlation function for stiff propagator
$\mathcal{C}_{Qz}(\theta,\theta')$	correlation function: stiff propagator and psychological time
R_Q	curvature orthogonal to normalized propagator's diagonal axis
$z_\pm$	combination of psychological future times $z(\theta)\pm z(\theta')$
η	scaling exponent for psychological future time

Chapter 9 Field theory of Treasury Bonds' derivatives and hedging

Δ,Γ,Θ	hedging parameters for Treasury Bond option
$\mathcal{V},\mathcal{V}_i,\mathcal{V}_\sigma,\mathcal{V}_\mu$	hedging parameter for variance of Treasury Bond option
$\mathrm{Cap}(t_0,t_*,\tau,X)$	interest rate cap
L_*	Libor $L(t_*,t_*,\tau)$
q_{cap}	volatility of interest rate cap
$\mathrm{Cap}_B(t_0,t_*,\tau,X)$	Black's formula for interest rate cap
F_{cap}	Forward contract for the interest rate cap
$n_1(t)$	hedging weight for delta hedging of bonds
$B(r;t;T)$	conditional expectation of a Treasury Bond, given r
Δ_i	hedging weights for minimization of variance
$\mathrm{Var}[\Pi(t_*)]$	variance of bond portfolio at future time t_*
V_0	variance of (an un-hedged) Treasury Bond
V	minimum of the residual variance of the hedged portfolio
V_*	minimum of the residual variance for finite time hedged portfolio
P_i	equal to $P(t,T_i)$
M_{ij}	hedging matrix for Treasury Bond
L_i	hedging vector for Treasury Bond
$\mathcal{F}_i$	equal to $\mathcal{F}(t,t_F,T_i)$
$X(t_*,T)$	function appearing in the expression for $B(r;t;T)$
$a(t_*,T)$	function appearing in the expression for $B(r;t;T)$

Chapter 10 Field theory Hamiltonian of forward interest rates

$f_{n,l}$	forward rates at discrete time n and future time l
F_n	forward rates excluding single (earliest) forward rate $f_{n,n}$
F_{n+1}	forward rates excluding single (latest) forward rate $f_{n+1,n+1+N_{FR}}$
$S(n)$	forward rates' action at time $t=n\epsilon$
$\mathcal{L}_n[\tilde{F}_n;F_{n+1}]$	forward rates Lagrangian connecting two time slices
$\mathcal{V}_n$	forward rates time-dependent state space at time n

$\mathcal{V}_{\mathrm{dual},n}$	forward rates time-dependent dual state space at time n
$\mathcal{F}_n$	forward rates physical state space at time n
$\mathcal{H}_n$	forward rates time-dependent Hamiltonian at time n
$\mathcal{I}_n$	the identity operator at discrete time n for the forward rates' state space
$\lvert f_t >, \lvert F_t >$	forward rates' continuum state space basis vectors at time t
$\lvert P(t,T) >, \lvert \mathcal{B}(t) >$	zero coupon and coupon Treasury Bond state vectors
$\mathcal{H}_f(t), \mathcal{H}_{f,z}(t)$	Hamiltonians for linear forward rates
$\mathcal{H}_\phi(t)$	Hamiltonian for nonlinear forward rates
$\mathcal{M}(x,x';t)$	kinetic operator for Lagrangian with stochastic volatility
$\mathcal{H}(t)$	Hamiltonian for linear forward rates with stochastic volatility
$\mathcal{H}_{\mathrm{eff}}(t)$	effective forward rates' linear Hermitian Hamiltonian
$\mathcal{O}(t), \beta(t,x)$	functions required for determining $\mathcal{H}_{\mathrm{eff}}(t)$
$< f_*\lvert$	basis of dual vector space for forward rates at future time t_*
$\mathcal{H}_f^*(t)\lvert_{\mathcal{R}}$	forward rates' Hamiltonian restricted to domain $\mathcal{R}$
W	time integrated forward rates' linear Hamiltonian

References

[1] K. I. Amin and V. K. Ng, 'Implied Volatility Functions in Arbitrage-Free term-Structure Models'. *Journal of Financial Economics*, **35** (1994) 141.

[2] K. I. Amin and V. K. Ng, 'Inferring Future Volatility from the Information in Implied Volatility in Eurodollar Options: A New Approach'. *Review of Financial Studies*, **10** (2) (1997) 333.

[3] B. E. Baaquie, 'Quantum Field Theory of Forward Rates with Stochastic Volatility'. *Physical Review E*, **65** (2002) 056122.

[4] B. E. Baaquie, 'Quantum Field Theory of Treasury Bonds'. *Physical Review E* **64** (2001) 016121-1.

[5] B. E. Baaquie, 'A Path Integral Approach to Option Pricing with Stochastic Volatility: Some Exact Results'. *Journal de Physique I*, **7** (12) (1997)1733.

[6] B. E. Baaquie and J.-P. Bouchaud, ' "Stiff" Field Theory of Interest Rates and Psychological Time', http://xxx.lanl.gov/cond-mat/0403713. *Wilmott Magazine* (2004) (to be published).

[7] B. E. Baaquie, C. Coriano, and M. Srikant, 'Hamiltonian and Potentials in Derivative Pricing Models: Exact Results and Lattice Simulations'. *Physica A* **334** (3) (2004): 531–57; http://xxx.lanl.gov/cond-mat/0211489.

[8] B. E. Baaquie, L. C. Kwek, and S. Marakani, 'Simulation of Stochastic Volatility using Path Integration: Smiles and Frowns'. http://xxx.lanl.gov/cond-mat/0008327.

[9] B. E. Baaquie and S. Marakani, 'An Empirical Investigation of a Quantum Field Theory of Forward Rates'. http://xxx.lanl.gov/cond-mat/ 0106317 (2001).

[10] B. E. Baaquie and S. Marakani, 'Finite Hedging in Field Theory Models of Interest Rates'. *Physical Review E* **69** (2004) 036130.

[11] B. E. Baaquie and M. Srikant, 'Comparison of Field Theory Models of Interest Rates with Market Data'. *Physical Review E* **69** (2004) 036129; http://xxx.lanl.gov/cond-mat/0208528 (2002).

[12] B. E. Baaquie, M. Srikant, and M. Warachka, 'A Quantum Field Theory Term Structure Model Applied to Hedging'. *International Journal of Theoretical and Applied Finance*, **6** (5) (2003): 443.

[13] C. Ball and A. Roma, 'Stochastic Volatility Option Pricing.' *Journal of Financial and Quantitative Analysis*, **29** (1994): 589.

[14] T. Bjork, Y. Kabanov, and W. Runggaldier. 'Bond Market in the Presence of Marked Point Processes'. *Mathematical Finance*, **7** (2) (1997): 211.

[15] F. Black and M. Scholes, 'The Pricing of Options and Corporate Liabilities'. *Journal of Political Economy*, **81** (1973): 637.

310

[16] Z. Bodie, A. Kane, and A. J. Marcus *Essentials of Investment*, Irwin (1995).

[17] J. N. Bodurtha and G. Courtadon, 'Empirical Tests of the Philadelphia Stock Exchange Foriegn Currency Options Markets'. *Working Paper* WPS 84-69, Ohio State University (1984).

[18] J.-P. Bouchaud and M. Potters, *Theory of Financial Risks*. Cambridge University Press (2000).

[19] J.-P. Bouchaud, N. Sagna, R. Cont, N. El-Karoui, and M. Potters 'Phenomenology of the Interest Rate Curve'. *Applied Financial Mathematics*, **6** (209) (1999) and 'Strings Attached'. *RISK* Magazine, **11** (7) (1998): 56.

[20] J.-P. Bouchaud and D.Sornette, 'The Black and Scholes Option Pricing Problem in Mathematical Finance: Generalization and Extension for a Large Class of Stochatic Processes'. *Journal de Physique I*, **4** (1994): 863.

[21] J.-P. Bouchaud and R. Cont., 'A Langevin Approach to Stock Market Fluctuations and Crashes'. *European Physical Journal B*, **6** (1998): 543.

[22] R. J. Brenner and R. A. Jarrow, 'A Simple Formula For Options on Discount Bonds'. *Advances in Futures and Options Research*, **6** (1993): 45.

[23] J. Y. Campbell, A. S. Low, and A. C. Mackinlay, *The Econometrics of Financial Markets*. Princeton University Press (1997).

[24] R. H. Cameron and W. T. Martin, 'The Wiener Measure of Hilbert Neighborhoods in the Space of Real Continuous Functions'. *Journal of Mathematical Physics*, **34** (1944): 195.

[25] C. Chiarella and N. El-Hassan 'Evaluation of Derivative Security Prices in the Heath–Jarrow–Morton Framework as Path Integrals Using Fast Fourier Transform Techniques'. *Journal of Financial Engineering*, **16** (2) (1996): 121.

[26] K. C. Chan, G. A. Karyoli, F. A. Longstaff, and A. B. Sanders, 'An Empirical Comparison of Alternative Models of the Short-Term Interest Rate'. *Journal of Finance*, **47** (3) (1992): 1209–1228.

[27] N. Chriss, *Black-Scholes and Beyond: Option Pricing Models*, Irwin Professional Publishing (1997).

[28] J. C. Cox and S. A. Ross, 'The Valuation of Options for Alternative Stochastic Processes'. *Journal of Financial Economics*, **3** (1976): 145–166.

[29] P. Damgaard and H. Huffel (eds.), *Stochastic Quantization*. World Scientific Publishers (1998).

[30] A. Das, *Field Theory: A Path Integral Approach*. World Scientific (1993).

[31] D. Duffie, *Dynamic Asset Pricing Theory*. Princeton University Press (1994).

[32] A. Etheridge, *A Course in Financial Calculus*. Cambridge University Press (2001).

[33] R. P. Feynman and A. R. Hibbs, *Quantum Mechanics and Path Integrals*. McGraw-Hill (1965).

[34] B. Flesker, 'Testing of the Heath–Jarrow–Morton/Ho–Lee Model of Interest Rate Contingent Claims Pricing'. *Journal of Financial and Quantiative Analysis*, **38** (1993): 483.

[35] M. Garman. 'A General Theory of Asset Valuation under Diffusion State Processes'. Working Paper No. 50, University of California, Berkeley (1976).

[36] H. Geman and M. Yor, *Mathematical Finance*, **3** (1999): 55.

[37] H. Geman, N. E., Khouri and J.-C. Rochet, 'Changes of Numeraire, Changes of Probability Measure and Option Pricing'. *Journal of Applied Probability*, **32** (1995): 443.

[38] P. Goldstein, 'The Term Structure of Interest Rates as a Random Field'. *Journal of Financial Studies*, **13** (2) (2000): 365.

[39] D. J. Griffiths *Introduction to Quantum Mechanics*. Prentice Hall (1994).

[40] J. M. Harrison and S. Pliska, Martingales and Stochastic Integrals in the Theory of Continuous Trading. *Stochastic Processes and their Applications*, **11** (1981): 215.

[41] E. G. Haug, *The Complete Guide to Option Pricing Formulas*. McGraw-Hill (1997).

[42] L. Haughton (ed.), *Vasicek and Beyond*. Risk Publications (1994).

[43] D. Heath, R. Jarrow, and A. Morton, 'Bond Pricing and the Term Structure of Interest Rates: A New Methodology for Contingent Claim Valuation'. *Econometrica*, **60** (1992): 77–105.

[44] D. W. Heerman, *Computer Simulations in Theoretical Physics*. Springer Verlag (1990).

[45] S. L. Heston, 'A Closed-Form Solution for Options with Stochastic Volatility with Application to Bond and Currency Options'. *The Review of Financial Studies*, **6** (1993): 327.

[46] T. S. Ho and S. Lee, 'Term Structure Movements and Pricing Interest Rate Contingent Claims'. *Journal of Finance*, **41** (1986): 1011–1028.

[47] J. C. Hull and A. White, 'Pricing Interest Rate-Derivative Securities'. *Review of Financial Studies*, **3** (1990): 573–592.

[48] J. C. Hull and A. White, 'Hedging the Risks from Writing Foriegn Currency Options'. *Journal of International Money and Finance*, **6** (1987): 131.

[49] J. C. Hull and A. White. 'The Pricing of Options on Assets with Stochastic Volatilities'. *The Journal of Finance*, **42** (2) (June 1987): 281.

[50] J. C. Hull and A. White, 'An Analysis of the Bias in Option Pricing Caused by a Stochastic Volatility'. *Advances in Futures and Options Research*, **3** (1988): 27.

[51] J. C. Hull, *Options, Futures and Other Derivatives*. Fifth Edition, Prentice-Hall International (2003).

[52] R. Jarrow and S. Turnbull, *Derivative Securities*. South-Western College Publishing (2000).

[53] J. C. Jackwerth and M. Rubinstein, 'Recovering Probability Distributions from Option Prices'. *Journal of Finance*, **51** (1998): 1611.

[54] N. L. Jacob and R. R. Pettit, *Investments*. Irwin (1989).

[55] P. Jaeckel, *Monte Carlo Methods in Finance*. Wiley Finance (2002).

[56] F. Jamshidian, 'Forward Induction and Construction of Yield Curve Diffusion Models'. *The Journal of Fixed Income Securities* (1991): 63.

[57] F. Jamshidian, 'An Exact Bond Option Formula'. *Journal of Finance*, **44** (1989): 205.

[58] R. A. Jarrow, *Modelling Fixed Income Securities and Interest Rate Options*. McGraw-Hill (1995).

[59] H. Johnson and D. Shanno, 'Option Pricing when the Variance is Changing', *Journal of Financial and Quantitative Analysis*, **22** 143.

[60] E. P. Jones, 'Option Arbitrage and Strategy with Large Price Changes'. *Journal of Financial Economics*, **13** (1984): 91.

[61] D. P. Kennedy, 'Characterizing Gaussian Models of the Term Structure of Interest Rates'. *Mathematical Finance*, **7** (1997): 107.

[62] H. Kleinert, *Path Integrals in Quantum Mechanics, Statistics and Polymer Physics*. World Scientific (1990).

[63] R. W. Kolb, *Futures, Options and Swaps*. Third Edition, Blackwell Publishers (2000).

[64] S. J. Kon, 'Models of Stock Returns – a Comparison', *Journal of Finance*, **39** (1984): 147.

[65] D. Lamberton, B. Lapeyre, and N. Rabeau, *Introduction to Stochastic Calculus Applied to Finance*. Chapman & Hill (1996).

[66] C. G. Lamoureux and W. D. Lastrapes, 'Forecasting Stock-Return Variance: Toward an Understanding of Stochastic Implied Volatilities', *Review of Financial Studies*, **6** (1993): 293.

[67] V. Linetsky, "The Path Integral Approach to Financial Modeling and Options Pricing", *Computational Economics* **11** (1998) 129; *Mathematical Finance* **3** 349 (1993).

[68] R. N. Mantegna, 'Levy Walks and Enhanced Diffusion in Milan Stock Exchange'. *Physica A*, **179** (1991): 232.

[69] R. N. Mantegna and H. E. Stanley, *Introduction to Econophysics*. Cambridge University Press (1999).

[70] S. Marakani, 'Option Pricing with Stochastic Volatility'. *Honours Thesis*. National University of Singapore (1998).

[71] D. W. Marquardt, *Journal of the Society for Industrial and Applied Mathematics*, **11** (1963): 431.

[72] A. Matacz, 'Path Dependent Option Pricing: the path integral averaging method'. http://xxx.lanl.gov/cond-mat/0005319v1 (2000).

[73] A. Matacz and J.-P. Bouchaud, 'Explaining the Forward Interest Rate Term Structure'. *International Journal of Theoretical and Applied Finance*, **3** (3) (2000): 381.

[74] A. Matacz and J.-P. Bouchaud, 'An Empirical Investigation of the Forward Interest Rate Term Structure'. *International Journal of Theoretical and Applied Finance*, **3** (3) (2000): 703.

[75] R. C. Merton, 'Option Pricing When Underlying Stock Returns are Discontinuous'. *Journal of Financial Economics*, **3** (1976): 125.

[76] R. C. Merton, *Continuous Time Finance*. Blackwell (1990).

[77] R. C. Merton, 'The Theory of Rational Option Pricing'. *Bell Journal of Economics and Management Science*, **4** (Spring 1973): 141–183.

[78] L. H. Mervill and D. R. Pieptea, 'Stock Price Volatility: Mean-Reverting Diffusion and Noise', *Journal of Financial Economics*, **24** (1989): 193.

[79] G. Montagna, O. Nicrosini, and N. Moreni, *Physica* **A310** (2002); *European Journal of Physics* **B27** (2002): 249.

[80] M. Musiela and M. Rutkowski, *Martingale Methods in Financial Modeling*. Berlin, Springer (1997).

[81] B. Oksendal, *Stochastic Differential Equations*. Springer-Verlag, NY (1998).

[82] M. Otto, 'Using path integrals to price interest rate derivatives', http://xxx.lanl.gov/cond-mat/9812318.

[83] A. M. Polyakov, 'Fine Structure of Strings'. *Nuclear Physics* **B268** (1986) 406. I thank Omar Foda for making me aware of this reference.

[84] W. H. Press, S. A. Teukolsky, W. T. Vetterling, and B. P. Flannery, *Numerical Receipes in C++* Cambridge University Press (2002).

[85] J. M. Poterba and L. H. Summers, 'The Persistence of Volatility and Stock Market Fluctuations', *American Economic Review*, **76** (1986): 1142.

[86] R. Rebonato, *Interest-Rate Option Models*. Wiley (1996).

[87] L. S. Ritter, W. L. Silber, and G. S. Udell, *Principles of Money, Banking and Financial Markets*. 10th Edition, Addison Wesley Longman (1999).

[88] B. M. Roehner, *Theory of Markets: Trade and Space-Time Pattern of Price Fluctuations: A Study in Analytical Economics*. Berlin: Springer-Verlag (1995).

[89] B. M. Roehner, *Patterns of Speculation*. Cambridge University Press (2002).

[90] G. Roepstorff, *Path Integral Approach to Quantum Physics*. Springer-Verlag (1994).

[91] M. Rosa-Clot and S. Taddei, 'A path integral approach to derivative pricing: Formalism and Analytical Results' http://xxx.lanl.gov/cond-mat/9901277.

[92] S. M. Ross, *Stochastic Processes*. Wiley (1983).

[93] M. Rubinstein, 'Nonparametric Tests of the Alternative Option Pricing Models Using All Reported Trades and Quotes on the 30 Most Active CBOE Option Classes from August 23, 1976 through August 31, 1978'. *Journal of Finance*, **40** (1986): 445.

[94] M. Rubinstein, 'Displaced Diffusion Option Pricing'. *Journal of Finance*, **38** (1983): 213.

[95] P. Santa-Clara and D. Sornette, 'The Dynamics of the Forward Interest Rate Curve with Stochastic String Shocks'. *Journal of Financial Studies*, **14** (1) (2001): 149.

[96] L. O. Scott, 'Option Pricing When the Variance Changes Randomly: Theory, Estimation and an Application'. *Journal of Financial and Quantitative Analysis*, **22** (1987): 419.

[97] M. Srikant, 'Stochastic Processes in Finance: A Physics Perspective" *M.Sc. thesis*, National University of Singapore (2003).

[98] J. Stein, 'Overreactions in the Options Market'. *Journal of Finance*, **44** (1989): 1011.

[99] E. M. Stein and J. C. Stein, 'Stock Price Distribution with Stochastic Volatility: An Analytic Approach', *Review of Financial Studies*, **4** (1991): 727.

[100] S. Sundaresan, *Fixed Income Markets and their Derivatives*. South-Western College Publishing (1997).

[101] H. Urbantke, 'Two-level Quantum Systems: States, Phases and Holonomy'. *American Journal of Physics* **59** (1991): 503.

[102] O. Vasicek, 'An Equilibrium Characterization of the Term Structure'. *Journal of Financial Economics*, **5**: 177.

[103] M. Warachka 'A Note on Stochastic Volatility in Term Structure Models' National University of Singapore preprint (2002).

[104] N. Wiener, 'Differential Space'. *Journal of Mathematical and Physical Sciences*, **2** (1923): 132.

[105] P. Wilmott, S. Howison, and J. Dewynne *The Mathematics of Financial Derivatives*. Cambridge University Press (1995).

[106] J. Zinn-Justin, *Quantum Field Theory and Critical Phenomena*. International Series of Monographs on Physics, Oxford Science Publications (1993).

Index

315

Printed in the United States
By Bookmasters